Silurian paulinitid polychaetes f

CLAES F. BERGMAN

Bergman, Claes F. 1989 11 30: Silurian paulinitid polychaetes from Gotland. *Fossils and Strata*, No. 25, pp. 1–128. Oslo. ISSN 0300-9491. ISBN 82-00-37424-6.

Silurian paulinitids constitute a large and diverse group of jawed polychaetes which flourished in the tropical epicontinental sea in the Gotland area. They are represented by five genera: *Gotlandites*, *Hindenites*, *Lanceolatites*, *Kettnerites* [with the subgenus *K.* (*Aeolus*)], and *Langeites*. Twenty species and nine subspecies and varieties from latest Llandovery to Late Ludlow are identified, based on apparatuses reconstructed by utilizing isolated elements (scolecodonts). A biological species concept is employed, incorporating taxa based on jaw elements as well as on apparatuses. The local temporal and geographic occurrences of the paulinitids on Gotland are established. Some lineages evolve very slowly throughout the sequence, and one group forms a complex of short lineages. A third category of taxa is found only in specific environments. It is concluded that at least some species are intracontinentally distributed. The study is based on several tens of thousands of jaws from more than 700 samples from 342 localities. □ *Polychaetes, eunicids, paulinitids, jaw apparatuses, scolecodonts, taxonomy, phylogeny, ontogeny, palaeogeographical distribution, Silurian, Gotland, Sweden, N5654 N5800 E1921 E1758.*

Claes F. Bergman, Department of Historical Geology and Palaeontology, University of Lund, Sölvegatan 13, S-223 62 Lund, Sweden; 1988 02 16.

Contents

Introduction

Polychaete jaw apparatuses of quite different types have developed in the widespread Recent errant taxa of the Phyllodocida and the Eunicida. In the Phyllodocida (including, e.g., Nereidae, Nephtyidae, and Glyceridae) the proboscis is axial. In the Eunicida (e.g. Eunicidae, Lumbrinereidae, Onuphidae, Dorvilleidae, and the extinct Paulinitidae) the proboscis is in the ventral position and armed with a more complex jaw apparatus than in the Phyllodocida.

It has been my ambition to employ a neontological species concept in the present work and to introduce a nomenclature which follows the ICZN rules. Thus, the two parallel taxonomical systems currently employed for fossil polychaetes, the form-taxonomic (usually based on isolated jaws) and what is known as the natural taxonomic (usually based on jaw apparatuses) systems are here merged into one.

The excellent conditions of preservation for the organic polychaete jaws in the Silurian sediments of Gotland greatly facilitated the present reconstruction work. Further, the Silurian marine, tropical, shallow shelf seas have offered a rich variation of habitats within different types of environments e.g. reefs, lagoons and open marine soft and hard bottoms. Thus, with dense sampling, it has been possible to study the evolutionary trends of populations as well as the preference of the polychaetes for different environments (i.e. the jaws have been found in a particular sediment).

The form taxonomic system has plagued studies of fossil jawed annelids ever since Hinde published his papers on annelid jaws towards the end of the 19th century. Based on a statement on Recent jaws by Claparède (1870, p. 24), Hinde assumed that the fossil jaws would also be useless in the taxonomical work and was conscious of the fact that his taxonomy was tentative (Hinde 1880, pp. 369–370). On the other hand, Hinde (1879) established the genus *Arabellites* to include several different jaws by analogy with the jaw apparatus of the Recent genus *Arabella*.

The state of affairs is similar to that of conodont taxonomy until the middle-late part of 1960's. Today most conodontologists have accepted the biological taxonomical concept, initiated by the works of Bergström & Sweet (1966), Webers (1966), Jeppsson (1969 and 1974), and von Bitter (1972), as the only practical solution.

Finds of complete polychaete jaw apparatuses are valuable, particularly for reconstructions, which include the identification of the type, number and arrangement of the elements. In spite of the large number of polychaete jaw apparatuses described, the polychaete taxonomy is in a bad state. One explanation may be that the scolecodonts are often more or less deformed, making identification of single elements and apparatuses difficult.

Among scolecodonts, a large number of names based on form-taxonomy of different jaw elements must be revised, and the apparatus-based names must be included in the revision. Several students of scolecodonts would probably agree on the desirability of this (e.g. Szaniawski & Wrona 1973) but only a few have been more or less successful in their attempts to deal with the problem (e.g. Männil & Zaslawskaya 1985a). The attempt by Kozur (1970) to combine the two taxonomical systems into one was premature. Still, it is possible to identify species on different single jaw elements, at least among the labidognatha (e.g. Kielan-Jaworowska 1966, pp. 40–42). However, scolecodontologists have not been convinced of the advantage of a merge of the two taxonomical systems.

The Gordian knot of the polychaete parataxonomy seems to be the fact that it might not be possible to identify all dispersed elements forming an apparatus. This obstacle may perhaps never be overcome but it must be much more important to distinguish the different taxa on the specific level even if only one element is identifiable in an apparatus, than to 'identify' broad form categories of no use for any other studies. Thus, the solution is simply to cut the knot and, at present, forget the identification of for example small anterior elements, since it is not of primary interest to identify all elements in an apparatus at the specific level.

A taxonomy based on populations is the platform for further work. The corresponding nomenclature includes names from both taxonomical systems. A sound species concept will without doubt facilitate stratigraphical, taxonomical, ecological, geographical and phylogenetical studies and make the polychaete jaws useful as tools for environmental interpretations and as index fossils.

An earlier version of the present monograph was printed and distributed in a limited edition (270 copies) in August, 1987, to meet the requirements for a Swedish Ph.D. thesis examination. Unfortunately, that version was not provided with a statement disclaiming the validity of the new systematic names (ICZN, Article 8b). Furthermore, it has been obtainable on request from the University of Lund. Consequently, the new taxonomic names must be treated as having been made available in 1987. Due to the limited availability of the thesis edition, full taxonomic information regarding these 1987 names is included in the present published version of the monograph.

Acknowledgements. – Much of my work has been carried out at the Department of Historical Geology and Palaeontology, University of Lund. During this time Gerhard Regnéll, Anita Löfgren, and Kent Larsson (present Head of the Department) have provided me with working facilities.

Sven Laufeld suggested the topic and introduced me to the geology of Gotland. Anita Löfgren has read the various drafts of the final version of this monograph and she has generously given me a never failing support in my work from strict editorial aspects to taxonomical considerations. Her comments have greatly improved this manuscript. Enrico Serpagli, Hubert Szaniawski and Stefan Bengtson have scrutinized the manuscript and suggested several most valuable improvements to it.

Grants to Anita Löfgren from the Swedish Natural Science Research Council (NFR) (G-GU 4746-100, 103, 104, 105, 110) and to Kent Larsson (G-GU 3676-102, 103, 107) have financed the main part of this project. Most of the extensive field work has been carried out under the auspices of Project Ecostratigraphy paid through the NFR. *Lunds Geologiska Fältklubb* and *Th. Nordströms Fond* supported part of the field work on Gotland. The Royal Swedish Academy of Science arranged, through the Polish Academy of Science, the possibility to study the extensive paulinitid collection deposited in the Palaeozoological Institute of the Polish Academy of Science, Warsaw. Travel costs to Poland have been paid by *Hans Emil Hanssons Fond*. Travel to Great Britain has been funded by grants from *von Beskows Fond*.

Gerhard Regnéll has given me information concerning the early Swedish literature. Birger Berg has corrected my latinized taxonomical names. Kristina Lindholm translated parts of the Russian papers.

Takeshi Miyazu gave technical assistance when the SEM equipment failed to follow my intentions. Sven Stridsberg coated the SEM preparates. It would have been almost impossible to process the samples without technical assistance, paid for through grants from NFR. Several of the sample residues derive from Lennart Jeppsson's samples, most of which have been processed by him and his staff. The following persons have carried out the laboratory handling of the samples, mostly involving the picking out of scolecodonts from the residues: Björn Olof Gustavsson, Marie Jönsson, Peter Mileson, Sara Nyman, Anja Rosenberg, Ewa Säll, and Cecilia Wieslander. Doris Fredholm has given me a helpful hand now and then by supervising my laboratory staff and also by checking my locality list. Three of my samples were processed by staff employed at the department.

The following colleagues have generously given me access to their samples: Lennart Jeppsson, Doris Fredholm, Kent Larsson, Sven Laufeld, Louis Liljedahl and Sven Stridsberg.

Material from the Vattenfallet section at the Swedish Museum of Natural History, Stockholm was lent by Valdar Jaanusson.

Access to the Hinde material at the British Museum (Natural History), London, was given by D. L. F. Sealy, that of Snajder at the Narodny Museum, Prag, by Rudolf Prokop and that of Kielan-Jaworowska and Szaniawski in the Palaeozoological Department in Warsaw by Hubert Szaniawski.

Finally, particular thanks are due to Lennart Jeppsson. He initiated the study, and has put considerable time and effort into this project. Recurrent discussions with him on the geology of Gotland, apparatus reconstructions and on various zoological aspects, from vertebrates to invertebrates have been most inspiring. His ideas and advice have been very valuable to me and his comments on the manuscript have greatly improved it.

To all these friends, colleagues, students, not forgetting the participants in the Project Ecostratigraphy and other persons who have supported and contributed to my study, I am deeply thankful.

Historical review

Among the earliest published accounts of jawed annelids are those by Eichwald (1854), Massalongo (1855), and Pander (1856), issued more or less simultaneously without the authors' being aware of each others' work. Both Eichwald and Pander described isolated annelid jaws from the Lower Palaeozoic of the Baltic region, referring to them as denticles of fish. Massalongo on the other hand described impressions of annelids with the jaws preserved in place, from Tertiary beds in Italy; thus he could identify the jaws as being those of polychaetes. His monograph was probably not widely circulated, and when Hinde worked with the

Lower Palaeozoic jaws decades later he was not aware of it. Thus, Hinde (1882, p. 4) and Thorell & Lindström (1885, p. 4) give N. P. Angelin, a Swedish palaeontologist, the credit for being the first to place fossil polychaete jaws in their proper taxonomic place. They refer to a letter written in 1864 by Angelin in which he has correctly identified annelid jaws. Like Massalongo, the German zoologist Ehlers (1864–68, 1868a, 1868b) described imprints of annelid bodies and jaws, but the jaws were in a bad state of preservation in Ehler's Solnhofen material. As opposed to Massalongo and Ehlers, Grinnell (1877) worked with isolated elements. He based a new annelid genus on a poorly preserved jaw fragment later regarded as a *nomen dubium* (Kozur 1970, p. 44).

Although not being the first to describe fossil annelid remains, Hinde must be regarded as one of the pioneers of scolecodont research with four contributions on Early Palaeozoic annelid jaws (1879, 1880, 1882, and 1896). Hinde was aware of the affinity of the jaws, and knew that in recent eunicid annelids the buccal armatures are differentiated with various elements composing an apparatus. In spite of this, Hinde worked both with an apparatus-based and with a morphological-based taxonomy. However, he notes (1880, pp. 369–370) that his work is of a tentative nature. The explanation of this treatment is found in a paper by Claparède, one of the leading authorities on Recent annelids at that time, who stated (1870, p. 24) that it was impossible to use the jaws of annelids for any taxonomic work. Thus, the myth that it is impossible to use isolated annelid jaws for the identification of 'true' species was born and the concept of morphological species was initiated which has since then been used by several later scolecodont workers.

Over the next decades, at the beginning of the twentieth century, very few reports on fossil worm jaws appeared. At the beginning of the 1930's however, the time seemed right for further studies. Still, each jaw type was treated as derived from a separate species. In an abstract Croneis & Scott (1933) denominated isolated polychaete jaws 'scolecodonts', and this term has won full acceptance and is widely used. According to Jansonius & Craig (1971, p. 252), Croneis & Scott supervised several doctoral theses on annelid jaws but published only the abstract of 1933 themselves. In the same year both Eller (1933) and Stauffer (1933) made their first contributions to scolecodont research. Eller continued to publish papers on scolecodonts until 1969, with more than 20 accounts dealing with Palaeozoic scolecodonts. Zebera (1935) described and discussed isolated scolecodonts and scolecodonts in clusters from the Palaeozoic strata of Bohemia.

Zebera's material was later revised by Snajdr (1951) who worked with a multielement species concept and by so doing is one of the first palaeontologists to apply a natural taxonomy to scolecodonts.

From Devonian strata of Paraná, Brazil, Lange (1947) recorded more or less complete polychaete jaw apparatuses on bedding planes. His penetrating study included a large number of apparatuses as well as comparative studies on Recent material. Lange concluded that there is a large variation in the morphology of the jaws in apparatuses of Recent annelids. Furthermore, he also referred to the literature on fossil jaws and singled out a number of isolated scolecodonts resembling the ones in his material. The identifications were based on illustrations and because of the confused taxonomic situation of scolecodonts and the lack of knowledge of other fossil annelid apparatuses, Lange (1949, pp. 48–56) noted that his synonymy list was tentative. As he was uncertain of the relationship between the scolecodonts described earlier and his material, he found it safer to name a new genus based on the apparatuses. This has had a major impact on subsequent fossil annelid research, because the parataxonomic system evolved out of this work. This system included the morphologically based system initiated by Hinde for dispersed jaws (scolecodonts) on the one hand, and the new biological system for apparatuses on the other. It has since then been very tempting for students working with apparatuses to use this apparatus-based taxonomy only, as there is then no pressure to work with the confused form-taxonomic scolecodont system. Later, Jansonius & Craig (1971) tried to sort out the scolecodont taxonomy, i.e. they grouped the elements in form-taxonomic categories. Because of the lack of information on the isolated jaws (e.g. the specific variation of the elements and the composition of apparatuses, etc.), they did not attempt to make a true biological taxonomy that should also have included the apparatus-based taxa in their work. But Jansonius & Craig (1971, p. 253) noted that with additional information it could be anticipated that eventually all scolecodonts can be assigned to a genus in accordance with the rules of ICZN.

Reconstructions of fossil polychaete jaw apparatuses from isolated elements have been made and discussed in several other studies and by different students, e.g. Sylvester (1959), Kielan-Jaworowska (1966), Szaniawski (1968), Kozur (1971), Corradini & Olivieri (1974), Jansonius & Craig (1974), Szaniawski & Gazdzicki (1978), Bergman (1979, 1980, and 1981b), Männil & Zaslavskaya (1985a, b). Their approaches to achieve a correct reconstruction represent different and feasible ways based on the available material (see the chapter 'Descriptions and reconstructions of annelid jaw apparatuses').

The idea of working with the apparatus-based taxonomy, the natural taxonomy introduced by Lange (1947, 1949), was accepted and used by several polychaete workers e.g. Kozlowski (1956), Martinsson (1960), Kielan-Jaworowska (1961, 1962, 1963, 1966, and 1968), Szaniawski (1968, 1970, and 1974), Szaniawski & Wrona (1973), Corradini & Olivieri (1974), Jansonius & Craig (1974, 1975), Boyer (1975), Mierzejewski & Mierzejewska (1975), Mierzejewski (1978b), Szaniawski & Gazdzicki (1978), Edgar (1984), and Colbath (1987b). Most of them have used isolated jaws together with apparatuses in their taxonomical work. For the most recent opinions on these questions, see the chapter 'Polychaete taxonomy'.

Accounts of fossil annelid jaws from Sweden

Silurian. – Though annelid jaws are fairly common in the Silurian strata of Gotland, reports on them are scarce. The first account of scolecodonts is the well-known one by Hinde (1882) on annelid remains from the Wenlockian Högklint and Slite Beds. Most of his material derived from the uppermost part, Högklint unit d, of the Vattenfallet section (Thorell & Lindström 1885, p. 4; Bergman 1979a). Lindström (1885, pp. 4–5) listed the species described by Hinde in his list of the Silurian fauna of Gotland, and the same species also appear in Lindström's list of the fossil faunas of Sweden (1888, pp. 5–6). In a study of the stratigraphy of the Visby district, Hedström (1910) noted the occurrence of annelid jaws and bristles from the same section but from lower strata, corresponding to Upper Visby Beds and Högklint Beds unit a and b. Bergman (1979) discussed the reconstruction, abundance and diversity of jawed polychaetes in a study comprising 32 samples from the uppermost part of the Lower Visby Beds to the Högklint Beds unit d of the Vattenfallet section.

In his first work on the geology of Gotland, Hede (1917) described the fauna from the transitional beds between the Wenlockian Slite Marl and Slite Siltstone. His faunal list included four of Hinde's annelid taxa. Hede must have sampled his annelid jaw collection close to the second locality (south of Klintehamn) that Hinde used in his 1882 paper. Hede noted the presence of annelid jaws in a number of localities in his descriptions (1921, 1925, 1927a, 1927b, 1928, 1929, 1933, 1936, 1940).

Angelin, who died in 1876, had prepared a number of plates to be published in his series *Palaeontologica Scandinavica*, some of which were never published. In one of the unpublished plates, number 53, a number of more or less well preserved scolecodonts have been illustrated (figs. 20–26). Later Regnéll (1952, p. 623) made an attempt to identify the jaws and referred to them as *Arabellites* cf. *hamatus* Hinde, *A.* cf. *contractus* Hinde, ?aff. *Lumbriconereites* sp., and others. Regnéll used the paper by Hinde (1882) to make his identifications, and it does not seem unlikely that Angelin's material derived from one of the same localities used by Hinde (the locality south of the town wall of Visby). This is the uppermost part of the Vattenfallet section (*'Pterygotus'* marl) on Gotland.

Martinsson (1960) found two assemblages of annelid jaws from the Hemse Beds, Hemse Marl SE part and Mulde Marl, respectively, and used the apparatus taxonomy when he named the material.

Ecological studies of fossil jawed annelids are so far very rare. However, Laufeld (1975, pp. 804–805) discussed the abundance of annelids and chitinozoans with regard to benthic marine life zones (*sensu* Boucot 1975) on Gotland.

Eisenack (1975) described and discussed apparatuses and isolated elements on material partly from Gotland. He concentrated on the construction of denticles and possible resorption and growth lines in mandibles. Later Mierzejewski (1984) redescribed one of the apparatuses from the Silurian of Gotland described by Eisenack (1975).

I have earlier discussed the jawed polychaete fauna from the Lower Wenlock of Gotland in my first attempts to use both isolated jaws and apparatuses for a natural taxonomy (Bergman 1979a, 1980b, and 1981b). The base for these reconstructions was a material of more than 13,000 elements from one of the localities. At that time I had not studied the type specimens involved and therefore I tentatively used the same names as used by Hinde. A later paper (Bergman 1984) dealt with the occurrence of an annelid jaw that had been transported by a density current.

At present I have studied more than 700 samples from the Silurian of Gotland (Fig. 1A and B, and appendix), representing several metric tons of different types of limestone, marlstone and siltstone. The samples were collected in order to cover all sedimentary rock types of the island to include a total geographical as well as stratigraphical distribution. My aim is to study the total jawed annelid fauna of Gotland, though in this publication only a minor part of the fauna is presented.

Scolecodonts from Gotland have also been illustrated and discussed in a popular science book by Brood (1982), who also used the names introduced by Hinde.

Scolecodonts from erratic boulders from Gotland were pictured by Schallreuter (1982).

Ordovician. – The occurrence of supposed annelid jaws from Ordovician strata of Sweden has been reported in various studies, e.g., Wiman (1893), Westergård (1909) and Hadding (1913, 1915). The first two accounts only noted the occurrence of annelid jaws, but in the latter, Hadding described and illustrated a number of conodont elements from Scania (Skåne), believing they were annelid jaws. These taxa have later been shown to represent conodonts (Lindström 1955). It seems very probable that the observations from the lowermost Ordovician by Wiman and Westergård also were accounts of conodonts. However, scolecodonts are present in the Middle and Upper Ordovician on Öland (Yngve Grahn, personal communication).

Erratic boulders. – The source and the age of erratic boulders may be hard to verify. Apart from a few unique specimens it is probably impossible to state the locality of origin and exact stratigraphical level of the boulders, thus making them less useful for palaeontological studies (see also chapter on taxonomy).

Schallreuter (1982, p. 5) has listed papers on scolecodonts from erratic boulders of various stratigraphical ages, some of the boulders treated probably being derived from the Gotland area.

Geology of Gotland

The Silurian strata of Gotland hide magnificent, three-dimensionally preserved fossils. The sediments, now found above sea level, were formed in a tropical, shallow and repeatedly fluctuating intracratonic sea. The palaeobathymetry roughly had a NE–SW strike with a mean dip of 0.15–0.3° (Laufeld 1974a, p. 7) towards the SE (e.g. Martinsson 1958; Agterberg 1958; and Bergman 1984). The slight dip of the tectonically almost undisturbed layers exposes a stratigraphical sequence of about 500 m ranging from Late Llandovery to Late Ludlow in age. The strata are composed of various types of carbonate-rich sediments including biohermal limestone, stratified limestone, oolite, marlstone and siltstone. The sediments have been deposited in subparallel, south-easterly migrating facies belts with a shore-line to the N to NE.

The sequence has been subdivided into thirteen lithological mappable units (Hede 1921 and 1925a; see Figs. 5 to 11 for areal extension). These comprise more or less heterogeneous complexes of strata. Therefore the units have been referred to as Groups, Formations and Beds by various authors.

Hede did an impressive piece of field work when he carefully mapped almost the entire island. The results were published in eight geological descriptions (1925, 1927a, b, 1928, 1929, 1933, 1936, 1940). Only minor amendments of Hede's conclusions have been made since then (see Laufeld 1974a and Jeppsson 1983, p. 126). The map descriptions are in Swedish, but an English version of his stratigraphy has been published (Hede 1960, pp. 44–52). Based on Hede's map descriptions, Laufeld (1974a, pp. 7–13) made a very useful summary of the stratigraphy and sediments of Gotland, and he completed Hede's subdivision of the major units. Awaiting a much needed modern sedimentological description of the sediments I refer to Hede's 13 major units as 'Beds' (Laufeld 1974a).

A short summary of the sediments and the Silurian palaeogeography of Gotland was given by Laufeld & Bassett (1981). The magnificent bioherms have attracted several geologists and resulted in special studies (e.g. Manten 1971; Watts 1981). Riding (1981) has given a short review of the structures of the bioherms.

In his subdivision of the strata, Hede also used the fauna, with emphasis on some taxa of macrofossils, e.g. *Pentamerus gothlandicus*. Later, diachronism between some of the topostratigraphical units and the stratigraphical distribution of fossil assemblages was noted, first by Martinsson (1967). Other students followed and produced evidence of diachronism of several of the boundaries (e.g., Larsson 1979; Franzén 1983). Not all suggested diachronisms are accepted by all students (e.g., Jeppsson 1983, pp. 128–129; Stridsberg 1985, p. 5). It must be pointed out that most of the stratigraphical work done is based on benthic organisms which are more likely to be influenced by the bottom environments than pelagic organisms are. The conodonts do not support diachronism of more than a few boundaries (Lennart Jeppsson, personal communication, 1986).

Correlation of the Gotland strata with other areas has been carried out with, for example, graptolites (Hede 1919, 1942), and ostracodes (Martinsson 1967), and recently Jeppsson (1983) showed that conodonts might become the optimal instrument in the correlation of the Gotland sequence. The relatively limited influence that the bottom conditions have on the distribution of conodonts, and their high abundance, makes them very suitable for correlations.

Some students have concluded that in the Gotland area during the Silurian very shallow water conditions prevailed. Estimations of the depth have also been made (e.g. Hadding 1941; Gray *et al.* 1974). It has become more and more clear that the maximum water depth was less than earlier estimated (150–200 m), probably not more than 90 m (Riding 1979). During shallower episodes subaerial conditions occurred repeatedly (e.g., Watts 1981; Cherns 1982; Frykman 1985, 1986).

Stratigraphical and geographical distribution of paulinitids on Gotland

The Silurian paulinitids were most probably carnivorous, living in burrows in the sediment. The distribution of jawed paulinitids is influenced by the conditions on and probably in the sediment. A large number of physical and chemical parameters are involved. A sedimentological study of the different samples combined with information of the faunal distribution would be most interesting. Post-mortem transportation of the jaws, bioturbation, and the reworking of sediment have, of course, changed the original geographical and, probably to a much smaller degree, the stratigraphical distribution.

With the present knowledge of the distribution of the paulinitids on Gotland and the very fragmentary information from the surrounding areas, it is impossible to establish the full stratigraphical range of the different taxa. Thus, the ranges shown here (Fig. 4) must be considered as local ranges, controlled by the environment. Thus, even though some of the lineages seem to show evidence of diachronism, it is hazardous to use the polychaete distribution as an argument in the discussion of diachronism on Gotland.

The stratigraphical diagram (Fig. 4) does not account for the abundance of specimens encountered or the number of localities where the taxon is represented in each unit. The areal distribution is shown in Figs. 5, 6, 7, 8, 9, 10 and 11. The abundance of species in some localities of importance is reported in Figs. 2 and 3.

Lower Visby Beds. – These beds are composed of alternating marlstone and argillaceous limestone. The Lower Visby Beds were probably deposited at a water depth close to the photic zone (Riding 1979) and it belongs to one of the deepest deposited strata on Gotland (Gray *et al.* 1974).

Lennart Jeppsson (personal communication, 1987, manuscript in preparation) has subdivided the Lower Visby Beds into the subunits b–e for the supramarine exposures of sediment. *Kettnerites* (*A.*) *siaelsoeensis* is a subspecies found exclusively in the lowermost unit (b) where it totally dominates the paulinitid fauna and also the jawed polychaete fauna. This subspecies is probably a stenotopic type with preference for deep water. *Kettnerites* (*K.*) *abraham abraham* is also found in the Lower Visby Beds unit b, but it ranges throughout the Visby Beds. *K.* (*K.*) *martinssonii* and *Lanceolatites gracilis* are not common in the Lower Visby Beds, and both have a long range. A characteristic species of the uppermost unit (e) of Lower Visby Beds is *K.* (*K.*) *versabilis*, including its varieties. The species is uncommon

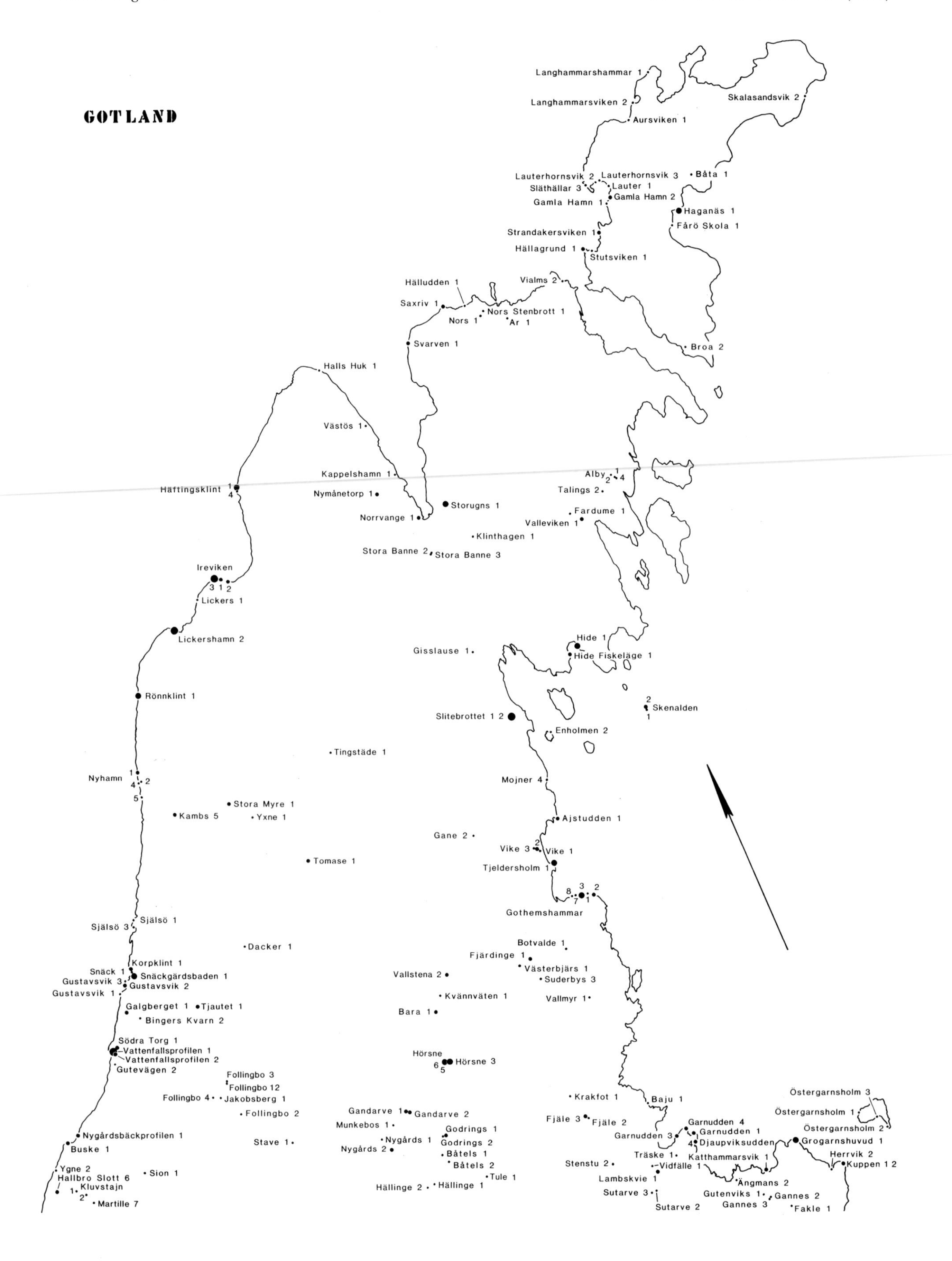

Fig. 1. Sketch maps of the northern and southern part of Gotland, showing the location of the sampled localities and their code names. The size of the dot indicates the number of samples at a locality without regard to sample size or fauna. Base map: the old topographical map sheets 1:100,000. For descriptions of the localities, see Appendix (herein).

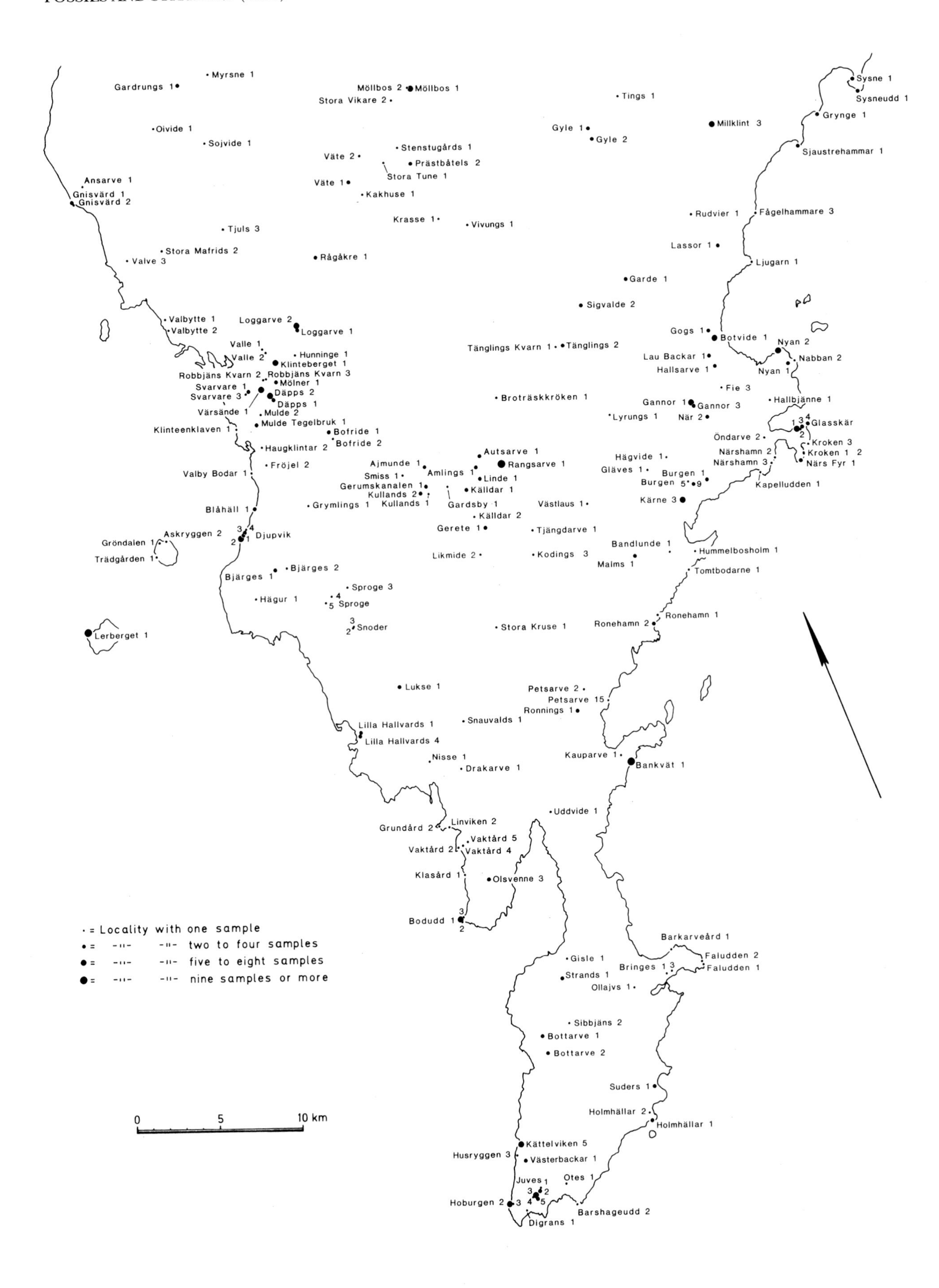
• = Locality with one sample
• = -"- -"- two to four samples
● = -"- -"- five to eight samples
● = -"- -"- nine samples or more
0 5 10 km

Fig. 1, cont.

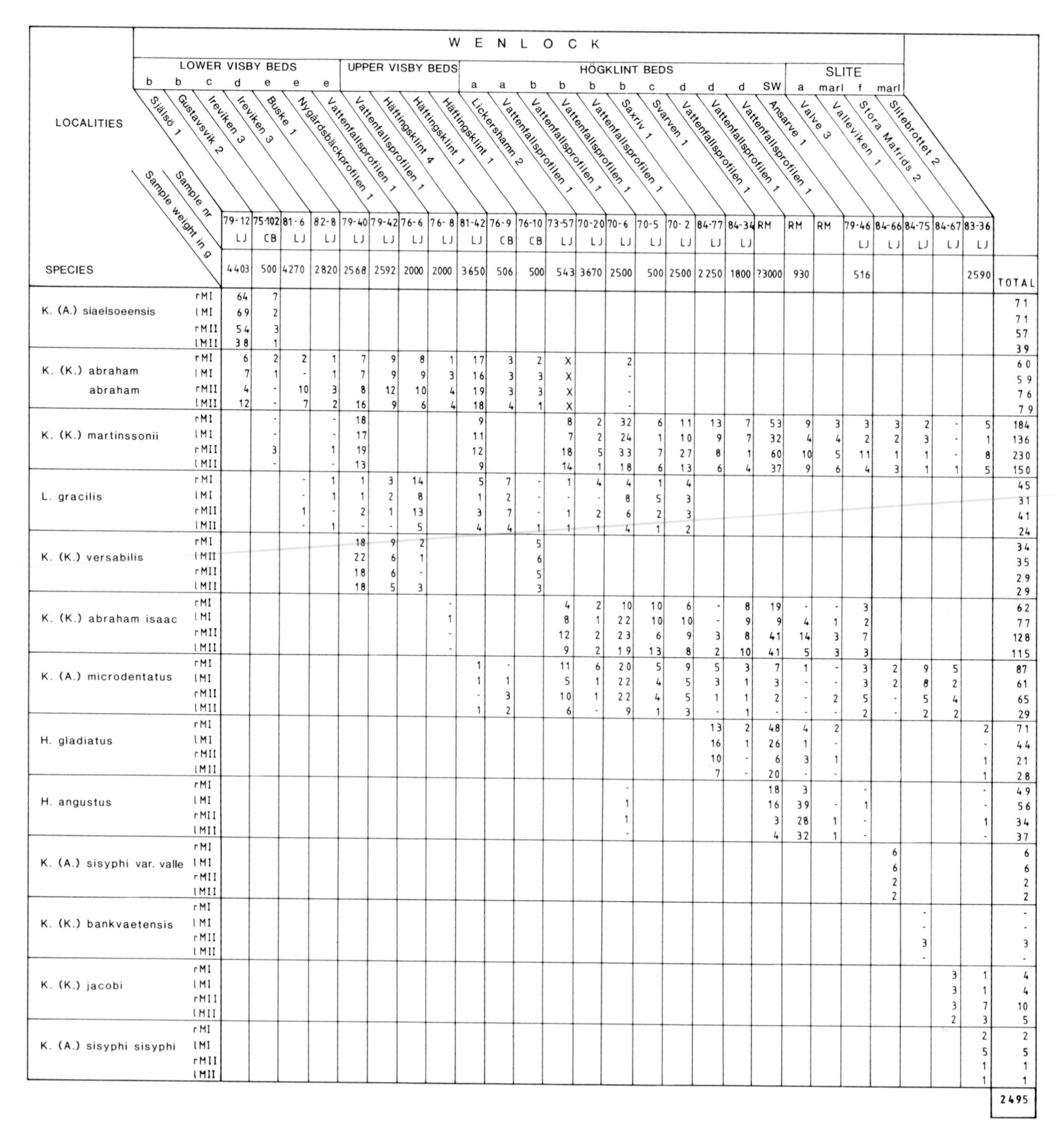

SPECIES	Sample nr	79-12	75-102	81-6	82-8	79-40	79-42	76-6	76-8	81-42	76-9	76-10	73-57	70-20	70-6	70-5	70-2	84-77	84-34	RM	RM	RM	79-46	84-66	84-75	84-67	83-36	TOTAL
WENLOCK		LOWER VISBY BEDS							UPPER VISBY BEDS				HÖGKLINT BEDS											SLITE				
		b	b	c	d	e	e	e					a	a	b	b	b	b	c	d	d	d	SW	a	marl	f	marl	
LOCALITIES		Själsö 1	Gustavsvik 2	Ireviken 3	Ireviken 3	Buske 1	Nygårdsbäckprofilen 1	Vattenfallsprofilen 1	Vattenfallsprofilen 1	Häftingsklint 4	Häftingsklint 1	Häftingsklint 1	Lickershamn 2	Vattenfallsprofilen 1	Vattenfallsprofilen 1	Vattenfallsprofilen 1	Vattenfallsprofilen 1	Saxriv 1	Svarven 1	Vattenfallsprofilen 1	Vattenfallsprofilen 1	Vattenfallsprofilen 1	Ansarve 1	Valve 3	Valleviken 1	Stora Mafrids 2	Slitebrottet 2	
		LJ	CB	LJ	LJ	LJ	LJ	LJ	LJ	LJ	CB	CB	LJ	LJ	LJ	LJ	LJ	LJ	LJ				LJ	LJ	LJ	LJ	LJ	
	Sample weight in g	4403	500	4270	2820	2568	2592	2000	2000	3650	506	500	543	3670	2500	500	2500	2250	1800	?3000	930		516				2590	
K. (A.) siaelsoeensis	rMI	64	7																									71
	lMI	69	2																									71
	rMII	54	3																									57
	lMII	38	1																									39
K. (K.) abraham abraham	rMI	6	2	2	1	7	9	8	1	17	3	2	X		2													60
	lMI	7	1	-	1	7	9	9	3	16	3	3	X		-													59
	rMII	4	-	10	3	8	12	10	4	19	3	3	X		-													76
	lMII	12	-	7	2	16	9	6	4	18	4	1	X		-													79
K. (K.) martinssonii	rMI		-		-	18				9			8	2	32	6	11	13	7	53	9	3	3	3	2	-	5	184
	lMI		-		-	17				11			7	2	24	1	10	9	7	32	4	4	2	2	3	-	1	136
	rMII		3		1	19				12			18	5	33	7	27	8	1	60	10	5	11	1	1	-	8	230
	lMII		-		-	13				9			14	1	18	6	13	6	4	37	9	6	4	3	1	1	5	150
L. gracilis	rMI			-	1	1	3	14		5	7	-	1	4	4	1	4											45
	lMI			-	1	1	2	8		1	2	-	-	-	8	5	3											31
	rMII			1	-	2	1	13		3	7	-	1	2	6	2	3											41
	lMII			-	1	-	-	5		4	4	1	1	1	4	1	2											24
K. (K.) versabilis	rMI					18	9	2				5																34
	lMII					22	6	1				6																35
	rMII					18	6	-				5																29
	lMII					18	5	3				3																29
K. (K.) abraham isaac	rMI								-				4	2	10	10	6	-	8	19	-	-	3					62
	lMI								1				8	1	22	10	10	-	9	9	4	1	2					77
	rMII								-				12	2	23	6	9	3	8	41	14	3	7					128
	lMII								-				9	2	19	13	8	2	10	41	5	3	3					115
K. (A.) microdentatus	rMI									1	-		11	6	20	5	9	5	3	7	1	-	3	2	9	5		87
	lMI									1	1		5	1	22	4	5	3	1	3	-	-	3	2	8	2		61
	rMII									-	3		10	1	22	4	5	1	1	2	-	2	5	-	5	4		65
	lMII									1	2		6	-	9	1	3	-	1	-	-	-	2	-	2	2		29
H. gladiatus	rMI																	13	2	48	4	2					2	71
	lMI																	16	1	26	1	-					-	44
	rMII																	10	-	6	3	1					1	21
	lMII																	7	-	20	-	-					1	28
H. angustus	rMI														-					18	3		-				-	49
	lMI														1					16	39	-	1				-	56
	rMII														1					3	28	1	-				1	34
	lMII														-					4	32	1	-				-	37
K. (A.) sisyphi var. valle	rMI																							6				6
	lMI																							6				6
	rMII																							2				2
	lMII																							2				2
K. (K.) bankvaetensis	rMI																								-			-
	lMI																								-			-
	rMII																								3			3
	lMII																								-			-
K. (K.) jacobi	rMI																									3	1	4
	lMI																									3	1	4
	rMII																									3	7	10
	lMII																									2	3	5
K. (A.) sisyphi sisyphi	rMI																										2	2
	lMI																										5	5
	rMII																										1	1
	lMII																										1	1
																												2495

Fig. 2. Distribution of paulinitid jaw elements in a few selected samples from the lower part of the Gotland strata i.e. Lower and Upper Visby Beds, Högklint Beds (no identifiable elements encountered from the Tofta Beds), and part of the Slite Beds (the uppermost part of the Slite Beds is reported in Fig. 3). Only the four main elements have been included in the count. The letter X denotes that the identification of elements is doubtful or that it has been impossible to separate the homologous elements of the different taxa.

outside this unit, but it is found in unit d and b and from below unit b (a Lennart Jeppsson sample, collected about 14 m below water level at Ireviken, not included in the locality list).

Upper Visby Beds. – Lithology similar to that of the Lower Visby Beds, but with the first sign of reef formation (e.g. tabulate-dominated mounds). The Upper Visby Beds were probably deposited in shallower water than the Lower Visby Beds.

The fauna is consistent within these strata. *Kettnerites* (*K.*) *abraham abraham, K.* (*K.*) *martinssonii*, and *Lanceolatites gracilis* and *L. gracilis* var. visby occur concurrently through the Upper Visby Beds. Their ranges are not exclusive for these strata, but their co-occurrence is typical of the Upper Visby Beds. *Lanceolatites gracilis* and *L. gracilis* var. visby are usually found in marl, which has been deposited in water of moderate to greater depth on Gotland. One subspecies which can be found throughout the succeeding parts of the Gotland sequence appears, viz. *Kettnerites* (*A.*) *microdentatus. Kettnerites* (*K.*) *versabilis* disappears in the lower part of the Upper Visby Beds.

Högklint Beds. – The most conspicuous rock types are large bioherms and argillaceous limestones intercalated with marlstone. The water energy varied considerably.

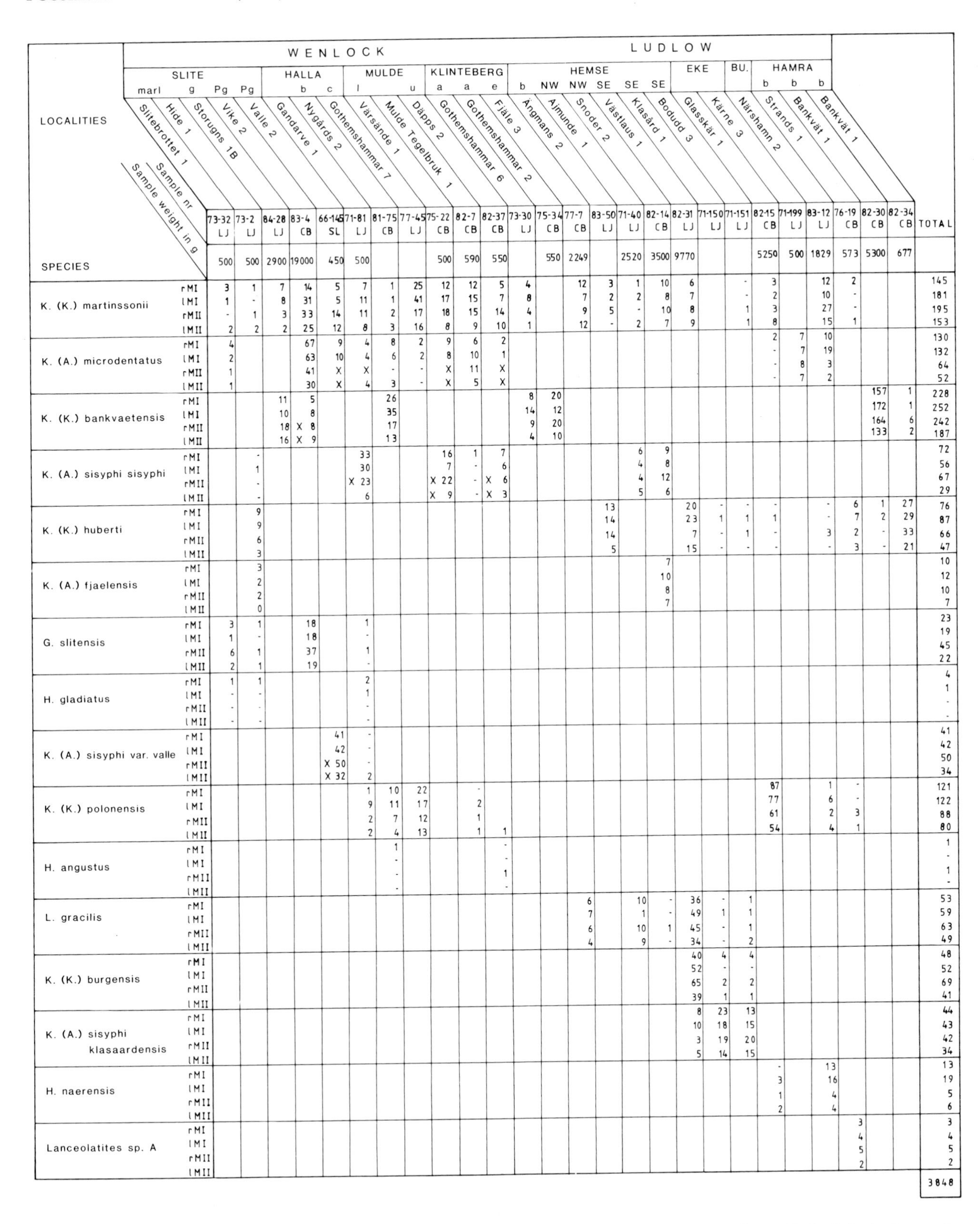

Series		WENLOCK														LUDLOW											
Beds		SLITE					HALLA			MULDE			KLINTEBERG			HEMSE						EKE		BU.	HAMRA			
Unit		marl		g	Pg	Pg		b	c	l		u	a	a	e	b	NW	NW	SE	SE	SE				b	b	b	
LOCALITIES		Slitebrottet 1	Hide 1	Storugns 1B	Vike 2	Valle 2	Gandarve 1	Nygårds 2	Gothemshammar 7	Värsände 1	Mulde Tegelbruk 1	Däpps 2	Gothemshammar 6	Gothemshammar 2	Fjäle 3	Ängmans 2	Ajmunde 1	Snoder 2	Västlaus 1	Klasård 1	Bodudd 3	Glasskär 1	Kärne 3	Närshamn 2	Strands 1	Bankvät 1	Bankvät 1	
Sample nr		73-32 LJ	73-2 LJ	84-28 LJ	83-4 CB	66-145 SL	71-81 LJ	81-75 CB	77-45 LJ	75-22 CB	82-7 CB	82-37 CB	73-30 LJ	75-34 CB	77-7 CB	83-50 LJ	71-40 LJ	82-14 CB	82-31 LJ	71-150 LJ	71-151 LJ	82-15 CB	71-199 LJ	83-12 LJ	76-19 CB	82-30 CB	82-34 CB	TOTAL
SPECIES / Sample weight in g		500	500	2900	19000	450	500			500	590	550		550	2249		2520	3500	9770			5250	500	1829	573	5300	677	
K. (K.) martinssonii	rMI	3	1	7	14	5	7	1	25	12	12	5	4		12	3	1	10	6		-	3		12	2			145
	lMI	1	-	8	31	5	11	1	41	17	15	7	8		7	2	2	8	7		-	2		10	-			181
	rMII	-	1	3	33	14	11	2	17	18	15	14	4		9	5	-	10	8		1	3		27	-			195
	lMII	2	2	2	25	12	8	3	16	8	9	10	1		12	-	2	7	9		1	8		15	1			153
K. (A.) microdentatus	rMI	4			67	9	4	8	2	9	6	2										2	7	10				130
	lMI	2			63	10	4	6	2	8	10	1										-	7	19				132
	rMII	1			41	X	X	-	-	X	11	X										-	8	3				64
	lMII	1			30	X	4	3	-	X	5	X										-	7	2				52
K. (K.) bankvaetensis	rMI			11	5			26					8	20												157	1	228
	lMI			10	8			35					14	12												172	1	252
	rMII			18	X 8			17					9	20												164	6	242
	lMII			16	X 9			13					4	10												133	2	187
K. (A.) sisyphi sisyphi	rMI		-				33			16	1	7					6	9										72
	lMI		1				30			7	-	6					4	8										56
	rMII		-				X 23			X 22	-	X 6					4	12										67
	lMII		-				6			X 9	-	X 3					5	6										29
K. (K.) huberti	rMI		9													13			20	-	-	-		-	6	1	27	76
	lMI		9													14			23	1	1	1		-	7	2	29	87
	rMII		6													14			7	-	1	-		3	2	-	33	66
	lMII		3													5			15	-	-	-		-	3	-	21	47
K. (A.) fjaelensis	rMI		3															7										10
	lMI		2															10										12
	rMII		2															8										10
	lMII		0															7										7
G. slitensis	rMI	3	1		18		1																					23
	lMI	1	-		18		-																					19
	rMII	6	1		37		1																					45
	lMII	2	1		19		-																					22
H. gladiatus	rMI	1	1				2																					4
	lMI	-	-				1																					1
	rMII	-	-				-																					-
	lMII	-	-				-																					-
K. (A.) sisyphi var. valle	rMI					41	-																					41
	lMI					42	-																					42
	rMII					X 50	-																					50
	lMII					X 32	2																					34
K. (K.) polonensis	rMI						1	10	22		-											87		1	-			121
	lMI						9	11	17		2											77		6	-			122
	rMII						2	7	12		1											61		2	3			88
	lMII						2	4	13		1	1										54		4	1			80
H. angustus	rMI							1				-																1
	lMI							-				-																-
	rMII							-				1																1
	lMII							-				-																-
L. gracilis	rMI														6		10	-	36	-	1							53
	lMI														7		1	-	49	1	1							59
	rMII														6		10	1	45	-	1							63
	lMII														4		9	-	34	-	2							49
K. (K.) burgensis	rMI																		40	4	4							48
	lMI																		52	-	-							52
	rMII																		65	2	2							69
	lMII																		39	1	1							41
K. (A.) sisyphi klasaardensis	rMI																		8	23	13							44
	lMI																		10	18	15							43
	rMII																		3	19	20							42
	lMII																		5	14	15							34
H. naerensis	rMI																					-		13				13
	lMI																					3		16				19
	rMII																					1		4				5
	lMII																					2		4				6
Lanceolatites sp. A	rMI																								3			3
	lMI																								4			4
	rMII																								5			5
	lMII																								2			2
																												3848

Fig. 3. Distribution of paulinitid jaw elements in a few selected samples from the upper Slite Beds to the Hamra Beds. Only the four main elements have been included in the count. The letter X denotes that the identification of the elements is doubtful or that it has been impossible to separate the homologous elements of the different taxa.

The evolution from *K.* (*K.*) *abraham abraham* into the younger subspecies *K.* (*K.*) *abraham isaac* occurs within the Högklint Beds, unit b (possibly also a), both subspecies being characteristic and easily identified taxa. *K.* (*K.*) *abraham isaac* continues through the Högklint Beds and is not firmly identified outside these strata. Only one occurrence is recorded from a sample of younger strata in the Halla Beds. The identification, based on a few poorly preserved jaws, is doubtful [*K.* (*K.*) aff. *abraham isaac*]. The genus *Hindenites* appears in the Högklint Beds with two lineages, *H. gladiatus* and *H. angustus*). The two species are often but not always found together. The Högklint Beds are the only strata on Gotland where *H. gladiatus* and *H. angustus* are fairly common. They probably lived in shallower water in the Gotland area. *Kettnerites* (*K.*) *martinssonii* is frequently found in every sample and *K.* (*A.*) *microdentatus* is also fairly common. Two new lineages appear, *K.* (*K.*) *bankvaetensis* and *K.* (*K.*) *polonensis*, but only a few jaws are recorded, and their identification is uncertain, based mainly on the second maxillary jaws (MII's).

Tofta Beds. – Only an unidentifiable jaw fragment has been encountered in the Tofta Beds. Unfavourable living conditions, high water turbulence, coupled with bad preservation conditions are probable explanations of the lack of fauna. The same environment seems to have persisted during the formation of the lowermost part of the Slite Beds.

Slite Beds. – The Slite Beds are composed of two main litho-facies: limestones (typical of the earliest units) and marlstones (typical of the succeeding units). There are a number of exceptions, however. The strata include a rich variety of sediments such as siltstone, reefal limestone, stratified limestone and marlstone. The earliest units, a to e, do not show any distinct faunal characters beside a very low abundance coupled with low diversity. A maximum of only two species are found in each unit. *Kettnerites* (*K.*) *martinssonii* is a typical representative of the fauna but must be regarded as an eurytopic species on Gotland. The species can be found both in typical marl levels and in limestone levels more or less all over the island.

A dramatic change of the fauna, probably inflicted by an environmental change, occurs in the undifferentiated part of the Slite Marl. A rich and varied fauna is encountered, particularly in the Slite Marl unit with *Pentamerus gothlandicus.* Typical species are *Kettnerites* (*K.*) *jacobi* and *Gotlandites slitensis*. *K.* (*K.*) *jacobi* is a conspicuous species, often with large jaws. It has a short stratigraphical range (Slite Marl, undifferentiated part to Lerberget Marl) on Gotland and is typical of unit g. *Gotlandites slitensis* is typical of the Slite Marl unit *Pentamerus gothlandicus* but ranges from the Slite Marl, undifferentiated part (Hide 1), to Halla Beds, undifferentiated part (Gandarve 1, where only two jaws are encountered). *Kettnerites* (*K.*) *huberti*, *K.* (*K.*) *bankvaetensis*, and *K.* (*K.*) *polonensis* lineages, all with long stratigraphical ranges, are first recorded with certainty in the Slite Beds. The latter two have been found in earlier strata but there only tentatively identified.

Halla Beds. – The strata consist of argillaceous limestone, small mounds and oolite, all deposited in a shallow-water environment.

The fauna in the Halla Beds has normally a fairly low diversity and abundance with only a few species per sample. *K.* (*K.*) *bankvaetensis* and *K.* (*K.*) *polonensis* become more common than in the Slite Beds. Together with *K.* (*K.*) *martinssonii* they totally dominate the fauna. *K.* (*A.*) *fjaelensis* appears.

Mulde Beds. – These are argillaceous limestone alternating with marlstone. The strata are more or less equivalent to the Halla Beds in time. The water energy was low but the water depth was probably fairly shallow.

The fauna is similar to that of the Halla Beds, but *Kettnerites* (*K.*) *martinssonii* becomes most conspicuous, with the large jaws of a new variety [*K.* (*K.*) *martinssonii* var. mulde]. This species dominates the fauna, together with *K.* (*A.*) *sisyphi*. The morphology of the latter species is variable, between the localities and often also within a sample. The latest recorded occurrence of *Hindenites gladiatus* and *K.* (*K.*) *polonensis* var. gandarve is recorded from these Beds.

Klinteberg Beds. – The lithology varies from well-washed white limestones to argillaceous limestones. Large bioherms are also typical.

The poor recovery of paulinitids from these beds is probably due to unfavourable living conditions and poor preservation. With few exceptions, the samples have yielded only a few fragmented jaws, but occasional extremely well preserved jaws are encountered. In the NE part (e.g. Gothemshammar) fairly rich faunas are recorded. A typical Klinteberg fauna of the eastern type is dominated by *Kettnerites bankvaetensis* with *K.* (*K.*) *polonensis* and *K.* (*K.*) *martinssonii* as normal accessory elements of the fauna. *K.* (*K.*) *bankvaetensis* is rare in the western area of the Klinteberg Beds. It is also rare in both the other younger and the older strata on the western side of Gotland (viz. in the more argillaceous sediments). The latest occurrence of the *Hindenites angustus* lineage is recorded from the lower–middle part of the Klinteberg Beds.

Hemse Beds. – The lithology is similar to that of the Slite Beds, with a large number of different types.

The rich and variable fauna is also reminiscent of the younger Slite fauna, particularly in some of the marly areas of the latter. *Kettnerites* (*K.*) *huberti*, earlier recorded from only two localities in the Slite Beds, is very common and a typical species in the Hemse Beds, except in the Hemse Marl, NW part. *Kettnerites* (*K.*) *polonensis* is another common species, particularly in sediments with a lower proportion of argillaceous particles. *K.* (*A.*) *sisyphi klasaardensis* and *K.* (*K.*) *burgensis* are almost endemic of the highly argillaceous sediments in the Hemse Marl SE and Marl Top. Both these lineages have a short range on Gotland. They appear in the Hemse Marl, NW part but very sparsely. Typically *K.* (*A.*) *sisyphi klasaardensis* dominates the paulinitid fauna in the Hemse Marl, SE part, particularly in the southeasternmost part along the present coast-line. *K.* (*K.*) *burgensis* seems to disappear in the latest Hemse Marl close to the boundary with the Eke Beds, while *K.* (*A.*) *sisyphi klasaardensis* continues through the boundary strata. *K.* (*A.*) *microdentatus* is

Fig. 4. Stratigraphical distribution of paulinitid jawed polychaetes on Gotland. Solid squares (■) represent fully identified taxa, solid circles (●) less confidently assigned ones. Open squares (□) represent identified taxa from localities with an imprecise stratigraphical position (e.g. lower–middle). The stratigraphical column does not reflect the true stratigraphical succession; Jeppsson (1983, Fig. 2) gives a more correct stratigraphical column but the coexisting units make it inappropriate to use in this connexion. The stratigraphical column is modified from Laufeld 1974a, Fig. 77, and Jeppsson 1983, Fig. 2.

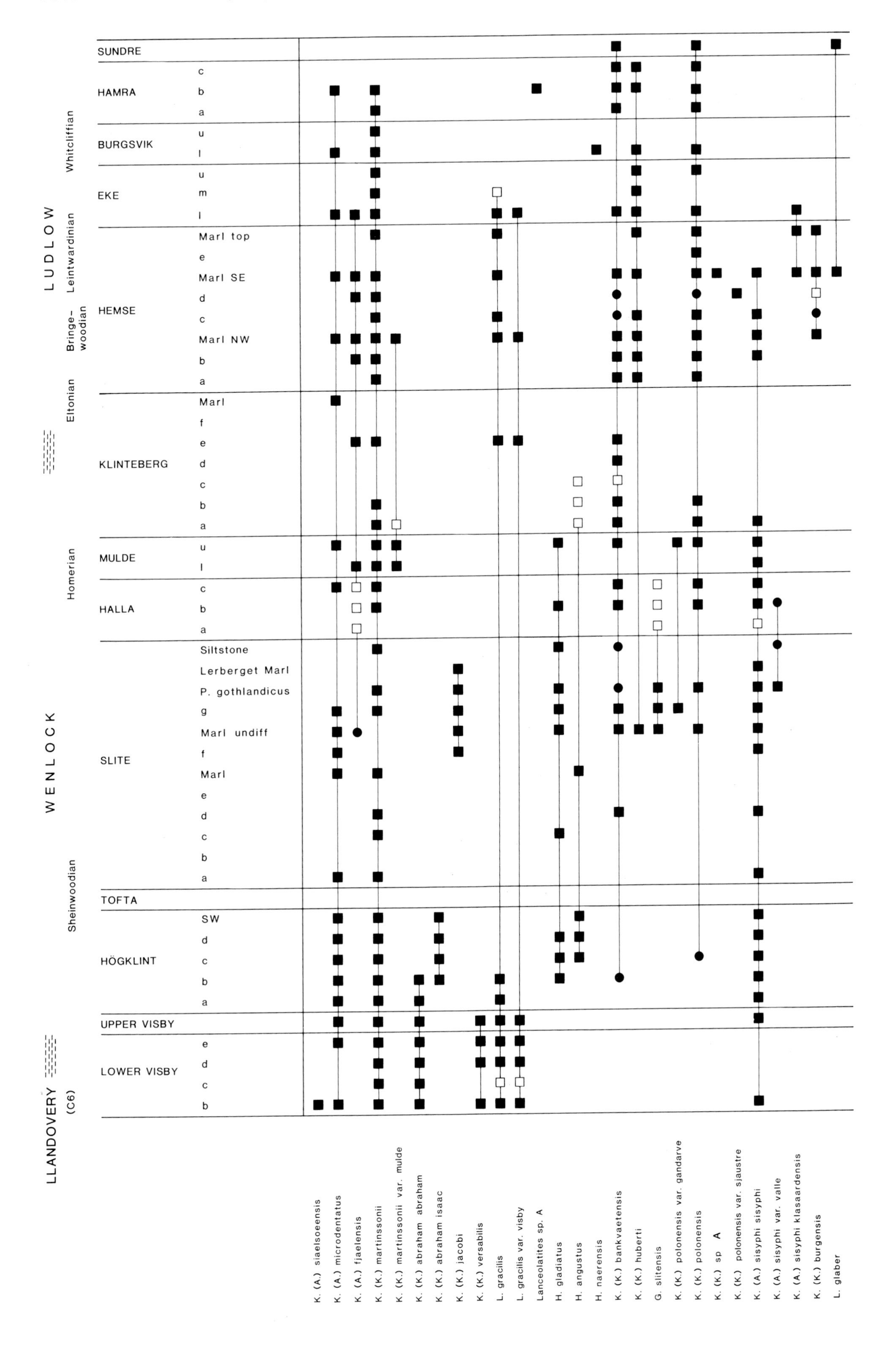
SUNDRE
HAMRA
BURGSVIK
EKE
HEMSE
KLINTEBERG
MULDE
HALLA
SLITE
TOFTA
HÖGKLINT
UPPER VISBY
LOWER VISBY
LUDLOW
WENLOCK
LLANDOVERY (C6)
Whitcliffian
Leintwardinian
Bringe-woodian
Eltonian
Homerian
Sheinwoodian
Marl top
Marl SE
Marl NW
Marl
Siltstone
Lerberget Marl
P. gothlandicus
Marl undiff
K. (A.) siaelsoeensis
K. (A.) microdentatus
K. (A.) fjaelensis
K. (K.) martinssonii
K. (K.) martinssonii var. mulde
K. (K.) abraham abraham
K. (K.) abraham isaac
K. (K.) jacobi
K. (K.) versabilis
L. gracilis
L. gracilis var. visby
Lanceolatites sp. A
H. gladiatus
H. angustus
H. naerensis
K. (K.) bankvaetensis
K. (K.) huberti
G. slitensis
K. (K.) polonensis var. gandarve
K. (K.) polonensis
K. (K.) sp A
K. (K.) polonensis var. sjaustre
K. (A.) sisyphi sisyphi
K. (A.) sisyphi var. valle
K. (A.) sisyphi klasaardensis
K. (K.) burgensis
L. glaber

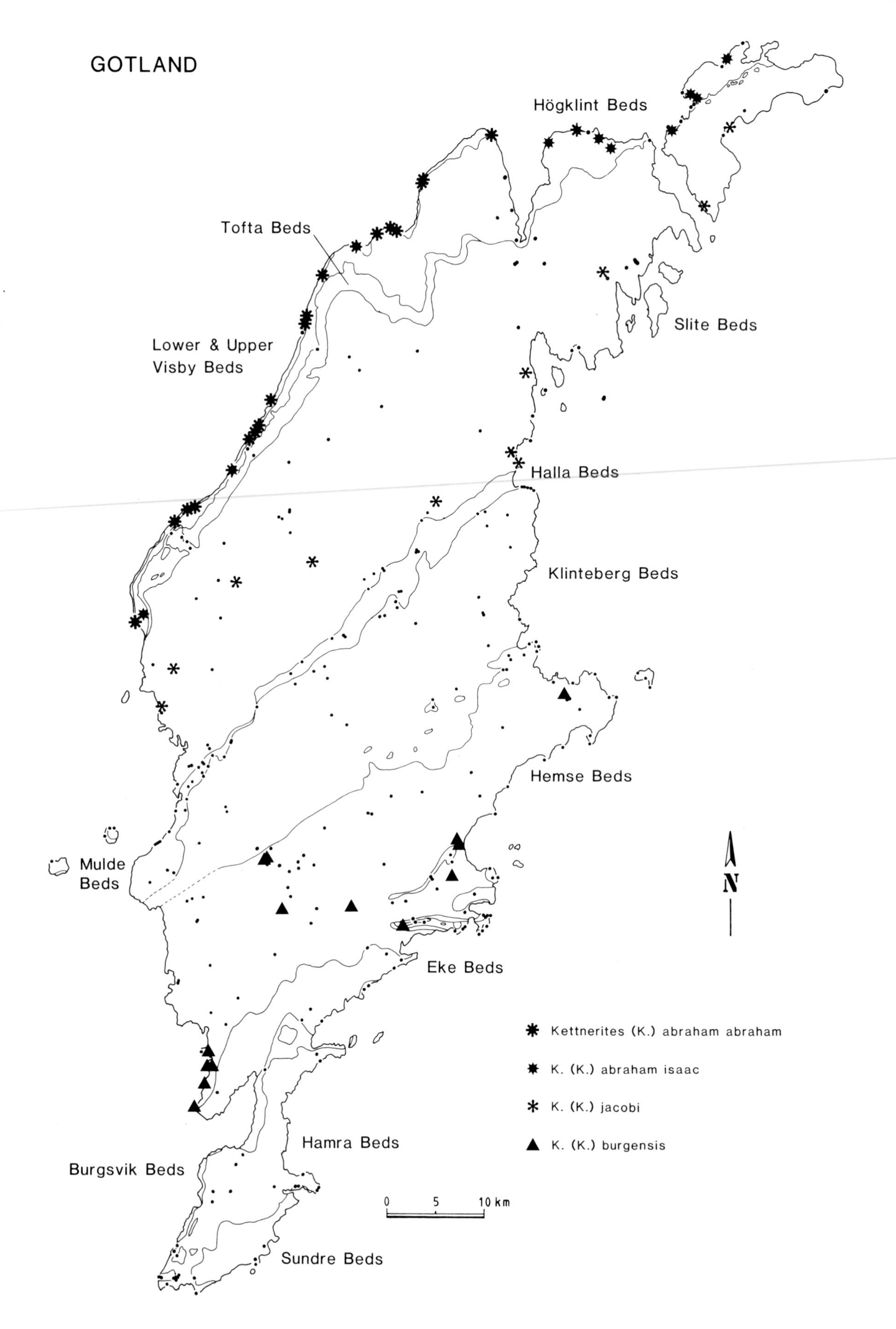

Fig. 5. Sketch map with the investigated localities represented by small dots and other symbols (see legend) indicating the identified taxa. Stratigraphical ranges: *Kettnerites* (*K.*) *abraham abraham*, Lower Visby Beds, unit b, to Högklint Beds, unit b. *K.* (*K.*) *abraham isaac*, Högklint Beds, unit b, to Högklint Beds, southwestern facies. *K.* (*K.*) *jacobi*, Slite Beds, Slite Marl, undifferentiated part, to Lerberget Marl. *K.* (*K.*) *burgensis*, Hemse Beds, from the Hemse Marl, NW part, to the Hemse Marl, uppermost part.

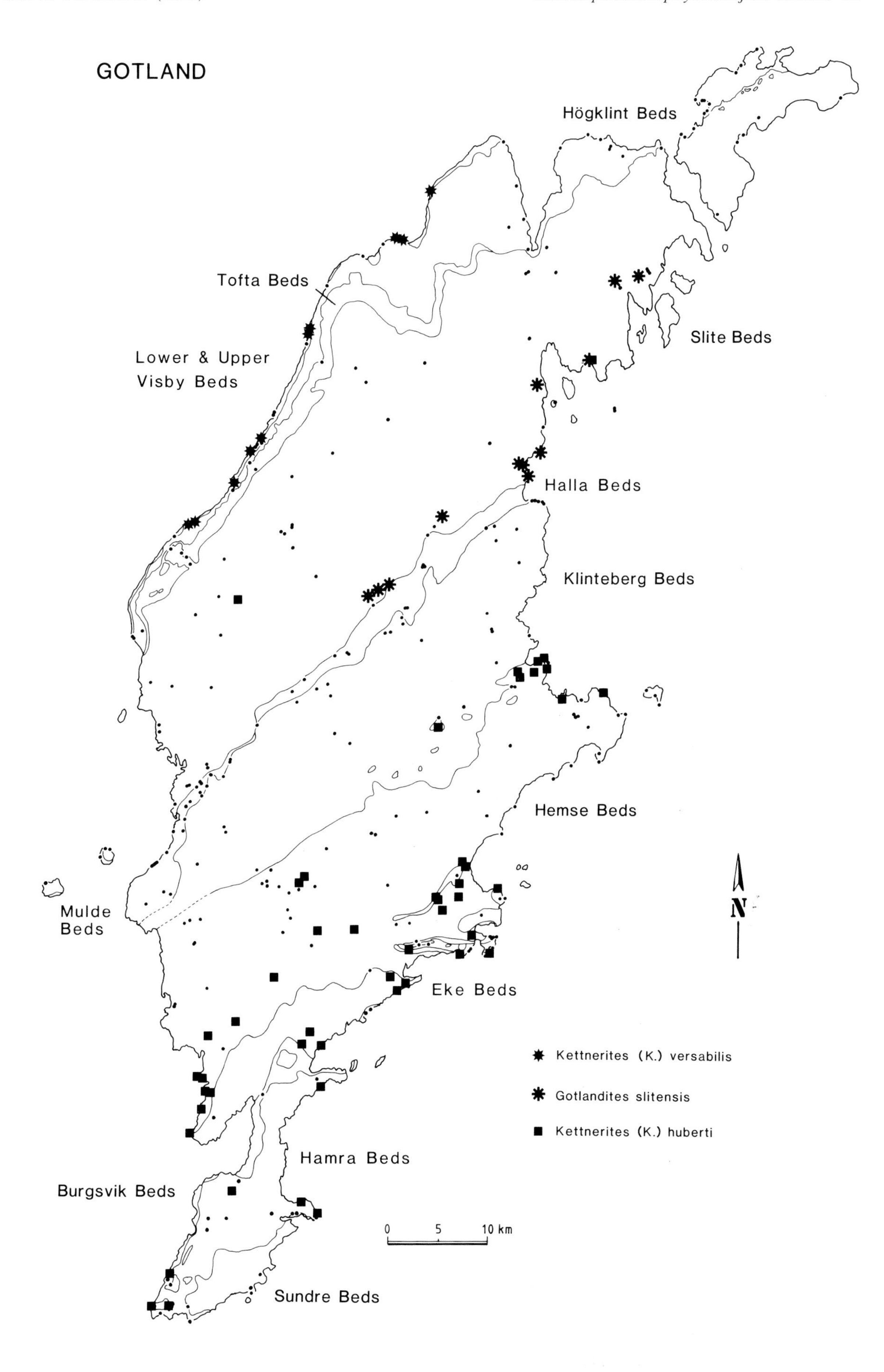

Fig. 6. Sketch map with the investigated localities represented by small dots and other symbols (see legend) indicating the identified taxa. Stratigraphical ranges: *Kettnerites* (*K.*) *versabilis*, Lower Visby Beds, unit b, to the lowest part of the Upper Visby Beds. *Gotlandites slitensis*, Slite Beds, Slite Marl, undifferentiated part, to Halla Beds. *K.* (*K.*) *huberti*, Slite Beds, Slite Marl, undifferentiated part, to Hamra Beds, unit c.

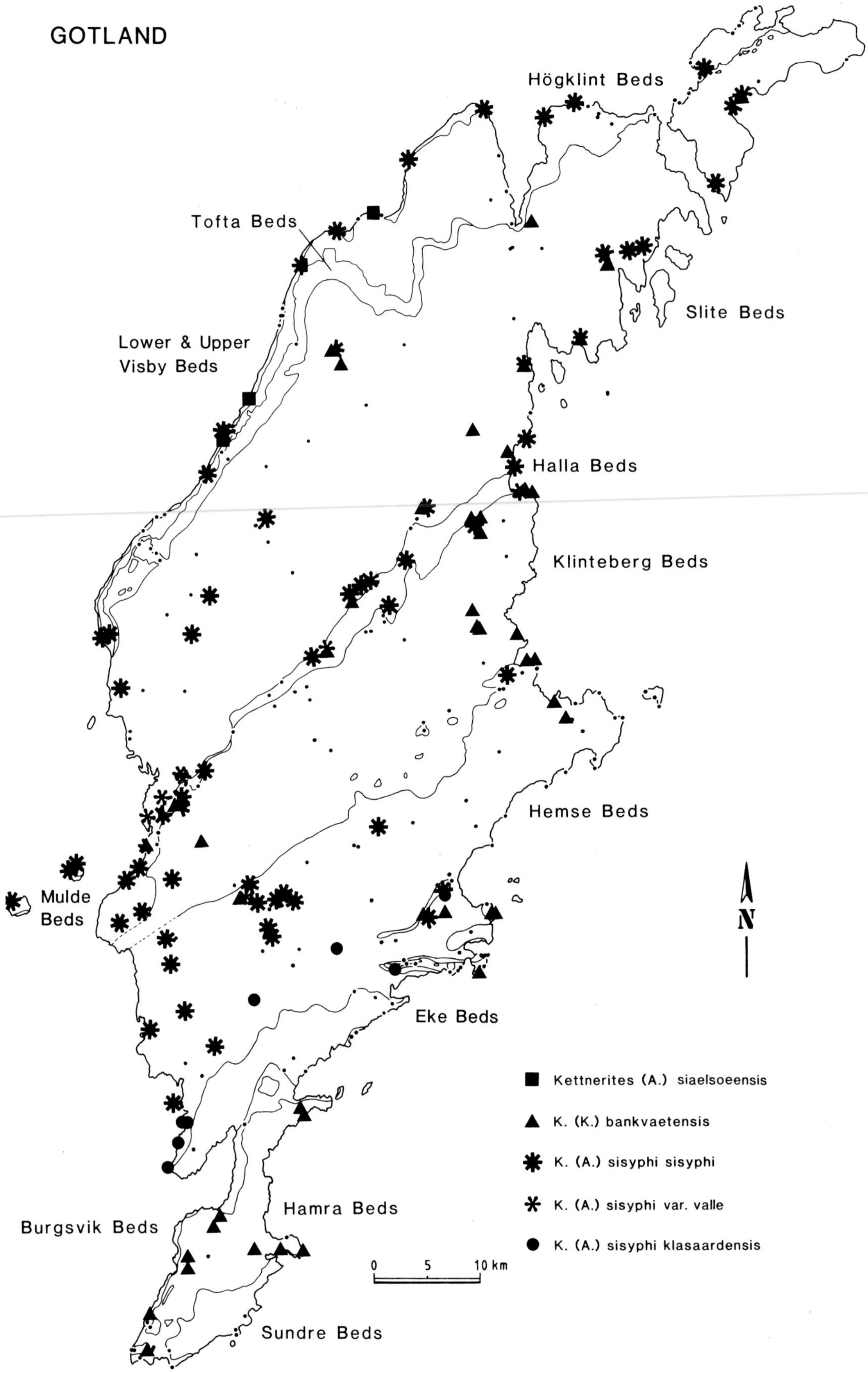

Fig. 7. Sketch map with the investigated localities represented by small dots and other symbols (see legend) indicating the identified taxa. Stratigraphical ranges: *Kettnerites* (*A.*) *siaelsoeensis*, Lower Visby Beds, unit b. *K.* (*K.*) *bankvaetensis*, fully identified from Slite Beds, unit d, to Sundre Beds. *K.* (*A.*) *sisyphi sisyphi*, Lower Visby Beds, unit b, to Hemse Beds, Hemse Marl, SE part. *K.* (*A.*) *sisyphi* var. valle, fully identified from the Slite Beds, unit *Pentamerus gothlandicus*, and probably ranging to the Halla Beds, unit b. *K.* (*A.*) *sisyphi klasaardensis*, Hemse Beds, Hemse Marl, SE part, to Eke Beds, lowest part.

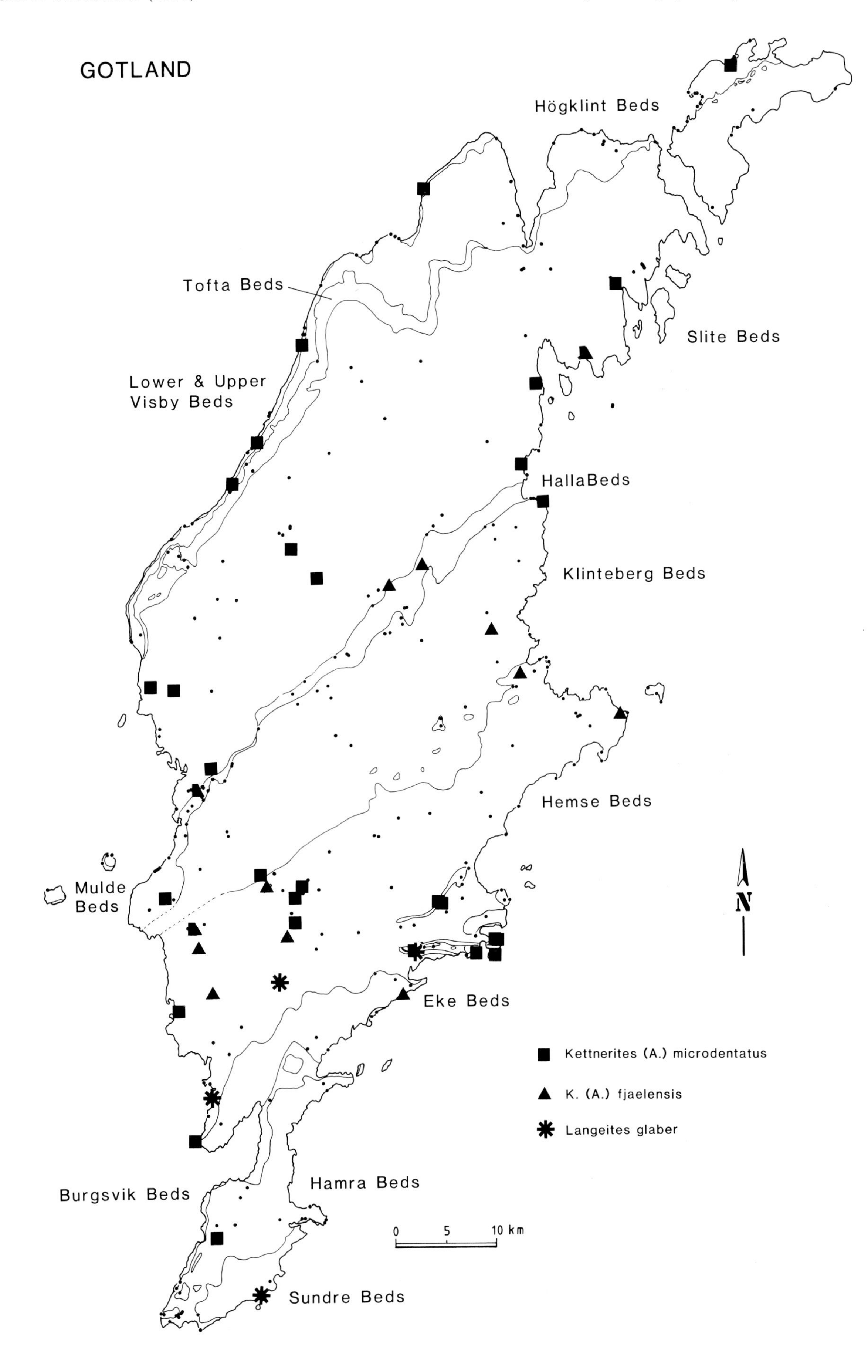

Fig. 8. Sketch map with the investigated localities represented by small dots and other symbols (see legend) indicating the identified taxa. Stratigraphical ranges: *Kettnerites* (*A.*) *microdentatus*, Lower Visby Beds, unit b, to Hamra Beds, unit b. *K.* (*A.*) *fjaelensis*, fully identified from the Halla Beds to the Lower Eke Beds. *Langeites glaber*, Hemse Beds, Hemse Marl, SE part, to Sundre Beds.

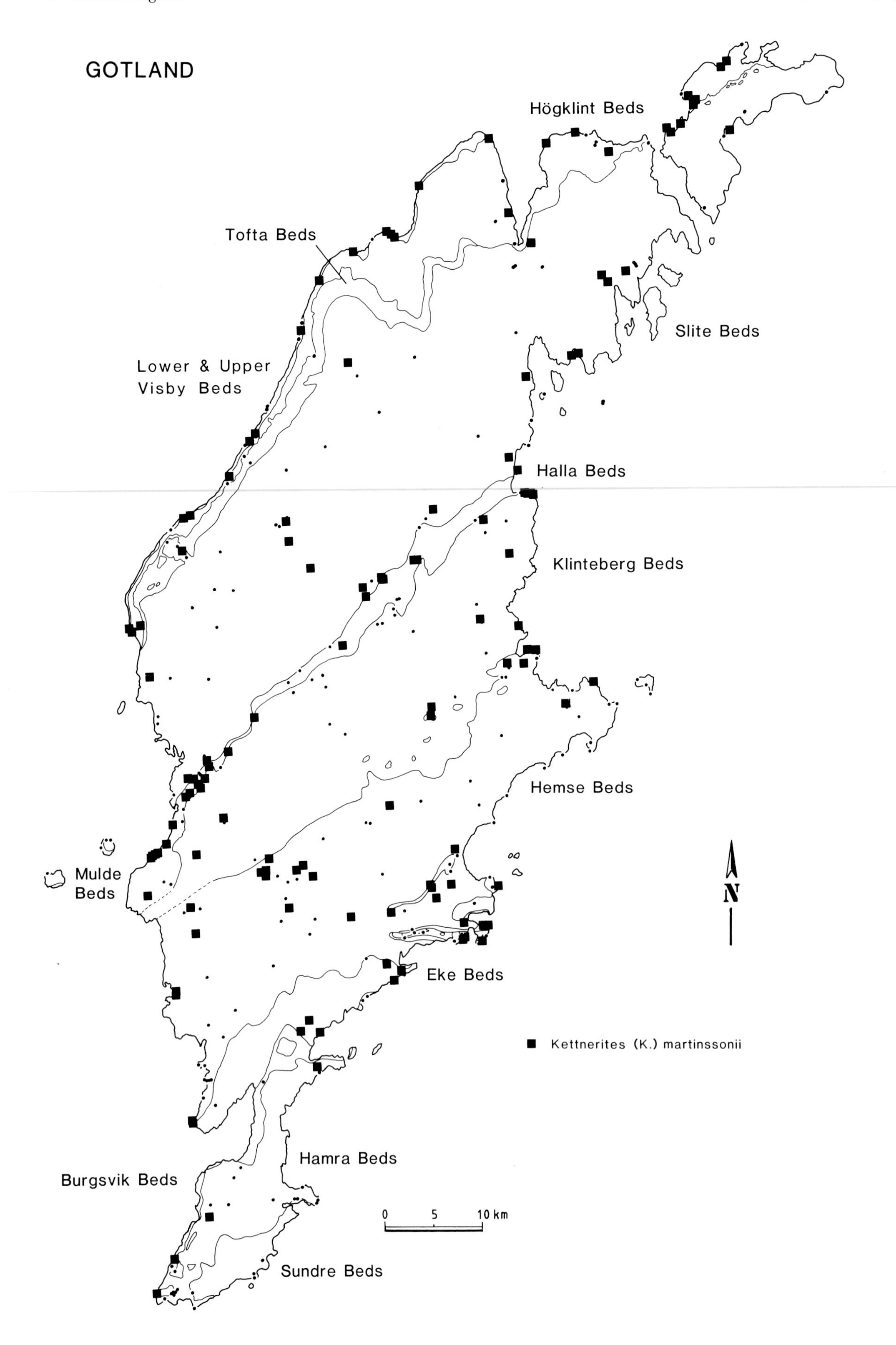

Fig. 9. Sketch map with the investigated localities represented by small dots and the filled square representing *Kettnerites* (*K.*) *martinssonii*, ranging from the Lower Visby Beds, unit b, to the Hamra Beds, unit b.

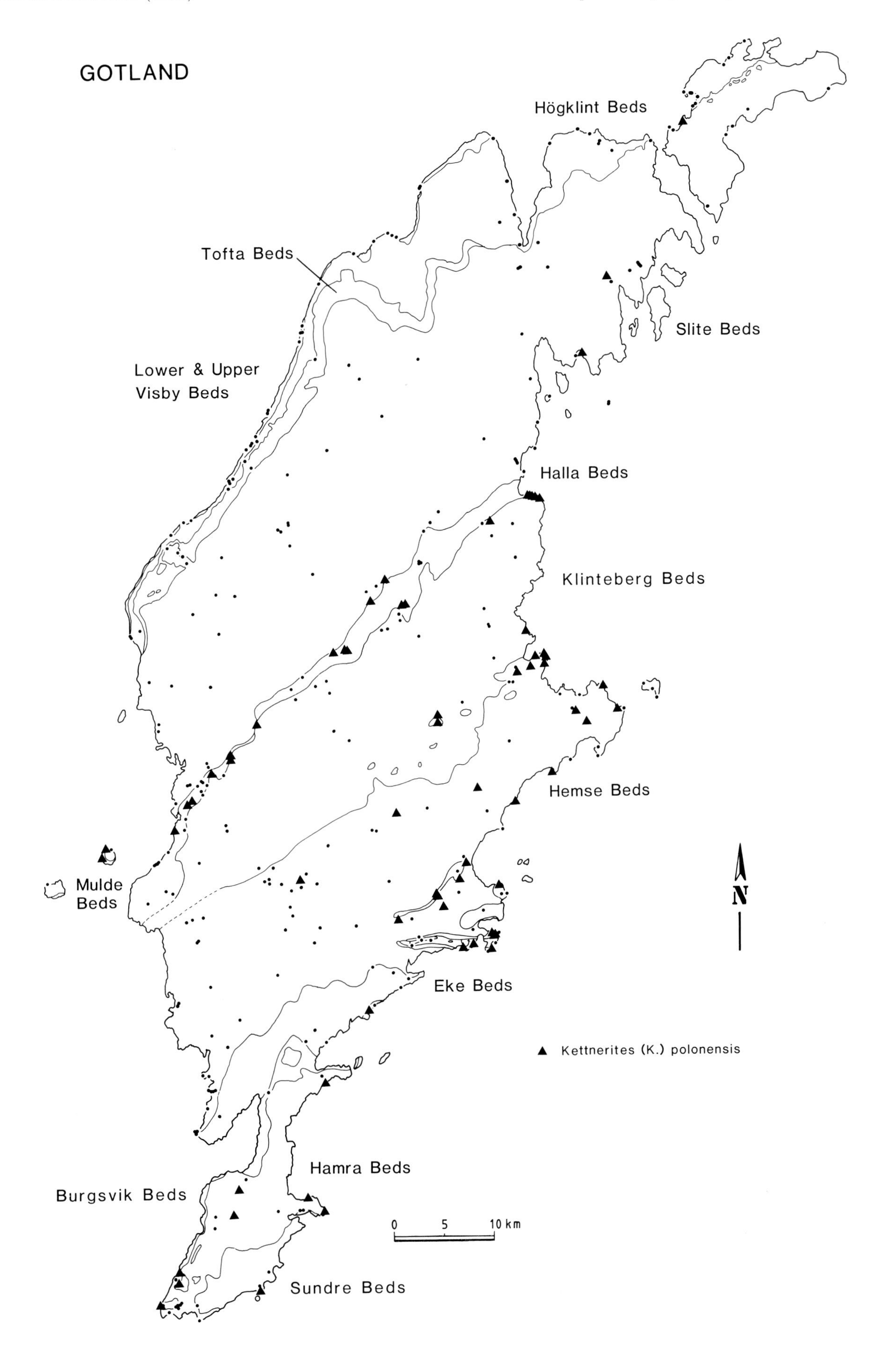

Fig. 10. Sketch map with the investigated localities represented by small dots and the filled triangle representing *Kettnerites* (*K.*) *polonensis*, typical populations of which range from the Slite Beds, Slite Marl, undifferentiated part, to the Sundre Beds.

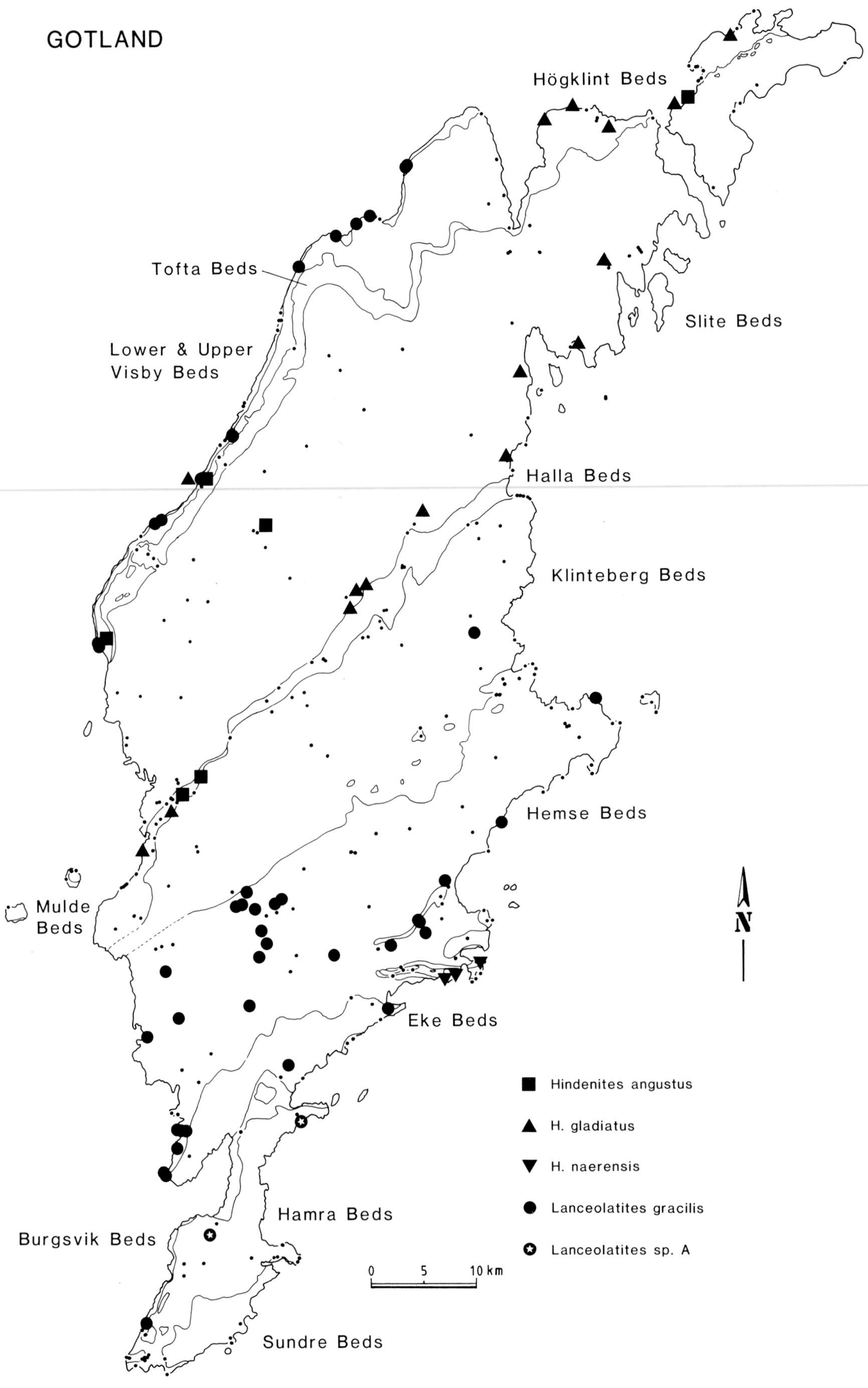

GOTLAND

Högklint Beds

Tofta Beds

Slite Beds

Lower & Upper Visby Beds

Halla Beds

Klinteberg Beds

Hemse Beds

Mulde Beds

Eke Beds

N

■ Hindenites angustus

▲ H. gladiatus

▼ H. naerensis

● Lanceolatites gracilis

✪ Lanceolatites sp. A

Hamra Beds

Burgsvik Beds

0 5 10 km

Sundre Beds

Fig. 11. Sketch map with the investigated localities represented by small dots and other symbols (see legend) indicating the identified taxa. Stratigraphical ranges: *Hindenites angustus*, Högklint Beds, unit c, to Klinteberg Beds, lower part. *H. gladiatus*, Högklint Beds, unit b, to Mulde Beds, upper part. *H. naerensis*, Burgsvik Beds, lower part. *Lanceolatites gracilis*, Lower Visby Beds, unit b, to Eke Beds, middle part. *L.* sp. A, Hamra Beds, unit b.

another example of a species which is found in sediments with a high argillaceous content. A very rare form, *Langeites glaber*, is recorded from two localities in the Hemse Marl, SE part, with only a few specimens. *Lanceolatites gracilis* is found in both the NW and SE parts of the Hemse Marl. In the localities where it is encountered it is often fairly abundant.

Eke Beds. – These are argillaceous limestones and mudstones, in places rich in calcareous algae, and in the northeastern part rich in small bioherms.

Paulinitids are normally less common in the Eke Beds than in the Hemse Beds. The change in the composition of the paulinitid fauna through the Hemse–Eke boundary is gradual. *Kettnerites* (*K.*) *burgensis* does not reach the base of the Eke Beds while *K.* (*A.*) *sisyphi klasaardensis* and *K.* (*A.*) *fjaelensis* pass through the boundary beds but disappear in the basal part of the Eke Beds. *K.* (*K.*) *bankvaetensis*, *K. huberti*, and *K.* (*K.*) *polonensis* are the dominating species in the Eke Beds.

Burgsvik Beds. – In the south, the Burgsvik Beds are dominated by clastic sediment (sand-, silt-, and claystone) and oolites to pisoolites, all with a low abundance of jaws. In the northeastern part the extension of the Burgsvik Beds is composed of an arenaceous limestone and biohermal limestone.

A polychaete fauna with a high abundance but fairly low diversity is encountered in the northeastern part of the areal extension of this unit. The most dominating species in the northeastern part of the boundary between the Eke Beds and the Burgsvik Beds is *Kettnerites* (*K.*) *polonensis*. The Burgsvik Beds, lower part, is characterized by the conspicuous *Hindenites naerensis*. This species is, however, fairly uncommon. *Kettnerites* (*A.*) *microdentatus*, which is usually found only in low numbers on Gotland, is here among the more common species, together with the eurytopic *K.* (*K.*) *martinssonii*.

Hamra Beds. – The argillaceous limestone intercalated with marlstone characterizes the middle unit b. Biohermal limestone is found in the upper part of the Hamra Beds.

Fig. 12 (p. 22). Hypothetic phylogenetic relationships between Gotland paulinitids. The connecting lines show probable or, if broken, possible relationships. Note that due to lack of space it has not been possible to place all the apparatuses in relation to each other in their proper stratigraphical orientation. The size of the drawings reflects roughly the mean size of the jaws (magnification about ×20). However, the smallest jaws have been enlarged and the largest jaws have been reduced in size in comparison with the medium-sized jaws. A comparison of size is also possible between Figs. 12 and 13. The jaws of Fig. 14 are, however, reproduced at a larger magnification (about ×30) than those in Figs. 12 and 13, due to the minute size of the jaws of some of the taxa included in that figure. Screening patterns represent the lithologies (bricks=limestone; broken ruling=marlstone) in which the illustrated specimens are found. □A. *Kettnerites* (*K.*) *martinssonii*, left and right MI and MII, Vattenfallsprofilen 1, Högklint Beds, unit b, sample 70-6LJ. □B. *K.* (*K.*) *martinssonii*, left and right MI and MII, Valle 2, Slite Beds, *Pentamerus gothlandicus* Beds, sample 66-145SL. □C. *K.* (*K*) *martinssonii* var. mulde, left and right MI and MII jaws from adult specimens, Snoder 2, Hemse Beds, Hemse Marl, NW part, sample 82-14CB. □D. *K.* (*K.*) *martinssonii*, left and right MI and MII, Snoder 2, Hemse Beds, Hemse Marl, NW part, sample 82-14CB. □E. *K.* (*K.*) *martinssonii*, left and right MI and MII, Sigvalde 2, Hemse Beds, lower–middle part, sample 71-115LJ. □F. *K.* (*K.*) *martinssonii*, left and right MI and left MII, Glasskär 3, Burgsvik Beds, lowest bed, sample 82-18CB, right MII, Glasskär 1, Burgsvik Beds, lowest part, sample 82-15CB. □G. *K.* (*K.*) *martinssonii*, left MI and right MII, Kauparve 1, Hamra Beds, lower–middle part, sample 76-13CB, right MI and left MII, Närshamn 2, Burgsvik Beds, lower part(?), 83-12LJ. □H. *K.* (*A.*) *siaelsoeensis*, left and right MI and MII, Själsö 1, Lower Visby Beds unit b, sample 79-12LJ. □I. *K.* (*A.*) *microdentatus*, left and right MI and MII, Häftingsklint 1, Upper Visby Beds, sample 76-9CB. □J. *K.* (*A.*) cf. *microdentatus*, left and right MI and MII, Vattenfallsprofilen 1, Högklint Beds, unit b, sample 70-6LJ. □ K. *K.* (*A.*) *sisyphi* var. valle, left and right MI and MII, Valle 2, Slite Beds, *Pentamerus gothlandicus* Beds, sample 66-145SL. □L. *K.* (*A.*) *sisyphi sisyphi*, left and right MI and MII, Däpps 2, Mulde Beds, upper part, sample 82-37CB. □M. *K.* (*A.*) *sisyphi sisyphi*, left and right MI and MII, Värsände, Mulde Beds, lowest part, sample 75-22CB. □N. *K.* (*A.*) *sisyphi klasaardensis*, left and right MI and MII, Vaktård 4, Hemse Beds, Hemse Marl, SE part, sample 81-35LJ. □O. *K.* (*A.*) *sisyphi sisyphi*, left and right MI and MII, Snoder 2, Hemse Beds, Hemse Marl, NW part, sample 82-14CB. □P. *K.* (*A.*) *fjaelensis*, left and right MI and MII, Fjäle 3, Klinteberg Beds, unit e, sample 77-7CB. □Q. *K.* (*A.*) *microdentatus*, left and right MI and MII, Stave 1, Slite Beds, Slite Marl, 75-11CB. □R. *K.* (*A.*) *microdentatus*, left and right MI, Närshamn 2, Burgsvik Beds, lower part(?), sample 82-12LJ.

Fig. 13 (p. 23). Hypothetic phylogenetic relationships between Gotland paulinitids. See Fig. 12 for explanation. □A. *Kettnerites* (*K.*) *polonensis* var. gandarve, left and right MI and MII, Gothemshammar 7, Halla Beds, unit c, sample 77-45LJ. □B. *K.* (*K.*) *polonensis*, left and right MI and MII, Glasskär 3, Eke Beds, sample 82-18CB. □C. *K.* (*K.*) *polonensis*, left and right MI, Glasskär 1, Eke Beds, sample 82-15CB. □D. *K.* (*K.*) *polonensis*, left and right MI and MII, Faludden 2, Hamra Beds, unit c, sample 76-16CB. □E. *K.* (*K.*) *polonensis*, left and right MI and MII, Sibbjäns 2, Hamra Beds, unit b, sample 82-32LJ. □F. *K.* (*K.*) *polonensis* var. sjaustre, left and right MI and MII, Sjaustrehammar 1, Hemse Beds, unit d, sample 82-19LJ. □G. *K.* (*K.*) sp. A, left and right MI and MII, Likmide 2, Hemse Beds, Hemse Marl SE part, sample 82-28LJ. □H. *K.* (*K.*) *burgensis*, left and right MI and MII, Västlaus 1, Hemse Beds, Hemse Marl, SE part, sample 82-31LJ. □I. *K.* (*K.*) *bankvaetensis*, left and right MI and MII, Vike 2, Slite Beds, unit *Pentamerus gothlandicus*, sample 83-4CB. □J. *K.* (*K.*) *bankvaetensis*, left and right MI and MII, Möllbos 1, Halla Beds, unit b, sample 77-28LJ. □K. *K.* (*K.*) *bankvaetensis*, left and right MI and MII, Fjärdinge 1, Klinteberg Beds, unit b, sample 77-5CB. □L. *K.* (*K.*) *bankvaetensis*, left and right MI and MII, Bankvät 1, Hamra Beds, unit b, sample 82-30CB. □M. *K.* (*K.*) *bankvaetensis*, left and right MI and MII, Kauparve 1, Hamra Beds, lower–middle part, sample 76-16CB. □N. *K.* (*K.*) *abraham abraham*, finely denticulated, left MI, Nygårdsbäckprofilen 1, Lower Visby Beds, unit e, sample 79-42LJ, right MI, Lickershamn 2, Lower Visby Beds, unit e, or lowest part of Upper Visby Beds, sample 73-53LJ. □O. *K.* (*K.*) *abraham abraham*, left and right MI and MII, Buske 1, Lower Visby Beds, unit e, sample 79-40LJ. □P. *K.* (*K.*) *abraham isaac*, left and right MI and MII, Vattenfallsprofilen 1, Högklint Beds, unit b, sample 70-6LJ. □Q. *K.* (*K.*) *jacobi*, left and right MI and MII, Slitebrottet 2, Slite Beds, unit g, sample 83-31LJ. □R. *K.* (*K.*) *huberti*, left and right MI and MII, Hide 1, Slite Beds, Slite Marl, sample 73-2LJ. □S. *K.* (*K.*) *huberti*, slender type, left and right MI and MII, Vaktård 4, Hemse Beds, Hemse Marl, SE part, sample 81-35LJ. □T. *K.* (*K.*) *huberti*, straight type, left and right MI and MII, Bankvät 1, Hamra Beds, unit b, sample 82-34CB. □U. *K.* (*K.*) *huberti*, coarse denticulated type, left and right MI, Bankvät 1, Hamra Beds, unit b, sample 81-39LJ, left and right MII, Bankvät 1, Hamra Beds, unit b, sample 82-32CB.

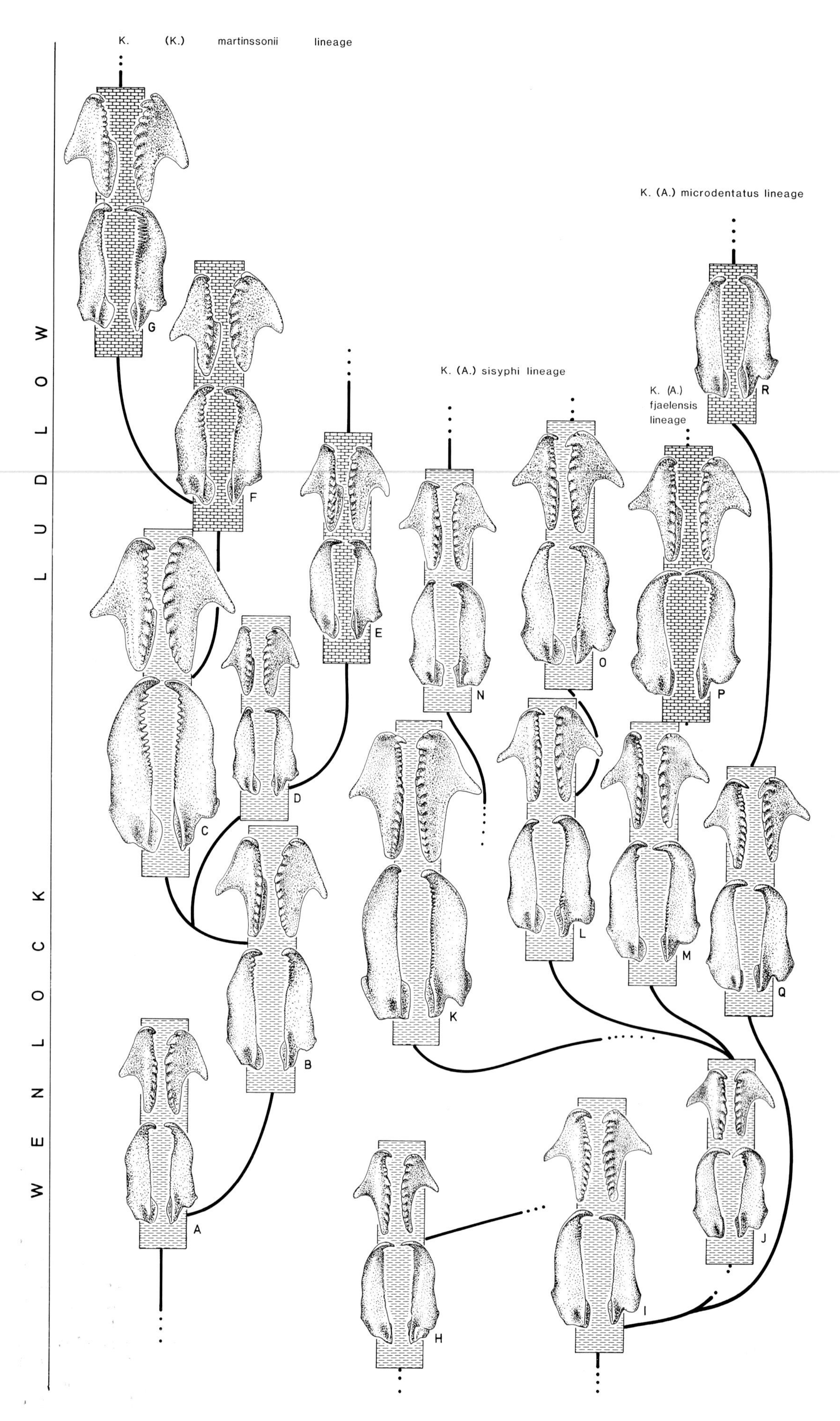

Fig. 12 (caption on p. 21).

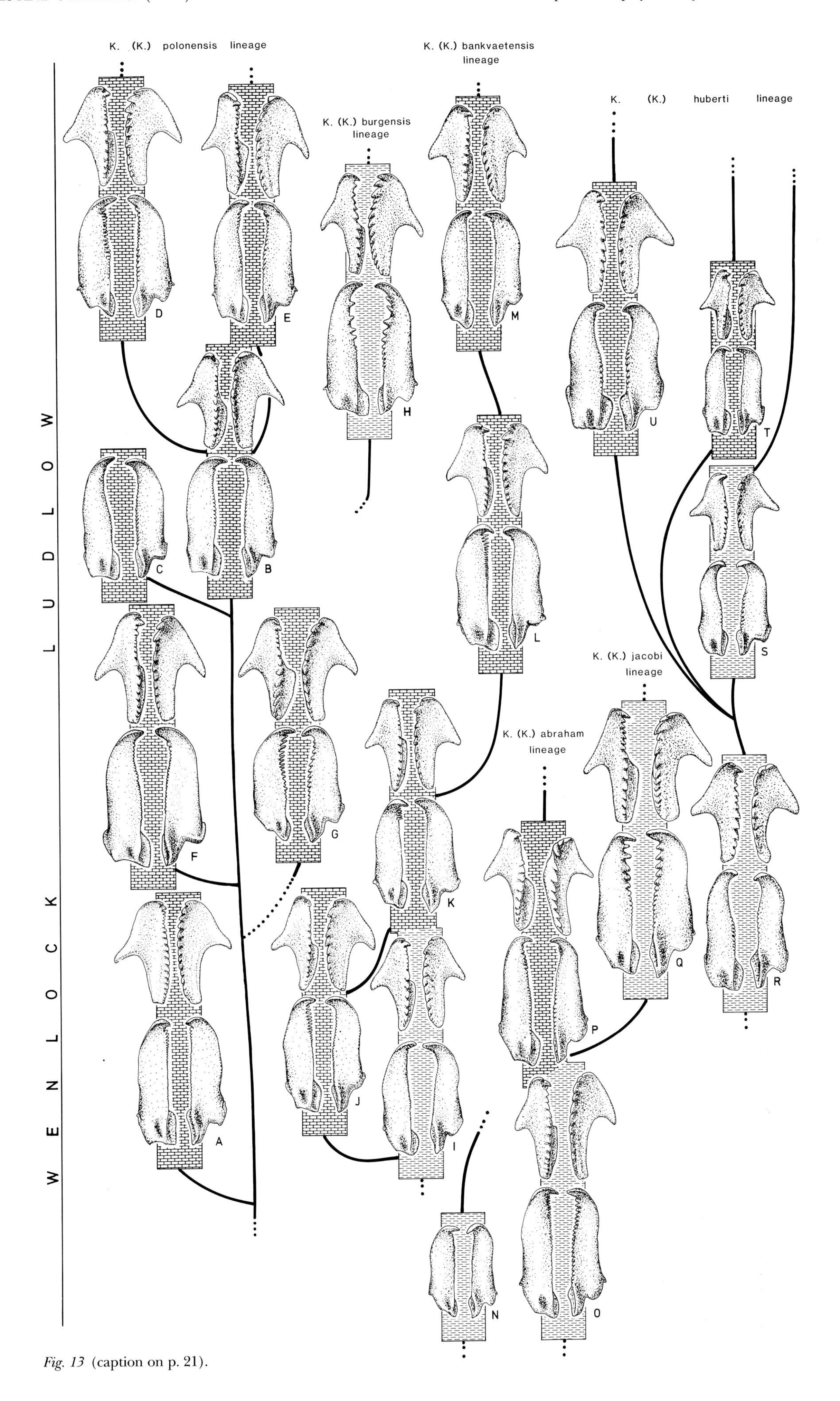

Fig. 13 (caption on p. 21).

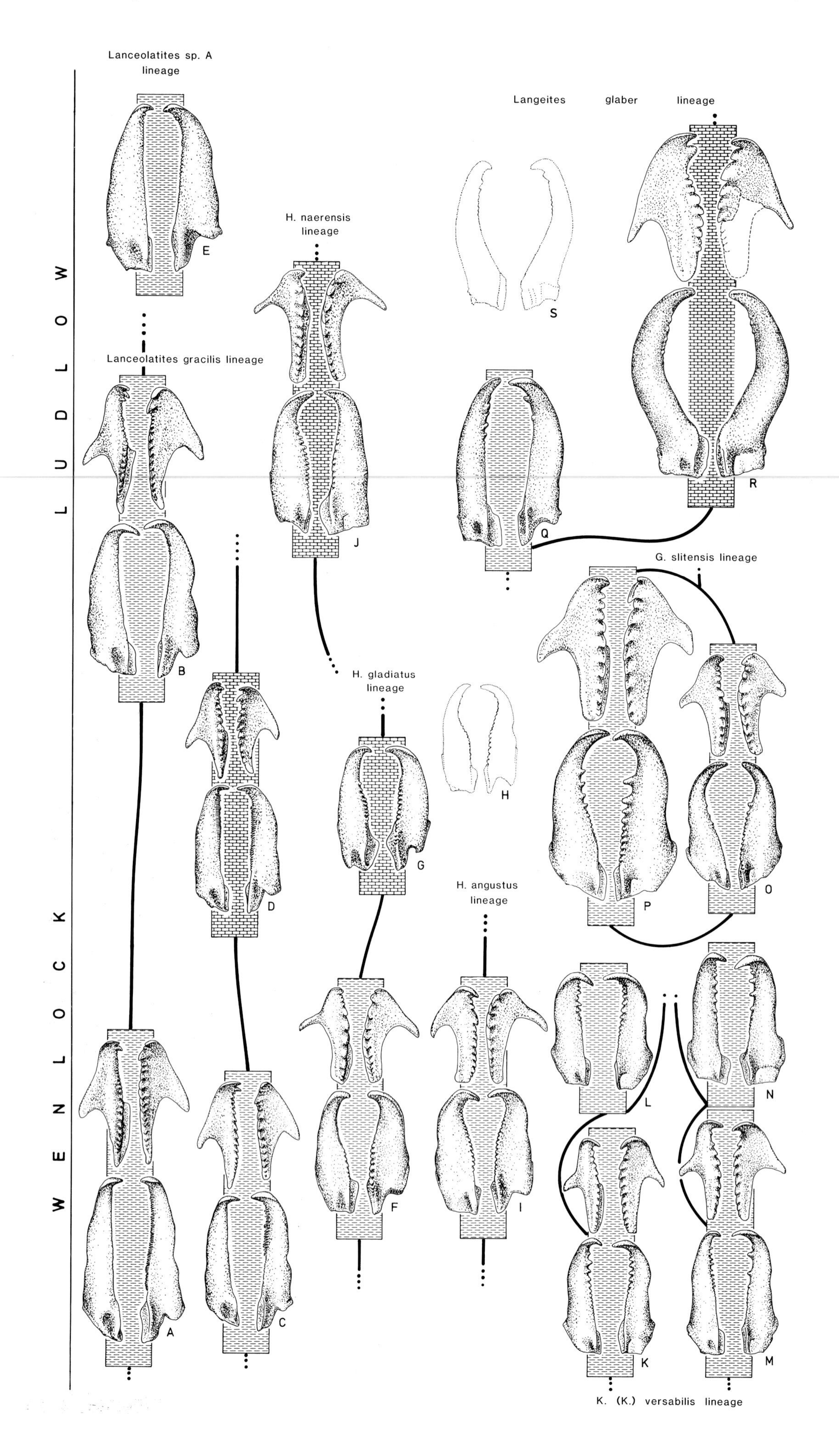
Lanceolatites sp. A lineage
Langeites glaber lineage
H. naerensis lineage
Lanceolatites gracilis lineage
G. slitensis lineage
H. gladiatus lineage
H. angustus lineage
K. (K.) versabilis lineage
L U D L O W
W E N L O C K
A
B
C
D
E
F
G
H
I
J
K
L
M
N
O
P
Q
R
S

Only unit b yields polychaete jaws in any great amount. The jaws are often very large, and from some samples (e.g. Bankvät 1) the fauna is totally dominated by one species. Several lineages seem to disappear in this unit. This is probably only an artifact, because the Hamra Beds, unit c, and the Sundre Beds are fairly poor in annelid jaws. Thus, the lineages may continue into younger strata exposed southeast of Gotland. One new species, *Lanceolatites* sp. A, appears and disappears in the Hamra Beds, unit b.

Sundre Beds. – Mostly crinoidal and biohermal limestones. Very few jaws, often fragmented, are encountered in these strata, and those found are often badly preserved. Shallow, highly turbulent water with unfavourable bottom conditions seems to be the explanation that the fauna is so limited. Very large specimens of *Langeites glaber* are encountered from a marl sample (collected by Anders Martinsson) at Holmhällar 1.

Ecological notes

The distribution of jawed polychaetes on Gotland seems to a large degree to be related to sediment types. Sediments on Gotland are often grouped into marlstone and limestone, but this subdivision is too imprecise to give any deeper knowledge of the distribution of the polychaetes. Without a study (including thin-sections) of the sedimentology, the ecological discussion (which is based, moreover, on only about one third of the polychaete fauna), can be only of a preliminary nature.

On Gotland, two to four paulinitid species is the normal diversity found in a sample yielding polychaete jaws. A low-diversity fauna, dominated by small specimens, and often coupled with a fairly low abundance, seems to indicate fairly deep water. Deeper water is usually inferred sedimentologically by a high content of argillaceous material, but the latter is only an indication of low water energy.

Faunas with a high diversity probably inhabited bottoms with low water energy, i.e. a marly substrate but not from the very deepest waters. Thus, lagoons and other restricted environments probably offered the most varied and optimal conditions for the paulinitid polychaetes.

Several samples from the limestone areas of, for instance, the Slite Beds, Klinteberg Beds and Hemse Beds have a very low abundance of polychaete jaws (Figs. 2–4 for the abundance and the stratigraphical distribution; Figs. 5–11 for the geographical distribution). This is probably partly due to unfavourable preservational conditions, but the faunal diversity was probably fairly low as well.

Most of the paulinitid species on Gotland show a preference for particular habitats. The following descriptions should be seen as sketches, and are not intended to cover the total variation of the fauna, only to indicate the trends.

- *Gotlandites slitensis*, stenotopic with a short range, found in marl, fairly deep water.
- *Hindenites gladiatus*, *H. angustus*, and *H. naerensis*, ranges of moderate length except the latter species which is found only at a few localities, found in shallow-water sediments, in the vicinity of reefs but not within them.
- *Kettnerites* (*K.*) *abraham*, probably eurytopic, with a short range, found in marl deposited in deep water.
- *Kettnerites* (*K.*) *jacobi*, more or less stenotopic with a short range, argillaceous limestone to marl.
- *Kettnerites* (*K.*) *bankvaetensis*, with few exceptions found in argillaceous limestones on the eastern side of the island.
- *Kettnerites* (*K.*) *burgensis*, stenotopic with a short range, found in marl deposited in both deep and fairly shallow water.
- *Kettnerites* (*K.*) *huberti*, eurytopic with a long range, found in marl deposited in deep water, and argillaceous limestone but not 'well washed limestone'.
- *Kettnerites* (*K.*) *martinssonii*, eurytopic with a long range, commonly found in most types of sediment except in marls deposited in deep waters.
- *Kettnerites* (*K.*) *polonensis*, long range, a typical limestone-loving form but also common in the Mulde Beds (marlstone). Taxon typical in sediments deposited in shallow waters.
- *Kettnerites* (*K.*) *versabilis*, stenotopic with a short range, found in marl deposited in deep water.
- *Kettnerites* (*A.*) *sisyphi sisyphi*, not very shallow or deep water.
- *Kettnerites* (*A.*) *sisyphi klasaardensis*, stenotopic with a short range, marl deposited in deep water.

Fig. 14. Hypothetic phylogenetic relationships between Gotland paulinitids. See Fig. 12 for explanation. □A. *Lanceolatites gracilis*, left and right MI and MII, Nygårdsbäckprofilen 1, Lower Visby Beds, unit e, sample 79-42LJ. □B. *L. gracilis*, left and right MI and MII, Ajmunde 1, Hemse Beds, Hemse Marl, NW part, sample 71-40LJ. □C. *L. gracilis* var. visby, left and right MI and MII, Vattenfallsprofilen 1, Lower Visby Beds, unit e(?), sample 76-6LJ. □D. *L. gracilis* var. visby, left and right MI and MII, Fjäle 3, Klinteberg Beds, unit e, sample 77-7CB. □E. *L.* sp. A, left and right MI, Strands 1, Hamra Beds, unit b, sample 76-19CB. □F. *Hindenites gladiatus*, left and right MI and MII, Vattenfallsprofilen 1, Högklint Beds, unit d, *Valdaria testudo* bed, sample from the Swedish Museum of Natural History (RM). □G. *H. gladiatus*, left and right MI, Gandarve 1, Halla Beds, sample 71-81LJ. □H. *H. gladiatus*, drawn from the type specimen (right MI) and the left MI from the same sample (sample O.308, cf. Kielan-Jaworowska 1966), locality and stratigraphic position unknown, probably late Wenlockian. □I. *H. angustus*, left and right MI and MII, Vattenfallsprofilen 1, Högklint Beds, unit d, *Herrmannina* bed, sample RM. □J. *H. naerensis*, left and right MI and MII, Närshamn 2, Burgsvik Beds, lower part(?), sample 83-12LJ. □K. *Kettnerites* (*K.*) *versabilis* form A, left and right MI and MII, Buske 1, Lower Visby Beds, unit e, sample 79-40LJ. □L. *K.* (*K.*) *versabilis* form A, left and right MI, Häftingsklint 1, Upper Visby Beds, sample 76-10CB. □M. *K.* (*K.*) *versabilis* form C, left and right MI and MII, Buske 1, Lower Visby Beds, unit e, sample 79-40LJ. □N. *K.* (*K.*) *versabilis* form B, left and right MI, Häftingsklint 1, Upper Visby Beds, sample 76-10CB. □O. *Gotlandites slitensis* juvenile specimen, left and right MI and MII, Ajstudden 1, Slite Beds, unit *Pentamerus gothlandicus*, sample 83-7CB. □P. *G. slitensis* adult specimen, left and right MI, Vallstena 2, Slite Beds, unit *Pentamerus gothlandicus* or slightly older, sample 77-2CB; left and right MII, Vike 2, Slite Beds, unit *Pentamerus gothlandicus* or slightly older, sample 83-4CB. □Q. *Langeites glaber*, left MI, Stora Kruse 1, Hemse Beds, Hemse Marl, SE part, sample 82-29LJ; right MI, Vaktård 4, Hemse Beds, Hemse Marl, SE part, sample 81-35LJ. □R. *L. glaber*, left and right MI and MII, Holmhällar 1, Sundre Beds, middle upper part, sample MS907AM. □ S. *L. glaber* drawn from the type specimen (right MI) and left MI from the same sample of unknown age (sample O.466, Kielan-Jaworowska 1966).

- *Kettnerites* (*A.*) *sisyphi* var. valle, short range, typical populations found in marl deposited in deep water.
- *Kettnerites* (*A.*) *microdentatus*, long range, usually found in marl but not in the deepest deposited sediment.
- *Kettnerites* (*A.*) *siaelsoeensis*, stenotopic(?) very short
- *Lanceolatites gracilis*, long range, normally found in marl deposited at moderate depth.
- *Langeites glaber*, range of moderate length, a very rare species found only in marl.

Comparison with paulinitid faunas from other areas

Due to the scarcity of reports on the jawed polychaetes and the confused taxonomy, the regional distribution of species is not known in any great detail. The question whether the paulinitid species were widespread or showed provincialism has so far been impossible to discuss since, at least, most of the Baltic species have been treated as one species. However, Taugourdeau (e.g. 1976) has tried to correlate Silurian and Devonian form taxa between North America, France and the Sahara. In my opinion it is dangerous to attempt comparisons based on badly preserved or partly buried elements, or with only one or a few elements of each type and taxon, or a combination of these features. The identification of biological species based on well-preserved elements of the four main types is the key to polychaete taxonomy, and thereby to reliable regional comparisons.

Sweden. – Only a few reports have been published on fossil jawed polychaetes from Sweden, and no paper has dealt with paulinitids outside the Silurian strata of Gotland. The Silurian fauna from Bjärsjölagård, south Sweden (Scania), is similar to that of Gotland but the morphology of the jaws is slightly different (courtesy of Fredrik Jerre, 1987).

The Baltic area. – A large number of polychaete jaws and apparatuses have been described by Polish students. Most of the material derives from erratic boulders of Baltic origin. However, the exact stratigraphic and geographic provenance of these boulders is not known and will probably never be possible to trace. The three species *Hindenites gladiatus*, *Kettnerites* (*K.*) *polonensis*, and *Langeites glaber*, described by Kielan-Jaworowska (1966) from erratic boulders, have all been found on Gotland. Her *H. gladiatus* resembles the Early to Middle Wenlock representatives of the species on Gotland. *K.* (*K.*) *polonensis* is very difficult to date due to the minute size of the type specimens; an Early Ludlow age is probable. *Langeites glaber* is very rare, both on Gotland and in the erratic boulders dissolved in Warsaw; the erratic material is transitory between the Gotland Sundre and Hemse material. The age of the erratic *Langeites glaber* fauna could be Late Ludlow. Beside these taxa, others, which have been described in this monograph, are present in the samples (e.g. samples described by Kielan-Jaworowska 1966) in Warsaw. For example, in boulder 187/32 jaws conspecific with *Kettnerites* (*A.*) *sisyphi* var. valle and in boulder 151/4 jaws similar to *Kettnerites* (*K.*) *abraham isaac* are found. The small number of 'marl loving' taxa found in erratic boulders is notable.

It is not improbable that at least some of the boulders found in Poland derived from, or close to, the Gotland area. Thus, a close resemblance of the faunas can be expected and will not increase our knowledge about the faunal distribution to any great extent.

Poland. – Polychaete jaws and apparatuses have also been described from borings in Silurian rocks in Poland, e.g. a paulinitid from the Mielnik boring (Szaniawski 1970). The paulinitid specimen described as *Paulinites polonensis* Kielan-Jaworowska is probably conspecific with *Kettnerites* (*K.*) *huberti* from Gotland.

Great Britain. – The jawed polychaete fauna from the Ludlow strata of the Welsh Borderland show a close similarity to the Gotland fauna. In Allison Brook's unpublished collections from this area I have seen some species which are also present on Gotland, e.g. *Kettnerites* (*K.*) *huberti* and *Lanceolatites gracilis* (courtesy of A. Brooks). The low diversity of the assemblages reported by Aldridge *et al.* (1979, p. 437) from the Whitcliffian of the British Isles is in agreement with the fauna of Gotland. I do not know whether this is caused by the local environment or whether it is due to a global impoverishment of the polychaete fauna.

Northern Siberia. – Männil & Zaslavskaya (1985a) have described Silurian polychaete faunas from northern Siberia, including the island Severnaya Zemlya in the Arctic Sea. The Llandovery right MI (738/51) from the Waterfall section on the island of Severnaya Zemlya is very similar to the corresponding element of *Kettnerites* (*K.*) *abraham abraham* in the Llandovery strata of Gotland. The left MI (738/50), illustrated together with 738/51 is less characteristic, and thus less easy to identify, but could possibly be the counterpart of the right MI. The two Wenlock MI jaws (left MI 738/54 and right MI 738/55) from the Srednij sequence of the same island seem to belong to different species, but the inferior preservation and the angle at which the photograph was taken make them difficult to assess. However, I believe the right MI to be conspecific with *Kettnerites* (*A.*) *sisyphi*. The left MII (738/56) from the Wenlock Srednij sequence is very similar to the corresponding element of *Kettnerites* (*K.*) *polonensis* from Gotland. The right MII (738/57) is older and from an other locality. They probably do not belong to the same species.

The fauna described by Männil & Zaslavskaya is very similar to the fauna from Gotland, but identifications based on single specimens are normally not enough for reliable conclusions.

Bohemia. – The type material of *Kettnerites kosoviensis* Zebera 1935 consists of very large jaws, flattened and in part crushed, partly buried in the sediment. It will not be possible to dissolve the matrix to free the jaws without breaking them. The jaws show a slight resemblance to *Kettnerites* (*K.*) *martinssonii* (described by Martinsson 1960 as *Paulinites* sp.).

France and Sahara. – Taugourdeau (e.g., 1968, 1976) has reported paulinitid jaws from the Silurian, Devonian and

Carboniferous of France and the Sahara. He has also identified them as conspecific with North American scolecodonts.

North America. – The North American polychaete fauna, from the Ordovician to the Devonian, is rich and varied. Eller (1934a, 1934b, 1936, 1938, 1940, 1941, 1942, 1945, 1955, 1963a, b, 1964, and 1967), Stauffer (1939), Sylvester (1959) and Boyer (1975) have described a large number of elements including apparatuses belonging to paulinitids. However, only few of the species could be placed within the 'Baltic genera' and no identical species seems to have been described, although a jaw from the Silurian of Wisconsin shown to me by Jeffrey J. Kuglitsch is very similar to *Kettnerites* (*K.*) *martinssoni* (Fig. 40B) from the Hamra Beds. Walliser (1960) described paulinitid elements from the Canadian Arctic Archipelago. Some of these jaws show a very slight similarity to species from Gotland (e.g. *Idraites* sp. and *Polychaetaspis? kozlowskii* described by Walliser 1960).

Material and methods

Localities and samples. – The locality name and geographical situation, stratigraphical level, sample number, sample level and recorded scolecodont fauna for each sample is given in the Appendix. New localities are described in accordance with Laufeld 1974b and localities described earlier, with reference to the original description(s).

In the field, each sample was given a unique sample code in the form initiated by Laufeld & Jeppsson 1976, e.g. G75-7CB. G stands for Gotland, but is omitted here since all samples (with sample number) discussed in this monograph derive from Gotland; 75 stands for the year of collection; 7 for the seventh sample of that year from Gotland; and the last two letters stand for the initials of the collector, in this case Claes Bergman. The collectors of other samples used in this publication are: Lennart Jeppsson (LJ), Doris Fredholm (DF), Sven Laufeld (SL), Sven Stridsberg (SS). I have also had access to a number of large scolecodonts picked out by Anders Martinsson and Kent Larsson from their ostracode and tentaculitid marl samples. The sample code of the museum material is abbreviated as follows: the Swedish Museum of Natural History, Stockholm as SMNH, and the Geological Survey of Sweden as SGU.

All illustrated specimens not already belonging to any museum collection are deposited in the type collection of the Department of Historical Geology and Palaeontology, University of Lund, under the designation LO + a number.

Collecting. – Standard micropalaeontological and stratigraphical field methods have been used. In order to get a proper view of the geographical and stratigraphical distribution, collecting was not restricted to highly productive lithologies, although during the later years of field work, I concentrated to lithologies that might be expected to yield a good fauna. Rock surfaces exposed to prolonged weathering in recent time were avoided, since an oxidizing environment is harmful to the preservation of annelid jaws. The positions of the samples within each section was measured relative to suitable reference points or levels, such as a bentonite layer, the base of a reef, or an artificial marker of more permanent type. In order to facilitate future collection at any sampled level, several of the localities were photographed with a polaroid camera and the sample levels marked on the print while still in the field. (The prints have been deposited at the Department of Historical Geology and Palaeontology, University of Lund.) The sampling interval within a section depends on the lithology. Normally, at least two samples were collected from a locality, except for small exposures with uniformly developed sediment. Sample weight varies from 3 to more than 50 kg. To start with, I collected samples of 2–4 kg depending on how difficult it was to collect. In later years normally about 8 to 10 kg of rock was sampled to get a better view of the fauna. It is now evident that samples of at least 50 kg will be necessary when dealing with the jawed polychaete fauna in many limestone areas on Gotland.

As a standard, in a section dominated by marly layers, the collected samples were taken from levels with the highest calcium carbonate content. In the limestone beds, most of the samples were taken from fine-grained sediment.

Laboratory technique. – The methods used in the laboratory have been improved considerably over the years, chiefly by Lennart Jeppsson (Jeppsson *et al.* 1985; Jeppsson & Fredholm 1987; Jeppsson 1987; Barnes *et al.* 1987). Methods invented elsewhere, published or available through oral or written communications, have also been adopted. Most of these changes do not effect the composition of the recovered scolecodont fauna, except that the increased efficiency has permitted larger samples to be processed. The increased number of methods have also permitted a wider selection of localities to be sampled.

The laboratory method now used is in brief as follows: First a slab, a pilot test, of at least 0.5 kg but not less than about one tenth of the total sample weight is cleaned mechanically with a wire-brush. If lichens, etc., need to be removed, the sample is etched in hydrochloric or acetic acid for a short time. The remaining part of the sample is stored. The uncrushed slab is dissolved in buffered 7% acetic acid (Jeppsson *et al.* 1985) and the insoluble residue is washed through 1.0 mm and 63 μm screens. All fractions (except those below 63 μm) are stored for future reference. The residues of very marly samples which do not fully disintegrate or are difficult to rinse free of clay, are heated to around 50°C for about 24 hours or until completely dry. Then the residue is immediately soaked in petroleum-ether for some hours, the excess fluid is poured off, but saved, and hot water poured over the sample. The marl starts to disintegrate within seconds and normally the sample is completely disintegrated after a few minutes. After sifting and rinsing in gently running water the sample is dried and the fossils picked out. Samples yielding abundant conodonts or fish scales are normally density separated and/or separated in a Frantz Isodynamic Magnetic Separator model L-1.

Most sample residues derive from Lennart Jeppsson's collections. The majority have been separated by heavy liquid and/or magnetic separation. Scolecodonts have normally been extracted from the light fraction, though most heavy fractions have been picked as well, chiefly in connection with the extraction of conodonts, phosphatic brachiopods and fish scales. The yield from the heavy fraction is very low as a rule, but exceptions do occur. I have so far avoided concentrating the scolecodonts from my own samples with heavy liquid, since the first maxillaes (MI) of the investigated taxa are hollow and very elongated with a strongly enclosed myocoele opening that may trap sediment. This would bias the result of the density separation as well as the magnetic separation. Experiments with magnetic separation of scolecodonts have so far been without success. An alternative method of concentrating the scolecodonts is to dissolve the sample in hydrochloric or hydrofluoric acid. This method has the disadvantages of being poisonous and destructive to phosphatic and siliceous material. Therefore, it is feasible only if the sample has first been broken down by buffered acetic acid and/or petroleum-ether and the phosphatic material removed by, for instance, heavy liquid separation. This treatment of samples will probably be routine when very large samples are to be dealt with in the future.

The dried residue is spread out on a metal picking tray with a grid system on a white background (available from Fema-Salzgitter as 'Hand-picking scales made of brass, Punched'. Address: Fa. Rudolf Stratmann, 3327 Salzgitter, Friedrich Ebert Str. 53, West Germany). The annelid jaws and other fossils are picked out from the residue by means of a single hair which has been charged electrostatically, and dropped through one of the holes into a microfossil slide below the tray (Barnes *et al.* 1985). The microfossil slide has a light coloured rectangular bottom, with dark lines in a grid system on it. The bottom of the slide is lightly pre-coated with a water-soluble glue based on gum arabicum (Jeppsson 1974, p. 5). The sorting of the annelid jaws is conveniently done with a fine wet sable brush (nr. 00), and the jaws are at the same time glued in place.

In order to facilitate comparison the jaws are arranged with each species in a horizontal row and homologous elements in vertical columns. Thus, the MI's are placed to the far left of the slide, the MII's to the right of the MI's of the corresponding species etc. A large quantity of jaws will of course violate this organization of the slide since some areas will be overcrowded, especially those intended for MI's and MII's. The jaws are normally placed with the dorsal side up, i.e. the myocoele opening facing down. This organization reveals the largest number of specific characters and will thus facilitate comparison between closely related taxa, since homologous jaws are similarly oriented and placed close to each other in the same column.

I have also worked with wet sample methods as described by Kielan-Jaworowska (1966, p. 15), with the intention to find scolecodont apparatuses. After more or less fruitless experiments I concluded that apparatuses of the genus *Kettnerites* and related genera must be fairly rare. Most of the partial apparatuses from Gotland studied were found in dried sample residues, because the majority of the sample residues have been dried. Apparatuses should be more common in wet samples, however, since drying will destroy any pellicle surrounding the apparatuses (Kielan-Jaworowska 1966, p. 11) of most, if not all, of the taxa. The majority of apparatuses found were more or less compressed and not in a good state of preservation.

The jaws forming an apparatus partly cover each other, and are often more or less flattened and distorted, making identification even more difficult than if single, well preserved specimens are used. The higher abundance of apparatuses expected in wet fractions is likely to be ascribed to small juvenile forms. I have not made much further effort in collecting apparatuses since most of the information they provide is also obtainable from isolated jaws.

The jaws may be bleached and become translucent by the use of an oxidation medium, e.g. sodium hypochlorite (Tasch & Shaffer 1961, p. 369) or hydrochloric acid and potassium chlorate (Kielan-Jaworowska 1966, p. 15). The bleaching will facilitate the study of inner structures while at the same time rendering the jaws fragile. I have bleached jaws of different sizes but were unable to bleach large, thick jaws to translucency – they become yellow but remain opaque.

Illustration of specimens

Scanning electron microscopy and photography. – A large material of well preserved elements, representing different populations from various stratigraphical levels was selected for scanning electron microscope (SEM) studies. The elements were mounted on specimen stubs with double-adhesive tape. In order to facilitate the orientation of an element on the stub, a small drop of water was added to the surface of the tape and the specimen was placed in the water. Thus, the element could easily be moved, but once outside the drop or when the water evaporated it was stuck to the tape. The specimens were oriented with the anterior part towards the centre of the stub, if possible in the same 'resting position' as on the slide. Thus, the 'light' will appear from the same angle on all scanning micrographs, and the micrographs are taken from the same angle at which the elements are normally studied, when using a normal stereoscopic microscope. I have not used the same specimen for the ventral and the dorsal view, since the risk of fracturing the specimens is very great when loosening them from the tape. The specimens were coated with an alloy of gold and palladium in the proportions 60:40. Initially, the hollow jaws tended to implode when air was let into the vacuum-chamber of the coating device. The very imprecise valve construction made it impossible to let in the air more gradually with slow compression. This obstacle was overcome by some ingenious work by Sven Stridsberg. The vacuum pump was allowed to work and the low pressure could then be decreased slowly with the help of the argon leak needle valve. After filling the chamber slowly with argon gas for a few minutes, (easily controlled on the torrmeter) the pump was stopped and after a light touch on the air valve the lid could easily be opened.

The SEM work was carried out using a Leitz AMR 1600T. Most specimens were photographed without tilt. A panchromatic roll film, Agfapan 25 professional, was used. This

24×36 mm fine-grain film will reproduce any fine detail on the 105×105 mm SEM screen, since only about two thirds of the screen height but all the width is used. This will not affect the total capacity since most objects investigated here are elongated. Further, no information is being lost using this 35 mm film compared to 120 films, since the film resolution gives about 50% overcapacity.

Agfa Rapidoprint paper was used for the prints. One standard of magnification was selected, to facilitate comparison. However, for practical reasons some other magnifications were also used. The background of the photographic figures was blocked with masks cut out from a semitransparent plastic film (Ulano Amber Clear; Amberlith; H6W 2655; cf. Bengtson 1986). Due to the small details (e.g. denticles) the innermost part were in some cases blackened with black ink.

Most of the drawings were made using a Nikon SMZ-10 stereoscopic microscope equipped with a drawing device.

Terminology

For ease of understanding, I have kept the number of technical terms as low as possible. The terminology is to a large extent based on Kielan-Jaworowska (1966) and on the glossary compiled and defined by Jansonius & Craigs (1971), Taugordeau (1978) and Wolf (1980). The jaws in dorsal position, named maxillary pieces (M), are normally numbered from posterior to anterior with Roman numerals (MI–MVI). The following glossary of descriptive terms is intended for paulinitid jaws (Fig. 15).

Basal angle. – The angle between the long axis of the undenticulated ridge and the outer margin of the basal portion of MI.

Basal furrow. – A fairly short furrow in the posterior part (basal portion) on the dorsal, outer side of MI, often parallel to the undenticulated ridge.

Basal plate. – A small to medium-sized right-hand jaw, closely fitted into a posterior bight or concavity of MIr (=right MI); in some species the basal plate is paired, the left-hand element being called the laeobasal plate.

Basal portion. – The posterior part of MI, which is widened compared with the denticulated anterior part of the jaw. The basal portion includes the inner wing, the undenticulated ridge, the basal furrow, the flange and sometimes a spur on the posterior outer part.

Bight. – A concavity in the margin of the outer and/or posterior face of a jaw, open to the posterior, especially in the right MI, where it provides a space into which the basal plate can fit; also used for ramal arch.

Bight angle. – The angle between a line tangential to the outer margin of the shank and: (1) a line tangential to the inner margin of the ramus (as in MII); (2) a line tangential to the posterior margin of the flange (as in MI).

Carriers. – Paired elements in the apparatus, situated behind the forceps and serving as a support for the posterior ends of the forceps or MI's.

Cusp. – The largest denticle, often the anterior one, in a series of denticles, esp. on the basal plate (if denticulated), MII, MIII, MIV. On the MII's, the cusp may be preceded by one, two or very rarely three pre-cuspidal denticles. On some MII's there are two anterior denticles of equal size forming a double cusp.

Cutting edge. – A small ridge, along the base to the apex on opposite sides of the denticle, oriented along the extension of the jaw. (Jeppsson 1979, in conodont terminology.)

Dentary. – Series of denticles along the inner (dorsal) margin; according to the density of denticles the dentary can be dense-spaced, normal-spaced, wide-spaced and paucidentate; in some forms part or all of the dentary is edentulate (=dentary without denticles), in which case the dentary is equal to the inner (dorsal) margin. Sometimes it is possible to subdivide the dentary into two parts, intermediate and posterior (=post-cuspidal denticles).

Denticulated ridge. – A ridge with denticles, here used in a restricted sense, on MI along the inner margin. The posterior part of the denticulated ridge may be without denticles. The term edentulate (Jansonius & Craig 1971) refers to such an undenticulated part of the ridge.

Depth. – The term is defined by Stauffer 1933. I have used the term *width* instead.

Double cusp. – A pair of cusps on MII, often with parts of equal size.

Falx. – The extension of the outer anterior part of the jaw, esp. in MI, often sickle-shaped, forming a hook or a fang.

Falcal arch. – The concave part of the anterior, inner margin, which may be denticulated or undenticulated (Jansonius & Craig 1971, emended herein.)

Fang. – The pointed, recurved termination of MI, less developed than a hook.

Flange. – A wing-like extension on the dorsal side of the posterior, outer part of the right MI. It is often more or less elevated, and if present the basal plate is fused to it. The term was used by Sylvester (1959) to include only the ligament rim.

Head. – The anterior, wider part of the carrier.

Hook. – The very large, recurved anterior termination of MI is denominated *hook*. The less developed anterior termination is named *fang*.

Inner margin. – A jaw seen in dorsal view has its inner margin along the side which is oriented facing the central axis of a paulinitid jaw apparatus, along the dentition and inner wing of the MI and MII elements.

Inner wing. – The posterior inner margin may be developed as a longitudinally elongated and laterally extended or downfolded and extended area.

Intermediate dentary. – A group of small, often equally sized denticles, usually two to eight in number, on the paulinitids. They are immediately posterior to the cusp of the left MII, and are less well-developed on the right MII.

Length of the element. – The largest dimension of a jaw parallel to its long axis (i.e. approximately parallel to the dentary) between the most posterior and the most anterior points.

Ligament rim. – A narrow structure (often a small ridge, groove or a combination of these) surrounding the myocoele opening along the anterior, the inner and the outer sides. Sylvester (1959) used the term *flange* for this structure.

Myocoele. – The space inside (pulpal or muscle cavity), and more or less enclosed by the jaw, extending to the tip of the fang, hook or cusp. (Partly from Kielan-Jaworowska 1961).

Myocoele opening. – The outline of the ventral margins of the jaw enclosing the myocoele.

Outer wing. – The posterior outer margin of the MI may be developed as a longitudinally elongated and laterally slightly extended area which is part of the margin of the myocoele opening.

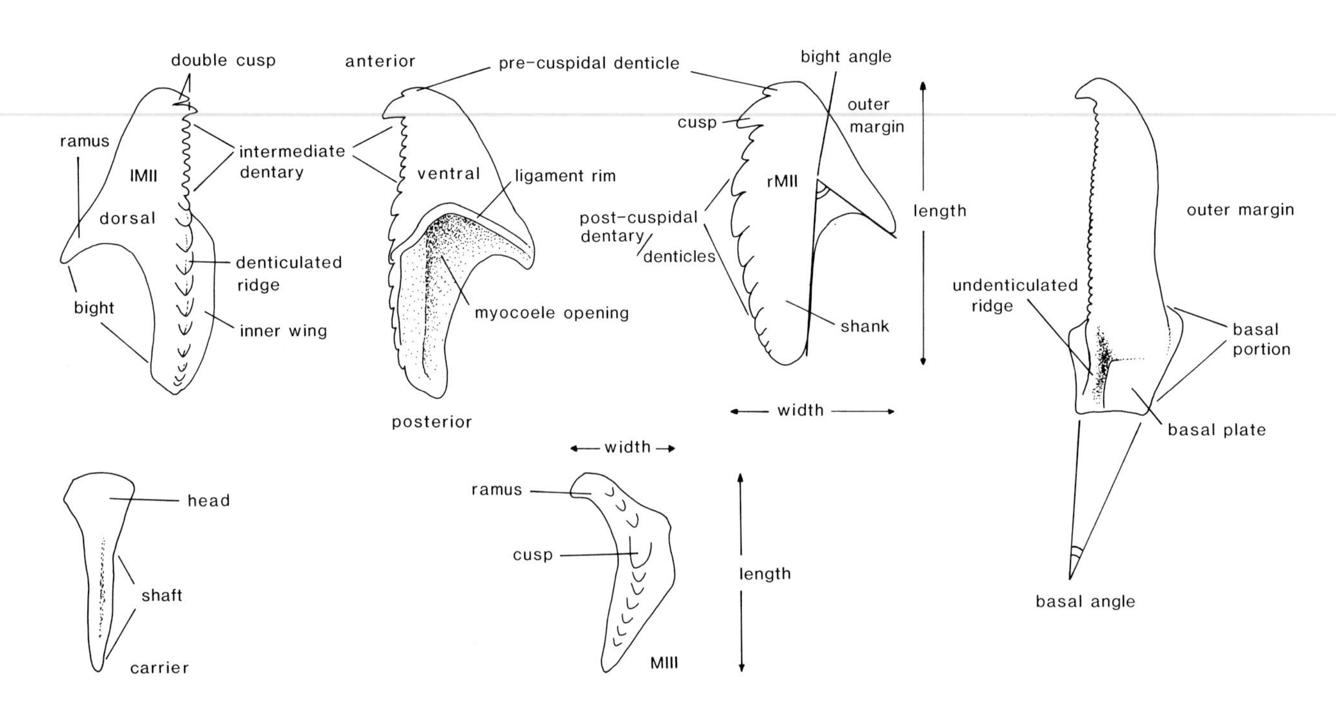

Fig. 15. Schematic drawings of paulinitid jaw elements to illustrate the morphological terms used. All are elements seen from the dorsal side, except the right figure of the left MII element (ventral view to show myocoele opening).

Paucidentate dentary. – With few and widely spaced denticles (see *dentary*).

Post-cuspidal denticles (*post-cuspidal dentary*). – The denticles on the posterior side of the cusp and posterior to the intermediate dentary. The term *posterior dentary* is also sometimes used (see *dentary*).

Pre-cuspidal denticle (*pre-cusp*). – Any denticle anterior to the cusp or a small denticle on the anterior side of the cusp (see *dentary*).

Ramus. – Any arm-like lateral extension of the face of a jaw projecting from the over-all outline, usually pointing posteriorly, esp. in MII, MIII, and MId (=right MI); in some forms pointing anteriorly, especially the basal plate. (Jansonius & Craig 1971.)

Scolecodont. – Any jaw piece of a polychaetous annelid; originally intended only for isolated fossil elements. (Croneis & Scott 1933.)

Shaft. – The posterior extension of a carrier or a mandible.

Shank. – The undenticulated posterior extension on the inner side of a jaw. The most posterior part of the shank can have a blunt or a sharp termination.

Slanting denticles. – Denticles pointing in the posterior direction.

Spur. – A short, ridge-like process which is a prolongation of the thickened margin of the myocoele opening on MI. The spur can often be seen in dorsal view on the outer margin on the anterior part of the basal portion.

Undenticulated ridge. – The ridge on the basal portion of MI, extending as a continuation of the denticulated inner margin (also referred to as edentulate by other authors).

Ventral side. – The side of the jaw to which the myocoele is open.

Width. – The largest dimension of a jaw perpendicular to its long axis and on a plane parallel to the myocoele opening. The width of the MI is measured anterior to the basal portion.

Structure, composition, and preservation of polychaete jaws

Chemical composition. – It has often been claimed that the jaws of polychaetes are composed of chitinous material (e.g. Schwab 1966, p. 416). This is, however, not the case according to Voss-Foucart *et al.* (1973), who demonstrated that Recent polychaete jaws are composed of scleroprotein. Olive (1980, pp. 572, 580) has published a thorough review of the much-varied composition of the jaw material in different taxa. The jaw material includes tanned protein, aromatic amino acids, glycerine, hystidine and metals such as iron, zinc and copper. Recent eunicids have a high concentration of calcium carbonate, while the phyllodocids have less. Colbath (1986) suggests that the heavily sclerotized jaws with little mineral reinforcement have a better prognosis for fossilization. Some of the fossil eunicid jaws might have had the same reinforcement of calcium carbonate as the Recent forms have (Colbath & Larson 1980, p. 485), though Boyer (1981) suggested that the calcium carbonate might have been replaced in the jaw *post mortem*. Some fossil eunicid jaws also include fluorapatite, but whether this is a reinforcement or of secondary origin is an open question according to Schwab (1966, p. 421). The Recent jaws studied lack fluorapatite (Schwab 1966, p. 421), although they are of different phylogenetic lineages than the fossil jaws investigated, and therefore give little information about the origin of the fluorapatite.

Preservation. – It has also been claimed that scolecodonts are very resistant to acids, oxidation, thermal alteration and to changes due to the recrystallization of the carbonate matrix (Jansonius & Craig 1971, p. 251). However, Kozur (1972, p. 753) states that the resistance towards oxidation and matrix recrystallization is very poor. From my material of the Lower Palaeozoic of Sweden it seems quite clear that scolecodonts are not common in any highly oxidized (red) sediment. Szaniawski (1974) and Colbath (1986) demonstrate that there is a considerable variation in the resistance to chemical oxidation among the jaws of different polychaete families.

The jaws have almost the same colour variation as amber: from brownish black through reddish brown to yellowish brown. The colour varies between some taxa but is consistent within the species apart from the smallest, often translucent, jaws. There is no sign of the jaws having been affected by high temperature. Dorning (1984) noted the possibility of using different types of fossils, including polychaete jaws, to enhance the precision of thermal metamorphosis indexes such as those based on conodonts (Epstein *et al.* 1977). Laufeld (1974a, p. 128) concluded that the Gotland chitinozoans had never been subjected to temperatures exceeding 100°C, and Jeppsson (1983, p. 122) found no sign of thermal alteration in the conodonts. This implies a maximum temperature of less than 80°C.

The preservation of scolecodonts from the Silurian of Gotland varies from oxidized, flattened, more or less unidentifiable fragments to beautiful three-dimensionally preserved elements without any sign of abrasion or compaction. Large elements, more than 1–2 mm in length, seem to be more easily oxidized and brittle, and thus more susceptible to fragmentation than smaller ones. In most soft marl samples the jaws are more or less flattened, whereas in the more competent calcium carbonate-rich layers, the jaws are normally preserved in full relief. Many limestones consist of well washed, more or less worn skeletal debris, e.g. the lower part of the Slite Beds and most of the Klinteberg Beds. The scolecodonts are normally very rare and fragmented in these limestones, although in some samples very well preserved specimens are encountered. In the oolitic beds only some undeterminable fragments of scolecodonts have been found.

The best preserved elements are found in fine-grained sediments with a high carbonate content. I have not noted nor searched for any effect of sediment recrystallization on the scolecodonts (cf. Kozur 1972, p. 753).

In several localities, soft marlstone is intercalated by competent argillaceous limestone. In the Lower and Upper Visby Beds, for instance, the limestone is often in the form of irregular nodules or thin lenses. This 'cyclic bedding' could be a result of a rhythmic change of the physical conditions and/or of mass transportation of either the clay- or the carbonate-rich component, but it could also be of diagenetic origin. The present work is not intended to interpret the sedimentology behind the 'cyclic bedding', but in samples from the Upper Visby Beds and Högklint Beds (e.g., Häftingsklint 1 and Vattenfallsprofilen 1) no difference in the composition of the jawed annelid fauna is noted between marlstone and limestone within a cycle. Transportation of any of the sediment components is not supported by the distribution of the scolecodonts, but bioturbation affecting only the annelid fauna could have occurred. In a similar type of bedding in the Hamra and Burgsvik Beds with dominating limestone sediments, a change of the jawed annelid fauna is noted, e.g. in Bankvät 1 (Hamra Beds), but not in Glasskär 1 (Burgsvik Beds).

Because no significant difference in the annelid diversity was found between limestones and adjacent marl beds, samples collected in the most recent years mainly derive from the competent limestone layers. Still, less than one sample in ten yields an abundant and well preserved fauna. Jeppsson (1983, p. 122) notes the same condition in conodonts, adding that the best conodont samples often yield excellently preserved scolecodonts without any sign of deformation.

Ultrastructure. – Only a few reports on the ultrastructure of the external and internal surfaces of annelid jaws have been published, such as Taugourdeau's (1972) study of the ornamented cuticle of Palaeozoic polychaete jaws. This limited interest is probably due to the usually smooth surface of the jaws. Light-microscopy studies by Tasch & Shaffer (1961) and Schwab (1966) on bleached jaws, were superficial in nature. The increasing use of transmission electron microscopy (TEM) and particularly scanning electron microscopy (SEM) has advanced the study of ultrastructures: almost simultaneously, several papers describing the ultrastructure of annelid jaws appeared, e.g. Strauch (1973), Corradini, Russo & Serpagli (1974), and Mierzejewska & Mierzejewski (1974). Strauch (1973) noted

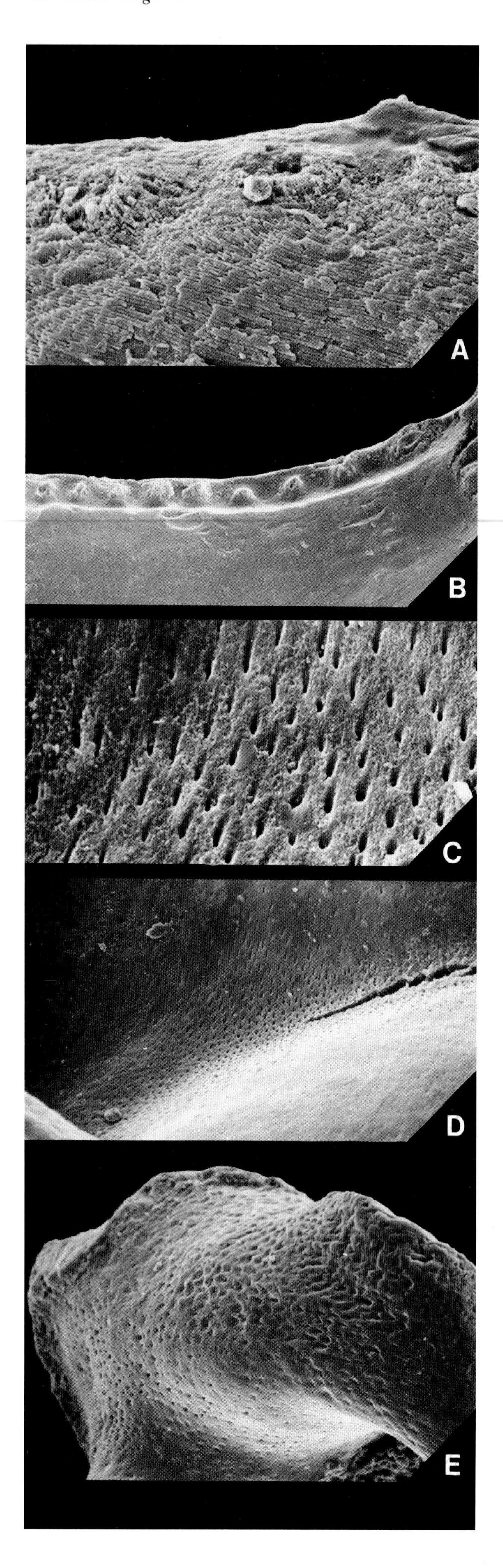

a layer with a smooth non-porous surface covering three different internal layers with pore-canals in different directions. He suggested that pore-canals could, among other functions, have increased the elasticity. The result was in general confirmed by Corradini *et al.* (1974), who worked with a larger material and made a more penetrating study. They distinguished a fourth layer and also studied the different pore-canals, attempting to interpret their function. Further, they noted exceptions to the smooth dorsal surface in some taxa, e.g. *Polychaetaspis tuberculatus*, which displays striation and very small pores. Shallow furrows were noted on the inner face of *Mochtiella*. Several different types of structures, including pores and what looks like small corroded pits, were recorded on the ventral side. Charletta & Boyer (1974) noted striation on denticles of *Arabellites auburnensis*. The taxonomic importance of the ultrastructure, hinted at by Corradini *et al.*, was demonstrated by Szaniawski & Gazdzicki (1978, pp. 7–10).

I have not noted any ultrastructural pattern on the smooth dorsal surface of unaltered paulinitid jaws from Gotland. Eroded denticles (Fig. 16B) show layers (Fig. 16A) similar to the ones described by, e.g., Corradini *et al.* (1974). However, on the ventral surface around the myocoele opening, pores similar to the ones illustrated by Corradini *et al.* (1974, p. 126, pl. 42) have been found (Fig. 16C, D, E).

Dental formula. – Several types of dental formula exist. The simplest one gives the total range of the number of denticles of the complete jaw. This number is usually not of major importance. A more significant type of dental formula was proposed by Taugourdeau (1968) and later modified by Jansonius & Craig (1971, pp. 261–264) and Taugourdeau (1978, pp. 14–17). Germeraad (1980, p. 5) noted that with simple or very complexly denticulated jaws it is easier to understand the dentation from an illustration than from a complex codification.

In my opinion, existing formulas are generally difficult to read and compare, and are typographically complicated. It is at least as simple to draw a profile of the dentition as to try to work out the formula for a complex denticulated jaw. Dental formulas are potentially useful for computerized comparisons, but with the present Silurian material the formula would have to express the number, direction, size and shape of the cusp, precusp(s), intermediate denticles and posterior denticles, not only small–large, but also in a more differentiated form. For the present purpose, formu-

Fig. 16. □A. Three deeply eroded denticles from the middle part of the denticulated inner margin of the left MI, *Kettnerites* (*K.*) *bankvaetensis*, LO 5841:1, Glasskär 3, Eke Beds, 82-18CB, ×1700. □B. Deeply eroded anterior part of the denticulated inner margin, right MI, *K.* (*K.*) *bankvaetensis*, LO 5841:2, same sample as A, ×325. □C. Close-up of the pores of the ventral surface within the posterior part of the pulp cavity of the left MI of *K.* (*K.*) *bankvaetensis*, LO 5829:8, Bankvät 1, Hamra Beds, unit b, 82-30CB, ×2000. □D. ventral view of part of the basal portion shown in C, same specimen as in C, ×510. □E. ventral view of the outer half of the basal portion of right MI, *K.* (*K.*) *bankvaetensis*, LO 5829:7, same locality as C, ×500.

las would be an unnecessary complication, and thus I have not used any type of dental formula besides stating the range of the number of denticles of the jaw.

Molting of jaws and ontogeny

Jaws may grow continuously (only possible with simpler jaw forms, such as a conical type), or by shedding. It is also possible that worn jaws were shed, although this is less probable. Continuous growth is demonstrated by growth lines in some jaws, e.g. mandibles (Paxton 1980). Growth lines have not been reported from eunicid maxillae, but are common in those of the Phyllodocida, e.g., Nephtyidae, Glyceridae and Nereidae (Herpin 1926, Schwab 1966 and Charletta & Boyer 1974).

Molting of jaws in Recent annelids has been described or suggested in a number of studies (summary with references in Kielan-Jaworowska 1966, pp. 49–51), for example that of Herpin (1926) on the recent phyllodocid *Odontosyllis ctenostoma*. Molting is also described from the intensively studied Recent eunicid genus *Ophryotrocha* (e.g., Bonnier 1893; Korschelet 1893; Heider 1922; Åkesson 1973; Jumar 1974). A closer study of molting and the morphological development of the jaws of *Ophryotrocha maculata* is given by Åkesson (1973), who studied a population of about 300 individuals for more than two years. Åkesson concludes that the jaws are replaced at intervals, and that an ontogenetic dimorphism occurs. Similarly, the jaw morphology of *O. puerilis* changes during ontogeny (Pfannenstiel 1977). Paxton (1980, p. 545) claims that among most Recent eunicids the jaws grow continuously during life and produce jaws without growth-lines. However, also eunicimorph jaws that are shed lack growth-lines. Recently Colbath (1987a) concluded that the internal structure of maxillae in eunicoid polychaetes is inconsistent with continuous growth, and he therefore rejects Paxton's theory.

Evidence for molting in fossil annelids is difficult to verify, however. Kielan-Jaworowska (1966, pp. 50–51) noted that there was no evidence of molting in fossil labidognath and prionognath types, but suggested that in placognaths, assemblages of morphologically identical jaws arranged as a jaw-within-a-jaw structure could indicate molting. Schwab (1966, p. 420) also reports nested placognath jaws belonging to the genus *Staurocephalites*, the second jaw being arranged within the first jaw, with the same alignment in every case. Schwab compared this with tooth replacement in mammals, but did not entirely exclude other explanations, such as growth abnormalities. Such phenomena have also been noted among the same type of jaws from the Silurian of Gotland. Mierzejewski (1978a) discussed different types of molting and evidence in support of it. He further stated that any increase in the size of eunicid jaws during ontogeny must occur by molting since a biochemical change occurs in the jaws which excludes the possibility of growth by accretion. Based on internal studies of Recent material, Colbath (1987a, p. 446) proposes that palaeontologists should tentatively assume that moulting occurred among the jawed fossil polychaetes.

The ontogenetic development shown in *Kettnerites* (*K.*) *martinssonii* and *Gotlandites slitensis* in this monograph, is also evidence of jaw replacement, if the possibility of resorption and accretion of the jaw material is excluded (see Mierzejewski 1978a).

Thus, in accordance with Colbath (1987a) I believe it is reasonable to assume that the jaws of many Lower Palaeozoic eunicids were shed and replaced during growth.

Size frequency classes. – The shedding of jaws could theoretically be proved by the identification of different, delimited size frequency classes, similar to the different instars, for instance, among the ostracodes (e.g. Martinsson 1962). Olive (1980) has stated the difficulty of obtaining correctly delimited size classes of complete worms in Recent material. According to Olive (personal communication, 1986) only the first and the second size classes of the polychaete group measured seem to be clearly delimited from the rest. The larger classes tend to merge, depending on individual differences in growth, caused by differentiated food supply and genetic differences.

The study of growth-lines of some Recent, potentially long-lived species within the family Nephtyidae permitted a size-independent estimate of age (Olive 1977). No growth-lines have so far been recorded from the eunicids, however.

A number of factors may affect a polymodal size-frequency distribution of the jaws. Three different reproductive strategies of Recent polychaetes have been recognized (e.g. Olive & Clark 1978). Reproduction may occur once during a lifetime, annually, or several times per year during an extended breeding season. How many times did the normal annelid replace its jaws? When did they replace them – cyclically or whenever needed? The shedding could be coupled to any regular phenomenon, such as lunar cyclicity or seasonal variations, for example in the abundance of prey. Is it then possible to identify the different stages, or do the stages partly overlap each other?

If shedding took place at the same periods in life, it would probably be possible to distinguish at least the different juvenile stages. Variable growth rates between different populations would obscure the pattern, however, and so would mixing of material of different stratigraphic age due to bioturbation or other kinds of reworking, and limits to sampling resolution. Transportation is also a possibility to be borne in mind, though Schäfer (1972, p. 176) noted that annelid jaws from the North Sea do not seem to be subject to any sorting by currents. However, accumulation of jaws may be caused by organisms preying on polychaetes. Transportation caused by density currents is likely to have occurred in the Silurian shallow shelf in the Gotland region (e.g. Bergman 1981a, 1984). Brenchley (1979) concludes that among species of Recent Eunicidae and Nereidae the fossilization potential of the jaws is highly variable depending on species and place of deposition.

Other obstacles hindering the clear identification of size classes could be a changing of the rhythm of shedding during the ontogeny, or a very large number of molt stages. Less probable is perhaps size-expressed polymorphism (e.g. male–female). In the present material no dimorphism

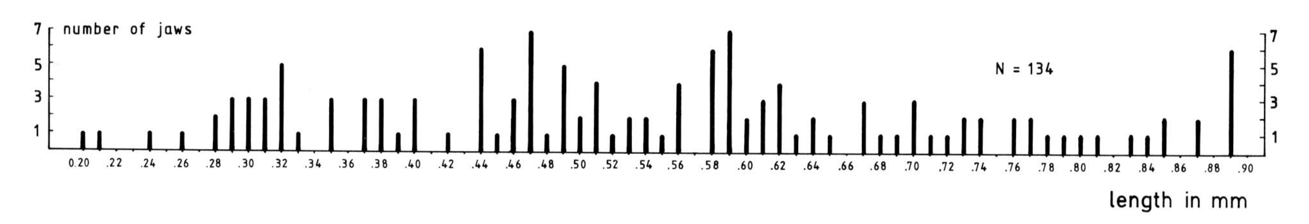

Fig. 17. Histogram showing the size-frequency distribution of 134 right MI elements of *Kettnerites* (*K.*) *bankvaetensis* from the locality Bankvät 1, Hamra Beds, unit b. The sample comes from a 30–40 mm thick bed representing a geologically short time interval. No obvious size-frequency classes can be observed. Studies from other beds and localities have given the same result.

implying sex-based differences in the apparatuses has been found. I have attempted to find polymodal size-frequency distributions at some localities (Fig. 17) yielding well preserved paulinitid faunas, but the result was inconclusive. Olive (personal communication, 1986) states that with Recent material it is virtually impossible to identify size classes beyond the second class.

General taxonomy

Morphological variation in a species exists in both time and space. Reconstructions of species distribution should be based on populations from different geographical areas and stratigraphical levels; scattered observations of one or a few specimens are of limited use. The study of populations can reveal variation within groups that exchanged genetic information and thus formed a species during a specific time. The imprecise dating of geological samples will give a blurred time resolution, which can increase the actual variation but may help us to follow the species through time. 'An evolutionary species is a lineage based on a sequence of populations, evolving separately from others with its own unitary evolutionary role and tendencies' (Simpson 1961, p. 153).

The taxa recognized in this study are by necessity based on jaws. The jaws are only a minor part of the animal, and many changes in the genetic code could take place without affecting the jaw apparatus. Further, a modification of the genetic code could affect the morphology of more than one part of the animal. But even so, some changes of connected characters would not necessarily mean that a new species has evolved. Species boundaries in time are largely subjective.

Variation within populations often leads to genetic polymorphism (Ford 1965), which in wide-spread species could increase and lead to polytypic species (Cain 1963). Studying Devonian conodonts of great variability, Murphy *et al.* (1981) replaced the previously used morphologic classification with a biologic one in which each taxon included a number of morphotypes. In my scolecodont material from Gotland, the lineages are not as complex as in the above-mentioned conodont example, perhaps as a result of the very limited geographical area sampled. Instead of introducing morphs, which might be premature concerning jawed polychaetes, I have chosen to use a concept of informal varieties when dealing with the more problematic polychaete populations. These varieties are to be regarded as infrasubspecific categories with no standing under the ICZN (Article 45e, g). Most varieties are described separately in order to facilitate transformation to a formal species-group taxon, if found necessary.

The criteria for delimiting the species could be specified for each element, i.e. for the four large posterior jaws which I am at present using. However, not every small change of the morphology of a jaw is used to base a new taxon on. Thus, the number of species are held at a lower level, which makes it easier to follow the stratigraphical, geographical, and ecological distribution of the species.

Criteria for delimiting genera are of course much more complicated since the genera involved are closely related and therefore have the same element arrangement.

The species identified can mostly be found in several levels through the stratigraphical sequence of Gotland. If in an evolutionary series there is a distinct change in the morphology of the jaws (e.g. *K.* (*A.*) *sisyphi sisyphi* and *K.* (*A.*) *sisyphi klasaardensis*) and the two forms do not exist at the same localities, then I treat the new form as a subspecies. If the two forms co-occur in the same sample I place the latter form as a variety of the first (e.g. *K.* (*K.*) *polonensis* and *K.* (*K.*) *polonensis* var. sjaustre).

Variability of jaws and apparatuses

The laboratory technique was designed to produce large numbers of jaw elements for population analysis, rather than to bring out the rare complete jaw apparatuses. Consequently the apparatuses described are reconstructed mainly on the basis of isolated jaws from a large number of localities. However, some more or less complete apparatuses were used in the reconstruction work.

The taxonomical value of Recent polychaete jaws has been under debate (e.g. Claparède 1870, p. 24; Treadwell 1921, p. 7; Wolf 1976, p. 55; Wolf 1980, pp. 97–99), probably due to morphological variation within some of the Recent polychaete species (e.g. Kielan-Jaworowska 1966; Fauchald 1970; Jumars 1974). The intraspecific variation is expressed in the position of elements in the apparatuses as well as in their symmetry (e.g. the symmetrical pair of the MIII; or the appearance of a basal plate in the left MI, normally developed only in the right MI). The number of elements may also vary within a species. Kielan-Jaworowska (1966, pp. 30–38) has reviewed Recent eunicid taxa exhibiting variation in number of denticles, often not correlated to jaw size. This common variability is found in *Arabella iricolor*, *Diopatra cuprea*, *Eunice floridana*, *Halla parthenopeia*, and *Onuphis eremita*. The number of dental plates and element shape are also subject to variation. This seems less common among Recent species but is reported from *Aglaurides fulgida* by Fauvel (1919, 1953) and Hartman (1944), *Arabella iricolor* by Kielan-Jaworowska (1966), *Arabella mutans* by Crossland (1924), and *Diopatra cuprea* by Kielan-Jaworowska (1966, p. 32). Barnes & Head (1977) note the variation in the number of paragnaths in *Nereis diversicolor*.

Change in jaw shape during ontogeny is reported from the genus *Ophryotrocha* (e.g., Bonnier 1893 on *Ophryotrocha puerilis* and Åkesson 1973 on *O. maculata*). Kozlowski (1956, pp. 183–185), in his pioneer study of fossil jawed annelids, demonstrated the ontogenetic development of the Palaeozoic *Polychaetaspis wyszgordensis* Kozlowski. Such dimorphism as shown by Åkesson (1973) must be very difficult to record on fossil material. However, Szaniawski & Wrona (manuscript in press, H. Szaniawski, personal communication) have been able to distinguish such an example on fossil material. Other examples of minor ontogenetic development are also shown in the present study, e.g. in *Kettnerites* (*K.*) *martinssonii* and *Gotlandites slitensis*. The change of shape occurs late in the individual growth; it probably represents sexual maturity.

A second type of denticle variation occurs in some of the Gotland taxa (e.g. *Kettnerites versabilis*, *K.* (*K.*) *martinssonii*, *K.* (*K.*) *burgensis*, and *Lanceolatites gracilis*). Within jaws of equal size, two different denticulation types occur, one with thin, densely spaced denticles and one with coarse denticles. The number of denticles varies within both types but the coarse type has fewer denticles. A similar type of variation is known from other animal groups, for example conodonts (Serpagli 1967, p. 54 described thinly and coarsely denticulated elements within the same genus). Kielan-Jaworowska (1966, p. 38) noted that the variability in jaw symmetry may cause taxonomic oversplitting.

Paulinitid MII jaws from Gotland have a very consistent dentition formula (Figs. 12, 13 and 14), though some variation exists. The right MII of *Kettnerites* (*K.*) *bankvaetensis* has normally one small and one large pre-cusp, but in about 2–3% of the jaws there are two (in one case even three) small denticles and one large. A geographically and stratigraphically dense sampling of well-preserved material makes it possible to identify several such cases of intraspecific variation.

Morphological features of taxonomical importance

MI and MII elements carry several taxonomically significant characters, although less so in the left elements. The left MI of two taxa may overlap morphologically, making identification based solely on this element almost impossible. This is not common in Gotland material, but it occurs, for instance, in some samples from the Halla–Klinteberg boundary at Gothemshammar, where *Kettnerites* (*K.*) *bankvaetensis* and *K.* (*K.*) *polonensis* occur together. In younger strata the morphology changes, and it becomes fairly easy to differentiate the two taxa solely on the left MI.

Jaw size in taxa and populations is useful for identification, though being also environmentally dependent it must be used with care.

Features that are usually specific at the species–subspecies level include:

Right MI. – Denticle number, size and form, extension of the dentary along the inner margin, transition from the posteriormost denticles to the undenticulated ridge, variation in denticle number versus jaw size, shape of the inner margin, form of the inner wing, sharp- or blunt-ended shank, fused or isolated basal plate, form, thickness and elevation of the flange, presence or absence of spur.

Left MI. – As for the right MI, except that the posterior part of the jaw (including the lack of basal plate) displays fewer traits of taxonomic importance.

Right and left MII. – The right MII's exhibit the most important characters (Fig. 18), including number, size, shape, position, and direction of precusp(s) (0–3) and cusp(s) (1 or 2). The cusp may be slender or swollen at the base. The most common number of intermediate denticles is 3–5, but varies. The delimitation of the intermediate dentary from the posterior dentary may sometimes be difficult. Size and form of the ramus and the shank, including the shape of the most posterior part of the shank are also important features.

Other elements which I have been able to identify in the Silurian eunicid apparatuses include an unpaired left MIII, left and right MIV, and carriers. However, this is only rarely possible, because of the few morphological characters on the jaws.

The mandibles are black, massive and often fragmented. Partly due to their usually bad preservation and thereby low number in the samples in relation to the maxillae, I have not assigned the mandibles to taxa.

Since most of the polychaete jaws are hollow they are often found more or less squeezed, thus creating identification problems. Some of the above mentioned criteria will still be possible to use, e.g. the dentition form of the anterior denticles of the MII and the dentition of the MI. However, I believe it is almost impossible to base an identification on flattened material without being familiar with a well preserved similar fauna (preferably preserved in full relief).

In the MII, the anterior part of the dentary, and in the MI, the dentary and the basal portion (especially the shape of the posteriormost tip of the shank) seem to be very persistent through time. These features might prove to be very important in phylogenetic studies.

Scolecodonts versus apparatuses

The natural taxonomy of fossil annelids is so far mostly based on jaw apparatuses. The scolecodont-based species described earlier are rarely illustrated in a way that permits comparison and identification from the photo or drawing. This has led several students to ignore these taxa without studying the type specimens. A large number of taxonomic

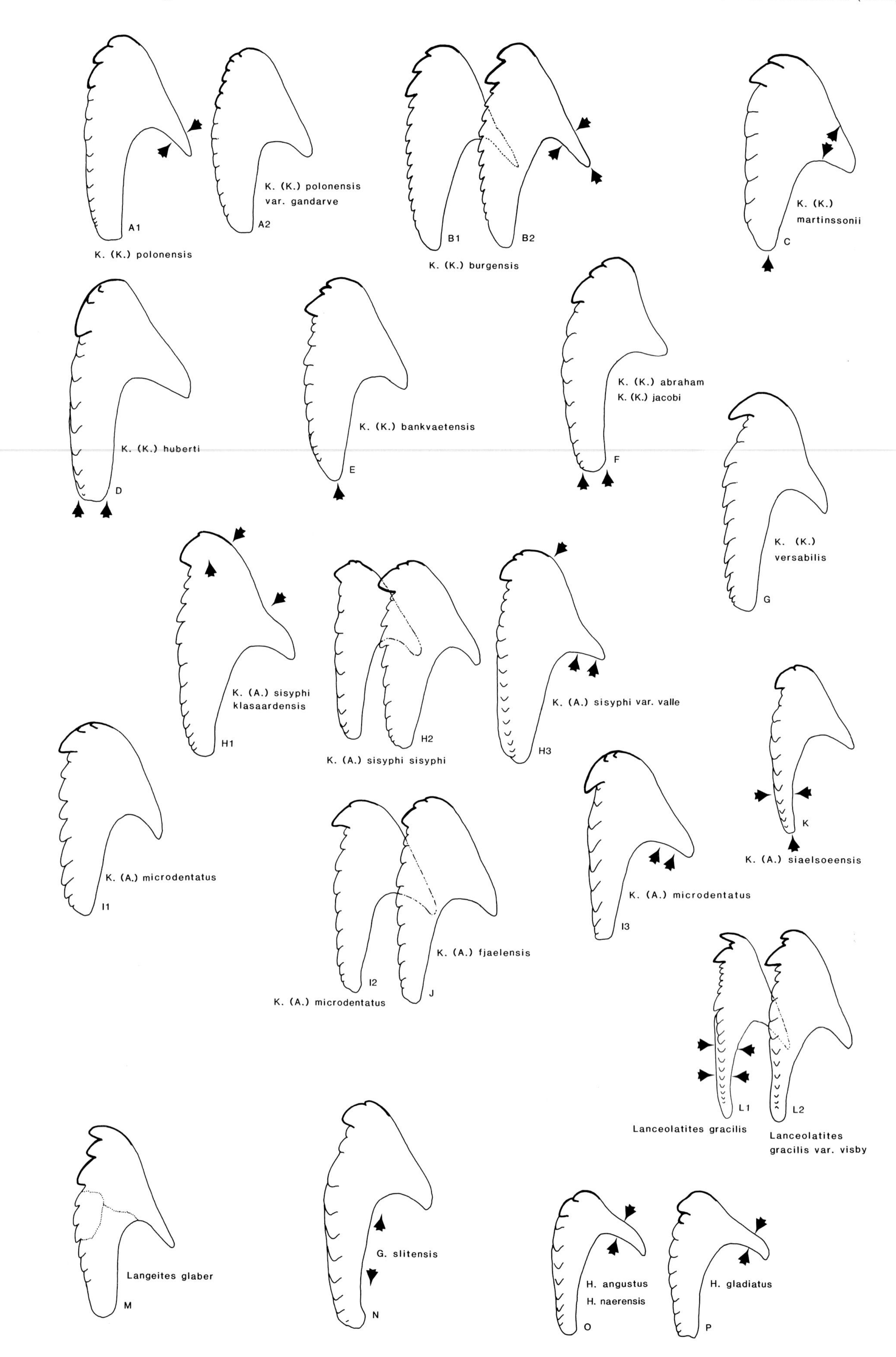

K. (K.) polonensis
K. (K.) polonensis var. gandarve
A1
A2
K. (K.) burgensis
B1
B2
K. (K.) martinssonii
C
K. (K.) huberti
D
K. (K.) bankvaetensis
E
K. (K.) abraham
K. (K.) jacobi
F
K. (K.) versabilis
G
K. (A.) sisyphi klasaardensis
H1
K. (A.) sisyphi sisyphi
H2
K. (A.) sisyphi var. valle
H3
K. (A.) microdentatus
I1
K. (A.) microdentatus
I2
K. (A.) fjaelensis
J
K. (A.) microdentatus
I3
K. (A.) siaelsoeensis
K
Lanceolatites gracilis
L1
Lanceolatites gracilis var. visby
L2
Langeites glaber
M
G. slitensis
N
H. angustus
H. naerensis
O
H. gladiatus
P

Table 1. A key to the identification of paulinitid jawed polychaetes from Gotland based on the right MII element. An asterisk (*) denotes that a species occurs in two different groups, i.e. they have a varied dentition formula.

1. With cusp(s) .... (see) 3
2. Without cusp or no pronounced cusp(s):
   A. *Kettnerites* (*K.*) *polonensis* var. gandarve
   B. *Hindenites naerensis*
3. Single cusp .... (see) 6
4. Double-cusp:
   A. *Hindenites angustus*
5. Double-cusp with small denticles in between (one, or rarely two):
   A. *Lanceolatites gracilis*
   B. *Lanceolatites gracilis* var. visby
   C. *Lanceolatites* sp. A
6. Single cusp with large pre-cuspidal denticle(s) ...... (see) 11
7. Single cusp with small pre-cuspidal denticle(s) .... (see) 9
8. Single cusp without pre-cuspidal dentary:
   A. *Kettnerites* (*K.*) *versabilis*
9. Single cusp with one small precuspidal denticle:
   A. *Kettnerites* (*A.*) *siaelsoeensis*
   B. *Kettnerites* (*A.*) *microdentatus**
   C. *Kettnerites* (*A.*) *sisyphi klasaardensis*
   D. *Hindenites gladiatus*
   E. *Gotlandites slitensis* (precuspidal denticle of fairly small to moderate size)
10. Single cusp with two small precuspidal denticles:
   A. *Kettnerites* (*A.*) *microdentatus**
   B. *Kettnerites* (*A.*) *fjaelensis*
   C. *Kettnerites* (*A.*) *sisyphi sisyphi*
   D. *Kettnerites* (*A.*) *sisyphi* var. valle
11. Single cusp with one large precuspidal denticle:
   A. *Kettnerites* (*K.*) *burgensis**
   B. *Kettnerites* (*K.*) *martinssonii*
12. Single cusp with two large precuspidal denticles:
   A. *Kettnerites* (*K.*) *polonensis*
   B. *Kettnerites* (*K.*) *polonensis* var. sjaustre
   C. *Kettnerites* (*K.*) sp. A
   D. *Kettnerites* (*K.*) *abraham abraham*
   E. *Kettnerites* (*K.*) *abraham isaac*
   F. *Kettnerites* (*K.*) *jacobi*
   G. *Kettnerites* (*K.*) *huberti*
   H. *Kettnerites* (*K.*) *burgensis**
   I. *Langeites glaber*
   J. *Kettnerites* (*K.*) *bankvaetensis* (the posterior denticle of the precuspidal dentary is usually smaller than the anterior one)

Fig. 18. Guide to the identification of the paulinitid jawed polychaetes from Gotland. The key element is the right MII, which is usually the most varied element, together with the right MI. The best features for identification down to species/subspecies level are the number, size and arrangement of the pre-cuspidal denticles and the cusp of the right MII. The arrows point out further details of importance for the identification, such as the convexity in the outer side of the ramus and sharp pointed shank. Two opposing arrows on opposite sides denote a narrow shank or ramus. Two parallel arrows on the posteriormost part of the shank indicate the almost parallel sides of the shank. Two triangles pointing away from each other within the ramus denote a wide ramus. Additional characteristics in the other element-types are sometimes important for proper identification (see Tab. 1 and the respective descriptions).

names based on a fragment or just a few jaw specimens, have thus been left out of consideration.

To erect a biological taxonomy of any organism parallel to an existing form taxonomy based on fragments is against the rules of the International Code of Zoological Nomenclature (1985 e.g. article 23). Eller (1964), Tasch & Stude (1965) were against the use of parataxonomy, and they argued for a combination of the two taxonomies. Kozur (1970, 1971) tried to combine the two different systems in accordance with the rules, but only to the generic level. His attempt was sketchy and was rejected as premature by Jansonius & Craig (1971). Szaniawski & Wrona (1973) also referred to Kozur's taxonomical idea as premature and criticized several of his interpretations.

After a review of the apparatus-based genera, Edgar (1984, pp. 256, 260) noted that posterior elements (e.g. MI) in the Palaeozoic apparatuses are identifiable at the genus level, both isolated and as part of an apparatus. In my opinion an identification at the specific level using the MI is also possible.

Descriptions and reconstructions of jaw apparatuses. – The first described apparatuses were found on bedding-planes (e.g., Massalongo 1855; Hinde 1896; Zebera 1935; Lange 1947; Martinsson 1960). The jaws are either fused together or lying on the bedding plane, like a cluster, more or less isolated. These finds are due to chance and/or a careful study of the sediment surface. When the etching method of searching for complete apparatuses was introduced (Kozlowski 1956; Kielan-Jaworowska 1966) it was possible to make more than just unique finds of apparatuses.

This technique avoids rinsing of the acid resistant residue through a sieve: instead the dispersed clay particles are poured out, leaving the larger particles, including the fossils, on the bottom of the vessel; the residue is studied when still under water (Kielan-Jaworoska 1966, p. 15).

Study of complete apparatuses makes it possible to appreciate the size relations between the elements in an apparatus, but the identification of single elements within the apparatus might be more difficult. Even the identification of the apparatus may offer problems, since the elements partly cover each other. For example, when I studied the type specimen of *'Paulinites' polonensis* Kielan-Jaworowska 1966, I had great problems in identifying the different elements composing the apparatus. It is a very small, probably juvenile, apparatus with indistinct morphological details. The elements are fused together, slightly distorted and fractured. The study was further complicated since the apparatus was kept in glycerine due to its fragile nature. The decision to base the taxonomy on apparatuses more or less forced Kielan-Jaworowska to choose a less good representative of the species as type, since apparatuses of paulinitid taxa are rare.

Thus, the description of fused jaws in an apparatus is complicated by the limited number of apparatuses and by the fact that the elements partly hide each other. I believe that it is an advantage to illustrate isolated elements as well.

Reconstructions based on isolated elements have not been very popular, yet there have been a number of more or less successful attempts utilizing different methods. Syl-

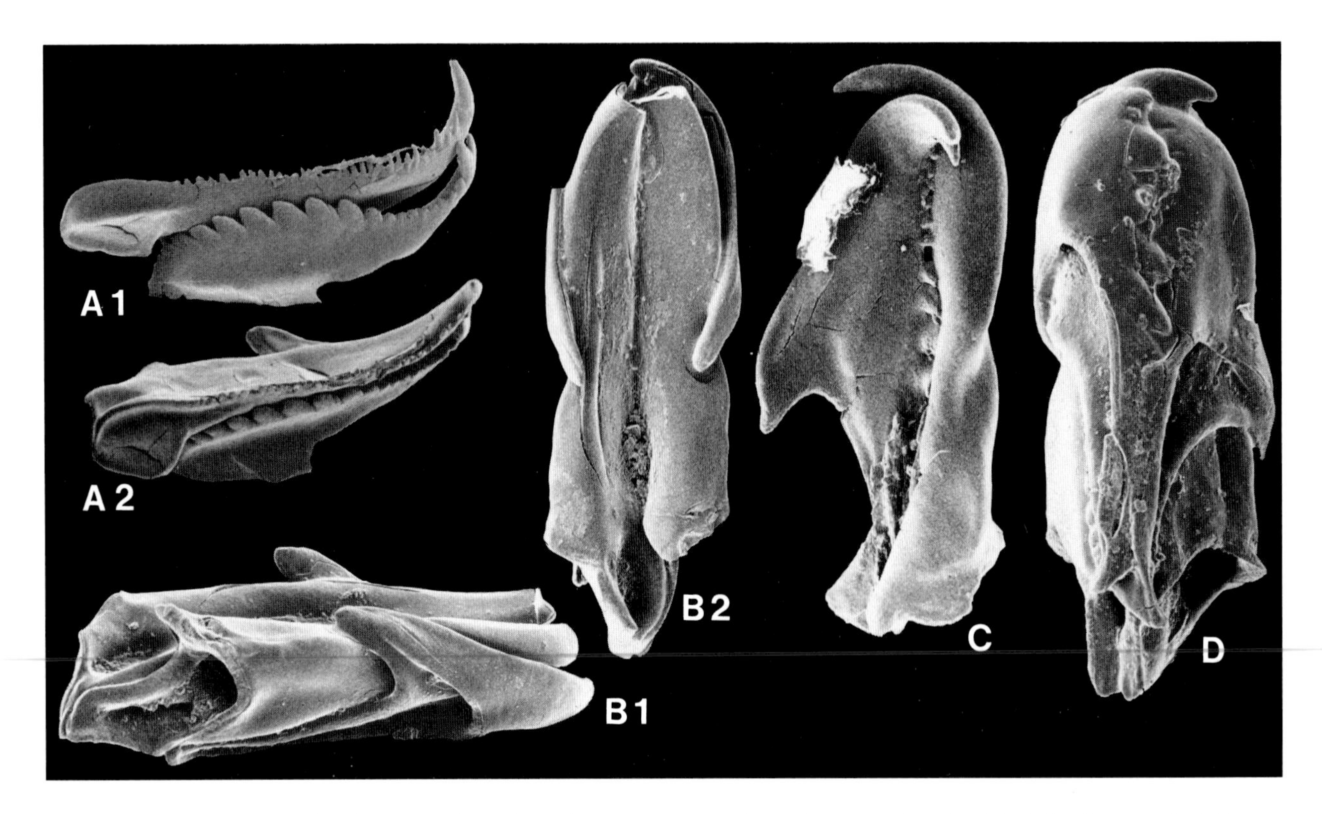

Fig. 19. □A. Left MI and MII fused together, *Kettnerites* (*K.*) *abraham*, Vattenfallsprofilen 1, Högklint b, 21.4–21.95 SL, ca. ×100; A1 lateral view, A2 lateral–dorsal view. □B. Left and right MI's and MII's fused together, *K.* sp., slightly squeezed and fractured, LO 5761:1, Vattenfallsprofilen 1; Högklint Beds, unit b, 70-6LJ, ×70; B1 lateral view, B2 dorsal view. □C. Dorsal view of right MI and left MII fused together, *Lanceolatites gracilis* var. visby, LO 5761:3, Vattenfallsprofilen 1, Högklint Beds, unit b, 70-6LJ, ×100. □D. Ventral view of left and right MI's and MII's fused together, *K.* sp., LO 5761:2, Vattenfallsprofilen 1, Högklint Beds, unit b, 70-6LJ, ×100.

vester (1959) attempted to reconstruct an apparatus on the basis of dispersed jaws. Although he did not support the reconstruction by, e.g., diversity or abundance of different elements, it might be correct. Later, Kielan-Jaworowska (1966) reconstructed some apparatuses from isolated elements, on the basis of co-occurrence of unusual element types (left and right MI).

Successful reconstructions of apparatuses have also been made by Szaniawski (1968), on the Permian species *Atraktoprion eudoxus*, and by Corradini & Olivieri (1974), on the Carboniferous–Permian species *Brochosogenys siciliensis* (Corradini & Olivieri) Colbath 1987b.

Reconstructing apparatuses from isolated jaws is a more delicate task than analyzing more or less complete apparatuses. However, dispersed jaws are easier to find and study, and the greater number of specimens makes the choice of a suitable type easier. As noted by Szaniawski & Gazdzicki (1978), reconstruction work of apparatuses is facilitated by similarities between Recent forms and the fossil ones, especially Mesozoic and Caenozoic. Ultrastructure also seems useful for apparatus reconstruction, and colour, surface texture (e.g. pits and grooves), and jaw size may be employed, at least within the same sample.

Taxonomy based on quantitative studies of single elements demands a large material, collected from localities of different geographical areas and stratigraphical levels. In theory, it is possible to encircle the jaws of a particular species by studying the co-occurrence of many elements in different environments. In practice, one usually has to work with samples from a limited range of environments and with a limited number of jaws. Reconstructions may also be complicated by the presence of several biological taxa in a sample. Nevertheless, in most cases it has been possible to reconstruct the apparatus by delimiting the taxa in 'favourable' samples. In comparison with the limited number of apparatuses, the large number of isolated elements makes it possible to choose types representing a mean within the morphological variation.

In my opinion, synonymization of names between the two taxonomical systems is inevitable, but must be done very carefully. Literature studies alone are insufficient for this purpose. A sound revision must be based on first-hand studies of fossil material to determine apparatus composition and variability of jaw morphology. The number and form of the jaws forming an apparatus can be derived from complete apparatuses or, at least in some taxa, numerical studies (e.g. Corradini & Olivieri 1974). No less important is the knowledge of the jaw variability within the population. Both ontogenetic and phylogenetic aspects are significant. A large number of specimens are required, and each element must be exposed so that a thorough study of every part of the element can be carried out.

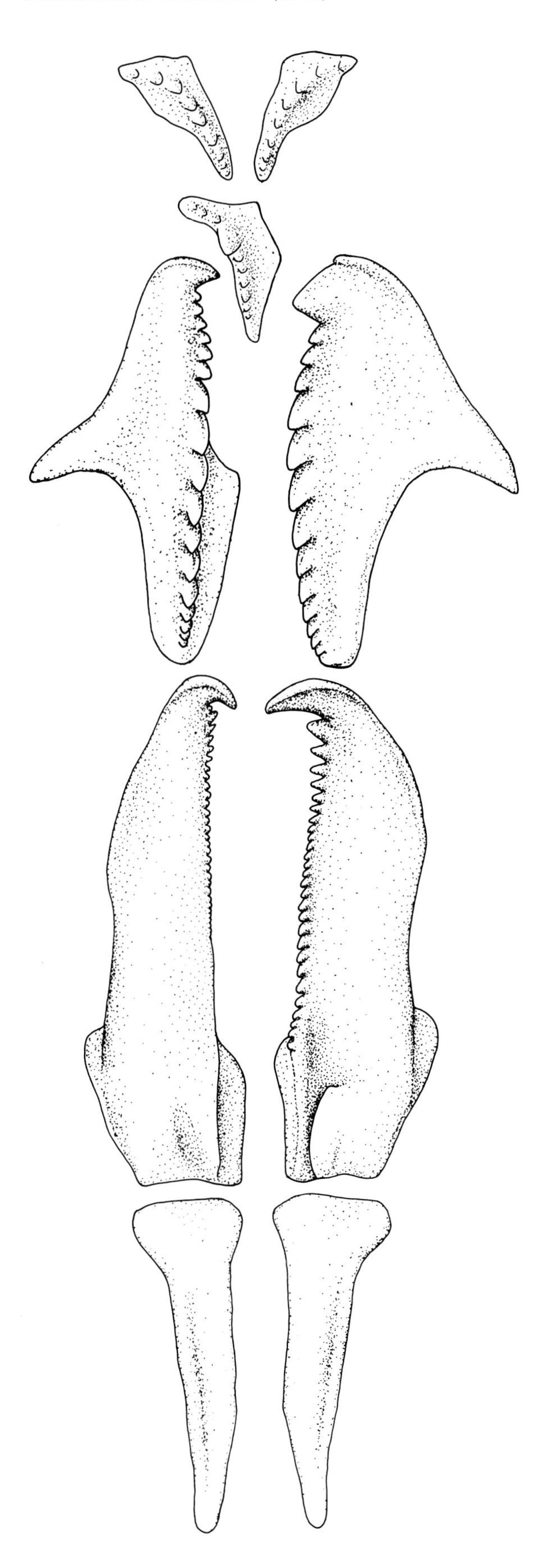

Fig. 20. Dorsal view of a reconstructed polychaete jaw apparatus of *Kettnerites* (*Aeolus*) *sisyphi klasaardensis.* The reconstruction with the elements separated includes the left and right of the following elements (from top): MIV, left MIII, MII, MI, and carriers. The reconstruction of the fused elements shows the position and the corresponding size of the left and right MI's and MII's in an apparatus.

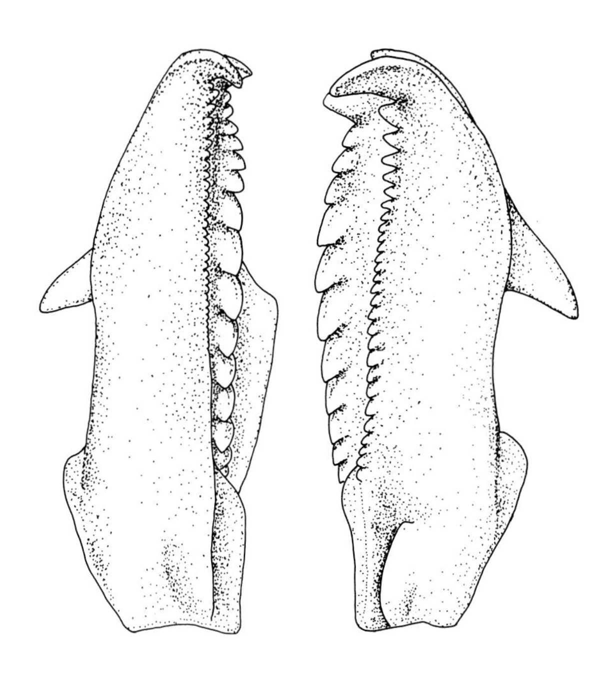

Classification

Phylum Annelida Lamarck 1809
Class Polychaeta Grube 1850
Order Eunicida Dales 1963
Superfamily Eunicea Grube 1852
Family Paulinitidae Lange 1947

Lange (1947) described one species (*Paulinites paranaensis*) within the family Paulinitidae. Species belonging to this family are reported from North and South America and Europe. The stratigraphical range of the Paulinitidae is not known in detail, Kielan-Jaworowska (1966, p. 125) gives the following range: Ordovician to Carboniferous (?Permian).

The paulinitid species from Gotland belong to the genera *Gotlandites*, *Hindenites*, *Kettnerites*, *Lanceolatites* and *Langeites*. The criteria for delimiting the genera are in short: The MI's of *Kettnerites* are fairly coarse with inner and outer margins almost parallel compared with *Lanceolatites*. The MI and MII of the latter genus are very slender, often with a large number of denticles, and the basal plate is normally fused to the MI. Elements (MI's) belonging to the genus *Langeites*, so far monospecific, have only very few denticles. The MI's are slender with almost parallel sigmoidal inner and outer margins ending anteriorly in a large denticulated hook. The MI's of *Gotlandites* and *Hindenites* have a wide posterior part and a more pronounced distinction between the hook and the rest of the jaw. *Gotlandites* differs from *Hindenites* in that the former has a larger hook and a more pronounced convex inner margin. The MII elements also differ between these genera.

A key to the species is given in Table 1.

Genus *Gotlandites* Bergman 1987

Derivation of name. – Latin *Gotlandites*, from the Island of Gotland.

Type and only species. – *Gotlandites slitensis* Bergman 1987

Diagnosis. – Right MI: Jaw approximately triangular, tapering strongly towards the large fang, inner margin slightly convex with paucidentate denticulation anteriorly. Denticles decrease in size and become less spaced posteriorly; large, wide basal portion, with a fairly small, fused basal plate.

Left MI: Jaw approximately triangular, tapering anteriorly, inner margin almost s-shaped, ending in a very large fang with paucidentate denticulation on anterior half of dentary.

Right MII: Fairly large cusp with a small to a fairly small pre-cuspidal denticle, followed by one or two minor denticles. Large shank and large denticles.

Left MII: Fairly large cusp without pre-cuspidal denticles, four large intermediate denticles. Large, wide ramus.

Gotlandites slitensis Bergman 1987

Figs. 14O, P, 18N, 21, 22

Synonymy. – □1987 *Gotlandites slitensis* n.sp. – Bergman, pp. 34–36, Figs. 14O, P, 18, 21, 22.

Derivation of name. – Latin *slitensis*, from the Slite Beds.

Holotype. – LO 5780:5, left MI, Fig. 21G.

Type locality. – Vike 2.

Type stratum. – Slite Beds, Slite Marl, *Pentamerus gothlandicus* Beds or slightly older.

Material. – Fig. 3; About 44 right MI, 42 left MI, 64 right MII, and 41 left MII.

Occurrence. – Figs. 4 and 6; Wenlock. Slite Beds, unit g and Slite Marl. Ajstudden 1, Fardume 1, Hide 1, Hide Fiskeläge 1, Munkebos 1, Nygårds 1, Slitebrottet 1, Talings 1, Tjeldersholm 1, Vallstena 2, Vike 1, 2, and 3.

Diagnosis. – As for the genus.

Description. – Right MI, dorsal side: Length 0.38–1.85 mm; width about 0.3 times the length. The jaw is approximately triangular and tapers towards the strongly developed fang. The inner margin is convex, denticulated from the falcal arch to the undenticulated ridge. The slightly slanting denticles occupy about 0.5–0.6 of the total jaw length in smaller specimens, and about 0.6–0.7 in larger ones (i.e. around 1.5 mm or longer). The anteriormost 4–5 denticles are moderately large to large, and paucidentate to widely spaced. The posterior remaining 5–10 denticles are closely spaced, ending abruptly at the undenticulated ridge. In smaller jaws the denticulation ends before reaching the inner wing; in larger ones it continues posteriorly, ending beside the basal plate. The length of the undenticulated ridge represents about 0.15–0.30 of the total length in smaller jaws and about 0.10–0.15 in larger ones. The undenticulated ridge forms the posterior sharp tip of the jaw, though in some cases the ridge ends without reaching the posteriormost part. The inner wing is almost rhomboidal, normal-sized; the inner margin parallel to the undenticulated ridge. The normal basal furrow is partly hidden under the skew, rectangular, fused basal plate. The posterior margin is almost straight, directed slightly anteriorly, and forming a smoothly rounded outer margin together with the outer margin of the large basal portion. The basal angle is about 30–35°. A more or less pronounced rim runs along the anterior outer margin of the basal portion. The outer margin anterior to the basal portion is smooth, shaped almost like a reversed s.

Ventral side: The crescent-shaped myocoele opening is narrow, strongly enclosed, representing about one third of the jaw length. The opening is surrounded by a narrow rim, and part of its interior surface forms a wide, smooth, rounded area along the outer margin.

Left MI, dorsal side: Length 0.35–1.93 mm, width ¼–⅓ of the length. The subtriangular, strongly tapering jaw ends in a slender, very large fang. The inner margin is s-shaped. The number of denticles is coupled to jaw size, the smaller jaws having 4–5 smaller, pronouncedly paucidentate denticles, and the larger ones up to around 20 denticles. The denticles are of normal size (cf. Fig. 15) and become subdued posteriorly, in larger jaws ending as crenulations. In larger jaws the size and spacing of denticles varies, with one or a few closely spaced, very small denticles interrupting a series of denticles of decreasing size and space. The anterior denticles are inclined slightly towards the anterior. The denticulation occupies around 0.3 to more than 0.6 of the length of the inner margin in smaller and larger jaws, respectively.

The transition from juvenile to adult is gradual. When the jaw attains a length of about 1.5 mm the number of denticles starts to increase, and the shape begins to change from sickle-shaped to more triangular.

The transition from the rounded denticulated inner margin to the undenticulated ridge is smooth. The inner wing is of normal size, about ¼ of the jaw length, and the rounded inner margin follows the orientation of the undenticulated ridge. The basal furrow is small, with a large outer face. A more or less pronounced furrow anterior of the outer face, dividing it from the rest of the jaw, is not uncommon; the structure resembles a very fused basal plate. The basal angle is much varied. The anterior part of the outer margin to the basal portion forms an outer wing. The outer margin anterior of the basal portion is smoothly rounded, often with a concavity in the lower middle part, ending in a large, sickle-shaped falx.

Ventral side: Almost a mirror image of the ventral side of the right MI.

Fig. 21. Gotlandites slitensis. All specimens except I and K from Slite Beds, Slite Marl, *Pentamerus gothlandicus* Beds or slightly younger. Specimens A, B, D, E, G, H, J, L and M are from Vike 2, 83-4CB. All specimens are in dorsal view. □A. Left MII, LO 5780:1, ×120. □B. Right MII, LO 5780:2, ×120. □C. Right MIV, LO 5779:1, Ajstudden 1, 83-7CB, ×170. □D. Left MII, LO 5780:3, ×80. □E. Right MII, LO 5780:4, ×80. □F. Vike 1, 75-30CB; F1 left MI, LO 5785:1, ×60; F2 right MI, LO 5785:2, ×60; F3 left MII, LO 5785:3, ×60; F4 right MII, LO 5785:4, ×60. □G. Holotype, left MI, LO 5780:5, ×80. □H. Right MI, LO 5780:6, ×120. □I. Right MI, LO 5774:2, Slitebrottet 1, Slite Beds, Slite Marl, 73-37LJ, ×120. □J. Right MI, LO 5780:7, ×80. □K. Left MI, LO 5774:1, same sample as I, ×150. □L. Left MI, LO 5780:8, ×120. □M. Right MI and MII fused together, LO 5781:3, ×170.

A
B
C
D
E
F3
F4
F2
I
J
H
F1
G
K
L
M

Fig. 22. Gotlandites slitensis. Dorsal (A–F) and ventral (G–J) views. □A. Left MI, LO 5776:1, Vallstena 2, Slite Beds, Slite Marl *Pentamerus gothlandicus* Beds or slightly younger, 77-2CB, ×80. □B. Left MII, LO 5779:2, Ajstudden 1, Slite Beds, Slite Marl *Pentamerus gothlandicus* Beds(?), 83-7CB, ×170. □C. Right MI, LO 5779:3, same sample as B, ×80. □D. Right MII, LO 5839:4, Gandarve 1, Halla Beds, 71-84LJ, ×120. □E. Left MI, LO 5778:3, Munkebos 1, Slite Beds, Slite Marl, *Pentamerus gothlandicus* Beds or slightly younger, 71-84LJ, ×120. □F. Right MI, LO 5776:2, same sample as A, ×120. □G–J. Vike 2, Slite Beds, Slite Marl, *Pentamerus gothlandicus* Beds or slightly younger, 83-4CB, ventral views, ×120; G, left MI, LO 5805:3; H, right MII, LO 5805:4; I, left MII, LO 5805:5; J, left MII, LO 5805:6.

Right MII, dorsal side: Length 0.32–1.50 mm, width about half the length. A small to moderately small pre-cuspidal denticle forms the fairly sharp anteriormost part of the jaw, somewhat variably situated on the cusp. The cusp is of normal size, slightly swollen basally, slightly slanted, and followed by one or two minor denticles forming the intermediate dentary. The post-cuspidal dentary is large, composed of 6–8 slanted denticles which decrease posteriorly in size. The shank is large, about 0.6 of the jaw length, posteriorly tapering, with inner margin almost straight to slightly convex, outer margin slightly concave, and posterior margin rounded. The bight angle is difficult to measure due to the variable ramus, which is fairly short, sigmoidal, wide, and more or less blunt-ended with a smaller sinus in the anterior outer margin.

Ventral side: The slightly enclosed myocoele opening represents about ⅔ of the jaw length. A fairly wide rim forms the anterior margin of the myocoele opening and continues along the ramus; it also extends along the shank. The shape of the margins varies considerably.

Left MII, dorsal side: Length 0.43–1.43 mm, width about half or slightly more than half of the length. The cusp is of normal size, and slightly swollen in its basal part. The intermediate dentary is composed of four denticles: three large denticles of equal size and a posterior, slightly larger one. There are 6–8 large, slanting, post-cuspidal denticles, decreasing in size towards the posterior. The shank occupies about half the length of the jaw, tapers posteriorly, and has a rounded ending. The almost triangular inner wing is widest anteriorly and occupies about half the length of the jaw. The bight angle is acute, the ramus is large, wide, and tapering, with a fairly rounded end having a distinct sinus at the anterior outer margin.

Ventral side: The slightly enclosed myocoele opening extends from slightly more than half to two thirds of the jaw length. The moderately crescent-shaped anterior margin of the myocoele opening extends along the ramus.

Comparisons. – The morphology of the MI is unusual, though the corresponding elements of the Late Palaeozoic *Brochosogenys siciliensis* (Corradini & Olivieri 1974) Colbath 1987b show a faint general resemblance in having a large fang and similar denticulation. The MII's are different in the two species. Due to the large time span separating *B. siciliensis* and *G. slitensis*, their relationship must remain hypothetical.

The left MII of *Gotlandites slitensis* could be mistaken for a left MII of *Hindenites gladiatus*. *G. slitensis* has a larger ramus and usually a smaller pre-cusp than *H. gladiatus*.

Genus *Hindenites* Bergman 1987

Derivation of name. – In honour of G. J. Hinde, one of the pioneers of the study of fossil polychaetes and the first to publish a paper on the polychaete jaws from Gotland.

Type species. – *Hindenites angustus* (Hinde).

Other species. – *H. gladiatus*, *H. naerensis*.

Diagnosis. – Right MI: Jaw tapers anteriorly, ending in a strong fang. Convex inner margin, fairly large denticles. Myocoele opening on ventral side extending for about 0.35–0.5 of length of jaw.

Left MI: More or less a mirror image of right MI except for basal portion. Inner wing almost triangular.

Right MII: Slender, very narrow ramus, long shank with almost parallel inner and outer margins. Cusp indistinct.

Left MII: As for right MII, except for cusp which is more pronounced but still fairly small.

Hindenites angustus (Hinde 1882)

Figs. 14I, 18O, 24B:1–7

Synonymy. – □1882 *Arabellites angustus* n. – Hinde, p. 19, Pl. 2:53, left MII. □1882 *Arabellites arcuatus* n. – Hinde, p. 19, Pl. 2:52, right MII. □1980 *Paulinites gladiatus* Kielan-Jaworowska 1966 – Wolf, pp. 86–87, Pl. 12:105–107. □1987 *Hindenites angustus* (Hinde 1882) – Bergman, pp. 37–40, Figs. 14I, 18, 24B:1–7.

Holotype. – Left MII, British Museum A 2221, illustrated in Hinde 1882, p. 19, Pl. 2:53.

Type locality. – Vattenfallsprofilen 1, Gotland.

Type stratum. – Högklint Beds, unit d, *'Pterygotus'* marl, *Herrmannina* Beds.

Material. – Figs. 2 and 3; at least 60 right MI, 59 left MI, 39 right MII, and 40 left MII.

Occurrence. – Figs. 4 and 11; Early to Middle Wenlock. Högklint, Slite, and Klinteberg Beds. Ansarve 1, Follingbo 12, Klinteberget 1, Loggarve 2, Strandakersviken 1, Vattenfallsprofilen 1.

Diagnosis. – Right MI: Tapering jaw with strong fang, convex inner margin, denticulated from middle part of falcal arch down to undenticulated ridge. Shank triangular and pointed, outer margin almost straight.

Left MI: Tapering jaw with convex inner margin, denticulated from extreme anterior end of falcal arch ending before reaching the short and almost rectangular inner wing. Basal furrow short and oriented parallel to outer margin of basal portion.

Right MII: Small double cusp, shank long and slender, ramus long, narrow and pointed.

Left MII: Cusp followed by three intermediate denticles and one large cusp-like denticle. Shank fairly long, ramus pointed.

Description. – Right MI, dorsal side: The jaw is 0.29–1.08 mm long, variable in width (usually about ⅓ of the length), tapering strongly from lower middle part. Anteriorly it ends in a large fang, usually bent upwards with regard to the fairly rounded dorsal surface. The inner margin is convex and 0.6–0.7 the length of the jaw. Denticulation extends from the anterior part of the falcal arch to a position anterior to the inner wing. The slightly slanting denticles increase somewhat in size towards their maximum in the posterior middle part; in some jaws the largest denticles are anterior. The posteriormost denticles usually gradually decrease in size and end abruptly. The number of denticles varies from 12, for a jaw length of 0.40 mm, to 19, recorded on 0.53–1.08 mm long jaws. The undenticulated ridge is long, about 0.3 of the jaw length, straight and fairly narrow; the transition to the denticulated ridge is smooth. The inner wing is nearly triangular with the anterior inner 'corner' sickle-shaped, of normal size (0.25 of the jaw length), strongly downfolded, almost perpendicular in relation to dorsal surface. The posteriormost part of the inner wing forms, together with the undenticulated ridge, the moderately sharpened posterior part of the shank. The basal furrow is long, fairly narrow, extending from the posteriormost denticles down to the posterior end of the shank.

Basal plate not recorded.

The flange is moderately thick-walled, highly elevated in the posterior, strongly downfolded along the long practically straight, outer margin. The ligament rim is exposed as a narrow straight ridge along the outer anterior margin of the basal portion. The bight angle is wide, about 100°, and the basal angle very sharp, about 5°. Thus, the ligament rim and the undenticulated ridge are almost parallel. The outer margin anterior to the basal portion is almost straight with only a minor median concavity, and ends in a large sickle-shaped falx.

Ventral side: The enclosed crescent-shaped myocoele opening extends anteriorly for about 0.4–0.5 of the jaw length. The opening is surrounded by a thickened rim along the anterior and outer margins.

Left MI, dorsal side: Length 0.39–1.14 mm, width ¼–⅓ of the length. The jaw tapers anteriorly, ending in a large fang, which is bent upwards in relation to the dorsal surface. The inner margin is convex, and the denticulation extends from the anterior part of the falcal arch over about 0.6 of the jaw length, ending short of the inner wing. Denticle number varies more or less regardless of size from, 14 denticles on a 0.48 mm long jaw to 24 on one of 0.64 mm length, but 18 and 20 denticles have been recorded on jaws 0.39 mm and 1.14 mm long, respectively. The anteriormost denticle is the largest, normally followed by some denticles decreasing in size towards the posterior part of the falcal arch, from where the denticle size increases to fairly large ones on the middle part of the inner margin. The posteriormost denticles are small. In some jaws the denticulation gradually decreases in size from the largest, anteriormost denticle. The anterior denticles point forward while most of the remaining denticles are slanting. The inner margin continues as a straight line onto the small and posteriorly narrowing undenticulated ridge. The moderately downfolded inner wing is angular, nearly rectangular, with an almost straight inner margin, representing about 0.2 of the jaw length. The basal furrow is short and narrow, parallel with the almost straight outer margin of the basal portion. The outer face of the basal portion is large, slightly more than 0.3 of the jaw length. The basal angle is narrow, about 15°. The outer margin anterior to the basal portion is slightly convex with a smaller concavity in the middle part.

Ventral side: The enclosed myocoele opening is crescent-shaped anteriorly, directed toward the anterior, surrounded by the fairly low ligament rim, which extends posteriorly at the inner and outer margins in the form of narrow spur-like ridges.

Right MII, dorsal side: Length 0.45–0.80 mm, width about ⅔ of the length. There is a very small double cusp, anterior cusp slightly smaller than the posterior one. They are followed by one, intermediate, fairly small denticle, in turn followed by seven to eight post-cuspidal denticles, only slightly smaller than the cusps. The denticles are slanting and decrease in size towards the posterior, ending before they reach the the posterior margin. The inner margin is slightly convex. The shank is extremely long, occupying more than two thirds of the length of the jaw, and the fairly slender and posteriorly pointed tip is somewhat bent to the outside. The margins of the shank are almost parallel (Fig. 24B:5) or taper slightly posteriorly (Fig. 24B:7). The small inner wing is normally not seen from the dorsal side. The bight is deep, the bight angle about right to acute. The ramus is very slender, with almost parallel margins and a pointed extremity. The outer margin is often slightly convex. A minor sinus is sometimes present on the anterior outer margin.

Ventral side: The slightly enclosed myocoele opening represents about ¾ of the jaw length. It is surrounded by a very narrow rim which forms a protruding corner on the anterior part of the inner margin.

Left MII, dorsal side: Length 0.38–0.85 mm, width about half the length. One cusp of normal size forms the sickle-shaped anterior margin. It is followed by three fairly small intermediate denticles which increase posteriorly in size. A conspicuous, large, cusp-like denticle divides the intermediate from the post-cuspidal dentary. There are eight to nine post-cuspidal denticles, about equal in size and slightly slanting towards the posterior. The almost parallel-sided shank is very large, about 0.6 of the jaw length, ending fairly bluntly. The posteriormost part extends to the outer ventral side. The bight is very deep, and the bight angle sharply acute. The ramus is fairly long and slender with a posteriorly oriented, pointed extremity. The anterior outer margin is almost straight.

Ventral side: The slightly enclosed myocoele opening represents about two thirds of the jaw length. The rim which surrounds the opening to the anterior (slightly curved) and along the inner margin of the jaw forms an almost right angle.

Discussion. – The holotype is the left MII; the right MII of the same species was described by Hinde (1882) from Gotland as *'A'. arcuatus*. A North American right MII with a similar denticle formula was described by Hinde (1879) as *'A'. similis* var. *arcuatus*. Although I have not seen the North American specimen, there seems to be a distinct difference between the Gotland form and the jaw from the North American Middle Devonian Hamilton Group. *'A'. similis* (Hinde 1879) from the Silurian Niagara Formation in North America also has a similar denticle formula, but judging from the crude illustration and description there is no close relationship to the Gotland form. Hinde did not compare this form with the one from Gotland (*'A'. arcuatus*).

Comparison. – The large myocoele opening of the MI is characteristic of the genus, distinguishing it from other paulinitid genera on Gotland. The MI differs from the corresponding element of *Hindenites gladiatus* in that the denticulation ends before reaching the inner wing. The MII has a unique dentition and form.

Hindenites gladiatus (Kielan-Jaworowska 1966)

Figs. 14F, G, H, 18P, 23, 24A, C–E

Synonymy. – □1882 *Arabellites hamatus* Hinde 1879 – Hinde, pp. 16–17 (pars), Pl. 2:42–43, left MI. □1966 *Paulinites gladiatus* n.sp. – Kielan-Jaworowska, pp. 129–131, Pl. 30:5A–C, 6A–C, left and right MI. □1979 *Oenonites* sp. a – Bergman, p. 99, Pl. 28:3A, B, left and right MI. □1987 *Hindenites gladiatus* (Kielan-Jaworowska 1966) – Bergman, pp. 40–42, Figs. 14F–H, 18, 23, 24A, C–E.

Holotype. – Right MI in Kielan-Jaworowska 1966, pp. 129–131, Pl. 30:6.

Type locality. – Loose boulder derived from the Baltic area.

Type stratum. – Silurian (?Upper Wenlock, ?Lower Ludlow).

Material. – Figs. 2 and 3; At least: 83 right MI, 55 left MI, 26 right MII, and 32 left MII.

Occurrence. – Figs. 4 and 11; Early Wenlock to Middle–Late Wenlock, Högklint Beds, unit b, to Mulde Beds. Ar 1, Aursviken 1, Däpps 2, Fardume 1, Gandarve 1, Hällagrund 1, Hide 1, Munkebos 1, Nygårds 2, Saxriv 1, Slitebrottet 1 and 2, Svarven 1, Valby Bodar 1, Vallstena 2, Vattenfallsprofilen 1, Vike 2.

Diagnosis (emended). – Right MI: Large denticles covering inner margin usually from posterior of the falcal arch, passing posteriorly to anterior part of the inner wing, ending slightly anterior of the bight. Myocoele opening covering 0.35–0.40 of the jaw length.

Left MI: Large fang, falcal arch with pronounced cutting edge but without true denticles. Large denticles along convex inner margin, end posterior of the anterior part of the inner wing.

Right MII: Fairly small pre-cuspidal denticle, cusp of normal size, very large shank, ramus long, narrow, with pointed extremity.

Left MII: Fairly large single cusp, very large shank, long, very narrow ramus with pointed extremity.

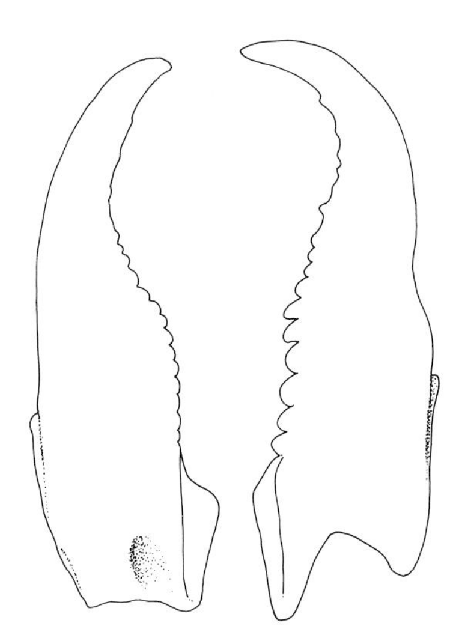

Fig. 23. Hindenites gladiatus (Kielan-Jaworowska 1966), camera-lucida drawing of the type specimen (right MI, Z. Pal. No. O.308/14*b*) and the left MI of the same species, ×84. Note that the denticulation of both right and left MI ends slightly to posterior of the anteriormost part of the inner wing.

Description. – Right MI, dorsal side: Length 0.55–1.20 mm, width 1/4–1/3 of the length. The jaw tapers strongly anteriorly, ending in a large, pointed, sickle-shaped fang, bent upward in relation to the jaw surface. The denticulation of the anterior part of the inner margin is formed by more or less indistinctly spaced crenulations, or as in the specimens from Saxriv 1 (Högklint Beds, unit b), by a cover of small denticles. The anterior denticles seem to be more developed in the smaller jaws, at least during Middle–Late Wenlock time. In the Slite Marl at Munkebos 1 an excellently preserved jaw exhibits large, thin, triangular denticles anteriorly. In the convex posterior part of the inner margin, the 11–13 denticles are large and slanting, passing along the anterior part of the inner wing, close to the bight. In the larger jaws from the Early Wenlock the posteriormost 3–5 denticles decrease in size towards the undenticulated ridge. In Middle–Late Wenlock specimens only the last denticle is smaller (e.g. Slitebrottet, Munkebos 1, and Gandarve 1). Denticles of the smaller jaws are of more or less equal size on the posterior half of the denticulated inner margin. Early Wenlock jaws show a prominent, short and straight undenticulated ridge. In Middle–Late Wenlock specimens the ridge is somewhat broader, the broadest part being at the posterior end. The strongly downfolded, almost triangular, inner wing occupies about 0.3 of the jaw length. The basal furrow has a smooth outer side and is wide and long, about twice the length of the undenticulated ridge.

Basal plate is not recorded.

The flange is thick-walled, bight angle about 90°. The basal angle is low, varying between 12° and 22°. The ligament rim is almost straight, occupying around one fifth to one quarter of the jaw length. The outer margin in front of the basal portion is convex with a more or less pronounced median concavity. The falx is large, wide, sickle-shaped.

Ventral side: The enclosed myocoele opening occupies around 0.35–0.40 of the jaw length. The opening is crescent-shaped anteriorly and surrounded by a sharp ridge along the innermost part of the inner wing.

Left MI, dorsal side: Length 0.53–1.26 mm, width about 1/4 of length. The jaw tapers strongly anteriorly, ending in a large, pointed, sickle-shaped fang. The fang has a prominent undenticulated cutting-edge along the large falcal arch, comprising 0.3–0.45 of the anterior part of the jaw, though 2–8 indistinct knobs may occur. The posterior part of the inner margin is convex, with 9–15 large to very large slanting denticles, occupying 0.34–0.50 of the jaw length. The largest denticles are in the middle part of the convex

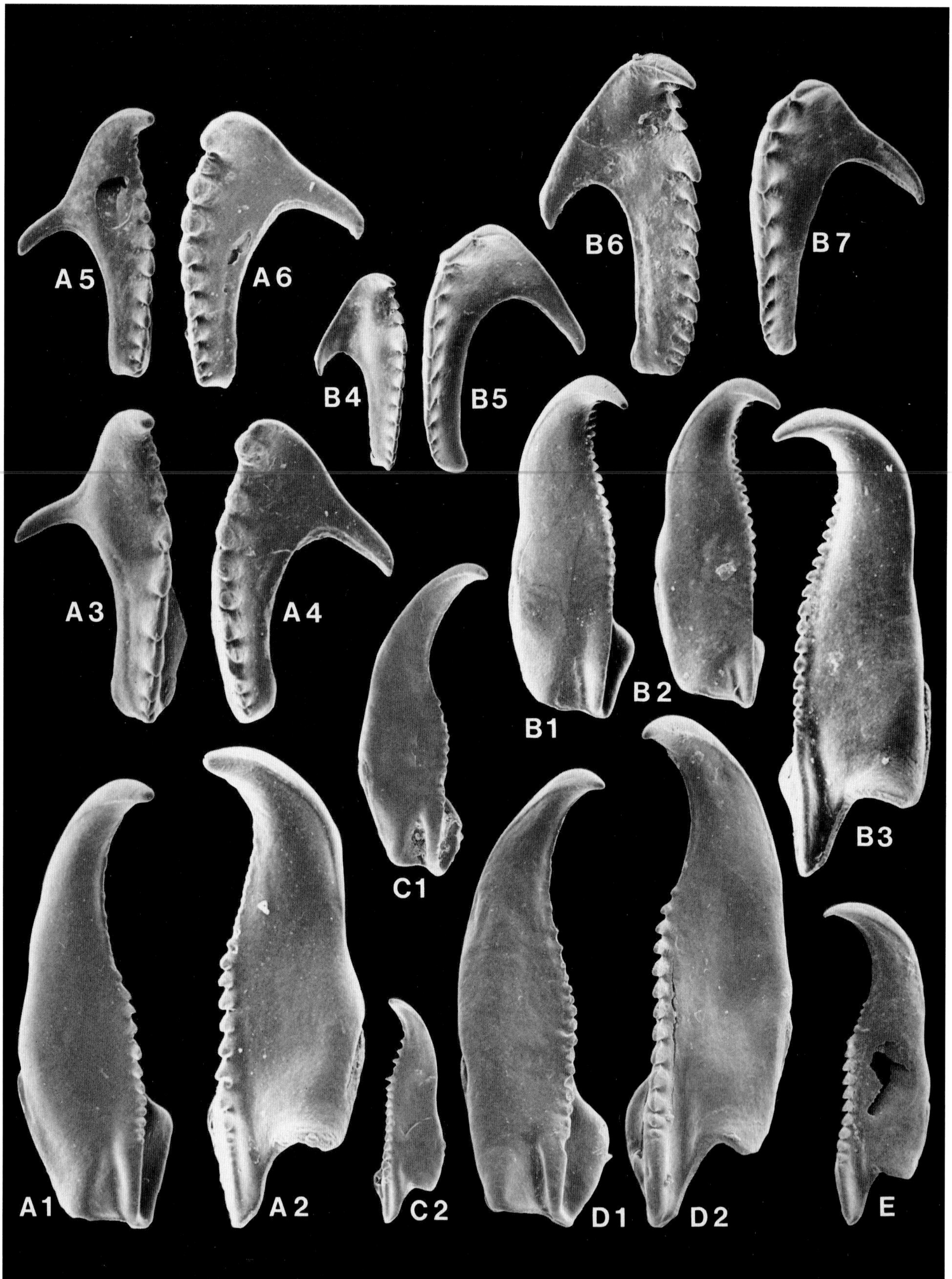

Fig. 24. All specimens in dorsal view. □A. *Hindenites gladiatus*, Vattenfallsprofilen 1, Högklint Beds, unit d, *Valdaria testudo* bed, RM sample; A1 left MI, AN 2702, ×80; A2 right MI, AN 2703, ×80; A3 left MII, AN 2704, ×80; A4 right MII, AN 2705, ×80; A5 left MII, AN 2700, ×80; A6 right MII, AN 2701, ×80. □B. *H. angustus*, Vattenfallsprofilen 1, Högklint Beds, unit d, *Herrmannina* bed, RM sample; B1 left MI, AN 2706, ×60; B2 left MI, AN 2707, ×80; B3 right MI, AN 2708, ×80; B4 left MII, AN 2709, ×80; B5 right MII, AN 2710, ×80; B6 left MII, AN 2711, ×120; B7 right MII, AN 2712, ×80. □C. *H. gladiatus*, Munkebos 1, Slite Beds, Slite Marl, *Pentamerus gothlandicus* Beds or slightly younger, 71-84LJ; C1 left MI, LO 5778:1, ×120; C2 right MI, LO 5778:2, ×120. □D. *H. gladiatus*, Gandarve 1, Halla Beds, 71-81LJ; D1 left MI, LO 5836:6, ×120; D2 right MI, LO 5836:7, ×120. □E. *H. gladiatus*, right MI, LO 5774:3, Slitebrottet 1, Slite Beds, Slite Marl, 73-37LJ, ×120.

inner margin. The undenticulated ridge is short, tapering anteriorly, forming a more or less straight continuation of the inner margin. At the posterior end it bends toward the base of the inner wing, which occupies about ¼–⅓ of the jaw length. Inner wing varies from nearly triangular to semicircular and, not so deeply downfolded as in the right MI, forms the posteriormost part of the jaw. The basal furrow is almost triangular and short, tapering anteriorly, its length about half the length of the inner margin. The almost straight ligament rim is notable from the dorsal view and occupies about one third of the jaw length. The basal angle is low. The outer margin in front of the basal portion is smoothly convex with a pronounced cutting-edge in the falx, though sometimes a shallow concavity in the middle of the jaw may occur.

Ventral side: The enclosed myocoele opening extends for 0.35–0.40 of the jaw length. The opening is crescent-shaped anteriorly and the ligament rim is thickened and surrounded by a narrow furrow on the anterior and inner part.

Right MII, dorsal side: Length 0.45–0.78 mm, width about ⅔ of the jaw length. A small pre-cuspidal denticle, on the cusp, forms the anteriormost, fairly rounded margin. The inner margin is slightly convex. The cusp is of normal size, followed by the post-cuspidal dentary. The anteriormost denticle is slightly smaller than the following, about 8, large denticles, which slant somewhat towards the posterior. The shank with almost parallel sides is very large, and occupies about 0.6 of the jaw length; the posterior extremity is fairly narrow and slightly rounded. The bight is deep, the bight angle acute to almost 90°. The ramus is long, very narrow, with a posteriorly oriented, pointed extremity. The anterior outer margin has a slight sinus.

Ventral side: The slightly enclosed myocoele opening represents about ¾ of the jaw length. It is delimited by the smooth rim along the almost straight anterior and inner margins.

Left MII, dorsal side: Length 0.36–0.88 mm, width approximately ⅔ of the jaw length. The inner margin is slightly convex. A single cusp of normal size is followed by the intermediate dentary containing 3–5 denticles of normal size, increasing slightly towards the posterior. The transition to the post-cuspidal dentary is more or less gradual. The size of the post-cuspidal dentary is varied, from normal to fairly large, posteriorly slanting denticles. The denticulation continues along the entire shank, which is very large, parallel-sided, and occupies about 0.6 of the jaw length; the posterior extremity ends bluntly. The inner wing is very short and has a rounded margin. The bight is fairly deep, the bight angle acute. The parallel-sided ramus is conspicuously long and slender, with a pointed extremity oriented towards the posterior. A pronounced sinus is found on the anterior outer margin.

Ventral side: Almost a mirror image of the ventral side of the right MII, except that the opening is more rounded and the inner wing extends to the rim on the left MII.

Remarks. – The most characteristic difference between *H. gladiatus* and *H. angustus* is that the denticles of the right MI of *H. gladiatus* extends, on the denticulated ridge, along the anteriormost part of the inner wing. The conspicuously large denticle of the left MII of *H. angustus* is also a distinct character.

Comparisons. – The type species *Arabellites hamatus* Hinde (1879, p. 377, Pl. 18:12) was found in the North American Cincinnati Group. The type specimen is squeezed and not well preserved, but it does not belong to the same species as the specimens described as *A. hamatus* from Gotland (Hinde 1882, pp. 16–17, Pl. 2:42–44).

Kielan-Jaworowska (1966, pp. 130–131) noted that *Nereidavus digitus* Eller (1963, p. 161, Pl. 1:14) from the Upper Devonian Sheffield Shale of Iowa and *Nereidavus exploratus* Eller (1961, p. 177, Pl. 1:20–21) from the Devonian of Michigan were 'reminding one in the shape of *Paulinites gladiatus* n.sp.'. In my opinion it is difficult to discern whether or not there is a closer relationship between the Baltic and the Laurentian taxa without studying populations from North America.

Hindenites naerensis Bergman 1987

Figs. 14J, 18O, 25

Synonymy. – □1987 *Hindenites naerensis* n.sp. – Bergman, pp. 42–44, Figs. 14J, 18, 25.

Derivation of name. – From the parish of När, referring to the type locality.

Holotype. – LO 5831:4, right MI, Fig. 25B.

Type locality. – Närshamn 2.

Type stratum. – Burgsvik Beds, lower part.

Material. – Fig. 3; more than 24 right MI, 26 left MI, right MII 5, left MII 6.

Occurrence. – Figs. 4 and 11; Early Whitcliffian, Ludlow. Burgsvik Beds. Glasskär 1, Kapelludden 1, Närshamn 2.

Diagnosis. – Right MI: Jaw smoothly rounded, tapering anteriorly, conspicuous almost triangular inner wing. Large, crescent-shaped, enclosed myocoele opening.

Left MI: Jaw smoothly rounded, tapering anteriorly, large crescent-shaped enclosed myocoele opening.

Right MII: No distinguishable cusp or intermediate dentary, shank extremely long, occupying about three quarters of the jaw length. Ramus very narrow and slender.

Left MII: Fairly small cusp, small post-cuspidal denticles. Shank narrow with parallel sides and extremely long, occupying about three quarters of the jaw length.

Description. – Right MI, dorsal side: Length 0.70–1.53 mm, width ¼–⅓ of the length. The jaw tapers unevenly anteriorly, and the fang is slightly bent upward in relation to jaw surface. The inner margin is somewhat convex, with crenulations of up to 5 small denticles along the inner margin of the falcal arch. Behind the falcal arch, there are 9–16 fairly large, slanting, and somewhat spaced denticles. The denticulated inner margin covers about 0.6 of the jaw length. The posteriormost denticles decrease in size posteriorly. The undenticulated ridge is long, 0.3 of the jaw length, narrow and low, forming a smooth arch. The ridge

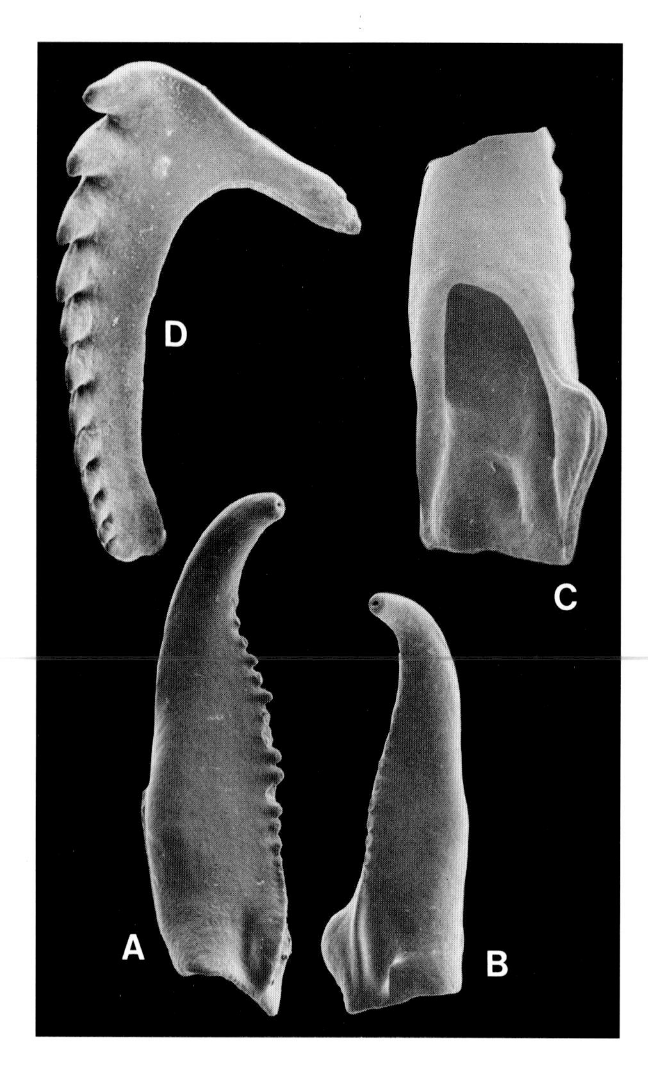

Fig. 25. Hindenites naerensis. Specimens A, B and D in dorsal view. Närshamn 2, Burgsvik Beds, 83-12LJ; □A. Left MI, LO 5831:3, ×80; □B. Holotype, right MI, LO 5831:4, ×80. □C. Right MI, ventral view, LO 5805:7, ×120. □D. Right MII, LO 5805:8, ×80.

disappears gradually without reaching the posterior margin. The inner wing is anteriorly rounded, short, about 1/4 of the jaw length, strongly downfolded in comparison with the dorsal surface, and forms the blunt posterior termination of the jaw.

The basal plate is small, nearly rectangular, without denticles. The plate and adjacent area of the outer face are highly elevated, forming the highest part of the jaw, the fang excluded, making the posterior margin undulating.

The outer margin of the basal portion is downfolded and almost straight in anterior–posterior direction. The basal furrow is wide, the deepest part being between the anteriormost part of the basal plate and the undenticulated ridge. The very long ligament rim extends almost 0.4 of the jaw length, along the outer margin. The basal angle is very low, the undenticulated ridge and the outer margin are almost parallel. The outer margin is almost straight, the basal portion included, with a minor convexity in the middle part of the jaw.

Ventral side: The elongated, enclosed myocoele opening represents about half of the jaw length. The opening is surrounded by a smooth, somewhat sunk ligament rim. A ridge, corresponding to the boundary between the basal plate and the flange on the dorsal side of the jaw, runs perpendicularly to the long axis of the jaw.

Left MI, dorsal side: Length 0.62–1.53 mm, width about 1/4–1/3 of the jaw length. The jaw tapers anteriorly, ending in a fang, bent slightly upwards in relation to the dorsal surface. The inner margin behind the undenticulated falcal arch is convex with some minute, and about 14–17 larger, slightly spaced, slanting denticles covering about half the length of the jaw. The denticle size increases from the anterior to the middle portion where it reaches a maximum and then decreases posteriorly. The undenticulated ridge is long, low and narrow with a small median inward bend, and normally does not reach the posterior margin. The inner wing is of varying but always small size, nearly rectangular and moderately downfolded in relation to the dorsal surface. The basal furrow is variable, but usually very short, almost rectangular, with its widest part posteriorly. The outer margin of the basal portion extends 0.3–0.4 of the jaw length, is slightly convex, deeply downfolded, almost hiding the long ligament rim. The outer margin in front of the basal portion is mostly occupied by the large sickle-shaped falx.

Ventral side: The enclosed, elongated myocoele opening extends for approximately 0.4 of the jaw length. The ligament rim is fairly thin, except at the outer anterior position where it forms a broad thick rim, narrowing posteriorly. The posteriormost part of the rim forms a narrow, sharp ridge.

Right MII, dorsal side: Length 0.77–1.00 mm, width slightly less than 2/3 of the jaw length. No cusp or intermediate denticles are distinguished. The number of denticles varies between 9 and 13 without relation to jaw size. They cover the entire length of the jaw and slant slightly towards the posterior. The size decreases slightly to the posterior, with the exception of the much smaller posteriormost denticle. The very narrow and extremely long (about 3/4 of the jaw length) shank has almost parallel sides and a slightly convex inner side. It occupies about three quarters of the jaw length. The posterior extremity is slightly widened and bent somewhat ventrally outwards. The bight is very deep, the bight angle acute, the ramus long and very narrow with a pointed extremity. The anterior outer margin is normally slightly convex to almost straight.

Ventral side: The slightly enclosed myocoele opening represents about 3/4 of the jaw length or more. It is boarded by a fairly narrow rim, gently crescent-shaped along the anterior and inner margins.

Left MII, dorsal side: Length 0.71–1.18 mm, width about half of the jaw length. A fairly small cusp forms the rounded anterior margin. The intermediate dentary has 1–3 small denticles increasing posteriorly in size. Ten to eleven fairly small post-cuspidal denticles decrease evenly in size from the base of the cusp to the posteriormost extension of the jaw. The shank, with almost parallel sides, is extremely long, occupying about 3/4 of the jaw length. The posterior extremity is slightly swollen, bent somewhat to the ventral outer side. The inner wing is of normal length, about 0.6 of the jaw length, and very narrow, the margin parallel with the denticulated ridge. The bight is deep, the bight angle acute, the ramus very slender with pointed extremity. The anterior outer margin has a slight sinus.

Ventral side: A mirror image of the ventral side of the right MII except that the opening represents about 2⁄3 of the jaw length.

Remarks. – The difference in the number of MI and MII encountered is probably due to the greater fragility of MII, but also to the fact that the MI is larger and therefore easier to recover. MI is also more easily identified by fragments.

Comparisons. – MI differs from the corresponding element of *H. gladiatus* by being more slender and straight, and in that the right MI has a fused basal plate. The denticle formula of the MII's is similar in the two species, but the latter is slightly more slender, particularly the ramus, than the former one. The denticulation of the MI's of *H. naerensis* and *H. angustus* are similar. *H. naerensis* is probably more closely related to *H. gladiatus* than to *H. angustus.* This assumption is based on the fact that the dentition formula of MII seems to be less changed through time. *H. gladiatus* and *H. naerensis* may even be in the same lineage.

Genus *Kettnerites* Zebera 1935

Jaw elements, found in Palaeozoic rocks from Europe and herein referred to *Kettnerites*, have been placed in the following genera: □*Oenonites* Hinde 1879 (*O. infrequens* Hinde 1879, p. 382 Pl. 20:2, left MI; *O. compactus* Hinde 1879, p. 384, Pl. 20:13, left MII; *O. aspersus* Hinde 1880, p. 373, Pl. 14:7,8, left and right MI; Hinde 1882, p. 13, Pl. 1:21, 22, 22A). □*Nereidavus* Grinnell 1877, (*Nereidavus solitarius* Hinde 1879, p. 385, Pl. 20:12, left MI). □*Kettnerites* Zebera 1935 (Snajdr 1951; Männil & Zaslavskaya 1985a and 1985b). □*Arabellites* Hinde 1879 (*A. perneri* Zebera 1935). □*Pronereites* Zebera 1935. □*Paulinites* Lange 1947; the type species from the Devonian of South America belongs to a different genus from *Kettnerites. Paulinites* has also been used by, among others, Martinsson 1960; Kielan-Jaworowska 1966; Szaniawski 1970; on species from Europe. However, in my opinion none of the European species belong to *Paulinites.*

The following scolecodonts from the Palaeozoic of North America might be placed in the genus *Kettnerites* (I have not noted any species in common between the North American and Baltic faunas): □*Nereidavus* Grinnell 1877 (*N. ontarioensis* Stauffer 1939, p. 507, Pls. 57:12, 13, 15; 58:26, 32, 34; *N. planus* Stauffer 1939, pp.507–508, Pl. 57:9, 10; *N. invisibilis* Eller 1940, pp. 16–17, Pl. 2:1–11; *N. harbinsonae* Eller 1941, pp. 325–326, Pl. 37:1, 2, 4, 5; *N. angulosus* Eller 1945, pp. 190–191, Pl. 7:12–16; *N. exploratus* Eller 1963a, p. 177, Pl. 1:20, 21; *N. incrassatus* Eller 1964, p. 231, Pl. 1:7–13). □*Arabellites* Hinde 1879 (*A. arcuatus* Hinde 1879 – Stauffer 1939, p. 501, Pl. 58:17; *A. priscus* Stauffer 1939, p. 504, Pl. 58:18; *A. dauphinensis* Stauffer 1939, p. 502, Pl. 58:4; *A. sinuatus* Walliser 1960, pp. 22–23, Pl. 5:3a–c). □*Oenonites* Hinde 1879 (*O. bidens* Eller 1940, pp. 26–27, Pl. 5:3–5). □*Ildraites* Eller 1936 (*I. geminus* Eller 1940, pp. 29–30, Pl. 6:1–5; *I. eminulus* Eller 1963a, p. 176, Pl. 1:18). □*Polychaetaspis* Kozlowski 1956 (*P.? kozlowskii* Walliser 1960, pp. 26–27, Pl. 6:1–3).

Type species. – *Kettnerites kozoviensis* Zebera 1935.

Subgeneric taxonomy. – Inside the genus *Kettnerites* two subspecies, *Kettnerites* (*Kettnerites*) Zebera 1935 and *Kettnerites* (*Aeolus*) Bergman 1987 can be recognized on basis of the following features: The MI's of *K.* (*Aeolus*) are fairly slender jaws often with a fairly wide basal portion and on the slender MII's the pre-cuspidal denticles, if any, are small.

Kettnerites (*K.*) *abraham* Bergman 1987

Figs. 13N–P, 18F, 26, 27,

Synonymy. – □1987 *Kettnerites abraham* n.sp. – Bergman, p. 45, Figs. 13N–P, 18, 26, 27.

Derivation of name. – As for the nominal subspecies.

Holotype. – As for the nominal subspecies.

Type locality. – As for the nominal subspecies.

Type stratum. – As for the nominal subspecies.

Material. – Fig. 2; more than 241 right MI, 205 left MI, 309 right MII, and 295 left MII.

Occurrence. – Figs. 4 and 5; Late Llandovery to Early Wenlock. Lower Visby Beds, unit b, to Högklint Beds unit d and southwest facies.

Diagnosis. – Right MI: Denticles normal to large size, slightly slanting, covering inner margin from falcal arch to undenticulated ridge. Shank ends in a sharp tip; almost rounded, large, thin-walled flange, with a spur or a spur-like structure.

Left MI: Denticles of normal to large size, covering inner margin from falcal arch to the undenticulated ridge. Almost rectangular inner wing. Spur or spur-like structure large to minute.

Right MII: Two large pre-cuspidal denticles of equal size, followed by a large cusp and a smaller denticle. Shank subparallel; ramus short and wide.

Left MII: Double cusp, parts arranged laterally almost side by side. Outer, slightly more anterior cusp, somewhat smaller. Intermediate dentary pronounced.

Description. – Right MI, dorsal side: Length 0.34–1.51 mm, width about 1⁄4 of the jaw length. The jaw is fairly rounded in cross-section. The inner margin increases in convexity from *K.* (*K.*) *abraham abraham* to *K.* (*K.*) *abraham isaac.* The dentary covers the total length between the falcal arch and the undenticulated ridge with moderate, equal-sized denticles in the nominal subspecies and larger unequal-sized denticles in the younger subspecies. The basal portion has a large flange with a distinct spur on the nominal subspecies and it becomes very conspicuously shaped and large but with a minor spur-like structure in *K.* (*K.*) *abraham isaac.* The shank ends sharply.

Left MI, dorsal side: Length 0.33–1.80 mm, width somewhat less than 1⁄4 of the jaw length. The inner margin is slightly convex on the nominal subspecies and becomes

A
B
C
D
E
F
G
H
I
J
K
L
M
N
O
P

s-shaped on *K.* (*K.*) *abraham isaac*. The denticles vary from moderate in size to large, covering the inner margin from the falcal arch to the undenticulated ridge. The inner wing is almost rectangular. In the nominal subspecies, the basal portion is fairly large with a prominent spur; in *K.* (*K.*) *abraham isaac* it becomes very large with a conspicuous outer side, but with a minor spur-like structure.

Right MII, dorsal side: Length 0.26–1.41 mm, width about half the jaw length. Two large pre-cuspidal denticles of equal size are present, with the anteriormost to the right of the posterior one. The cusp is large, slanting posteriorly, followed by one minor denticle. The denticulated ridge is composed of 9–10 large denticles, which slant and decrease in size towards the posterior. The shank occupies about half the jaw length, tapers slightly to the posterior and terminates in a rounded posterior extremity. The bight is shallow; the bight angle varies from acute to a right angle. The ramus is short and wide. The anterior outer margin is almost straight or has a small concavity.

Left MII, dorsal side: Length 0.33–1.48 mm, width about half the jaw length or slightly more. The parts of the double cusp are positioned almost side by side, the outermost being slightly smaller. The intermediate dentary is pronounced, composed of 4–7 denticles of equal size. Posterior to these, on the denticulated ridge, around nine fairly large denticles slant towards the posterior. The shank tapers slightly to the posterior, occupying about half of the jaw length. The inner wing is almost triangularly rounded, representing about half the jaw length. The bight is shallow. The ramus is almost triangular, fairly wide, with a convexity occupying the main part of the anterior margin.

Discussion. – The most characteristic feature to differentiate the two subspecies from each other is the shape of the basal portion. The long s-shaped outer margin of the basal portion of *K.* (*K.*) *abraham isaac* is very conspicuous.

Fig. 26. Kettnerites (*K.*) *abraham abraham.* All specimens except N and O are in dorsal view, ×120. □A. Left MI, LO 5364:1, Buske 1, Lower Visby Beds, unit e, 79-41LJ. □B. Right MI, LO 5363:2, Lickershamn 2, Lower Visby Beds, unit f, 73-53LJ. □C. Left MI, LO 5365:1, Nygårdsbäckprofilen 1, Lower Visby Beds, unit e, 79-42LJ. □D. Left MI, LO 5364:2, same sample as A. □E. Right MI, LO 5363:3, same sample as B. □F. Left MI, left MII and MIII fused together, LO 5360:9, Själsö 1, Lower Visby Beds, unit b, 79-12LJ. □G. Holotype, right MI, LO 5364:3, same sample as A. □H. Right MI, LO 5362:4, Buske 1, Lower Visby Beds, unit e, 79-40LJ. □I. Right MII, LO 5367:5, Häftingsklint 1, Upper Visby Beds, 76-9CB. □J. Left MII, LO 5760:1, Vattenfallsprofilen 1, Upper Visby Beds, lowermost part, 70-14LJ. □K. Right MII, LO 5368:1, Vattenfallsprofilen 1, Upper Visby Beds, 76-8LJ. □L. Left MI, LO 5363:4, same sample as B. □M. Left MI, LO 5362:5, same sample as H. □ N. Left MI, LO 5365:2, same sample as C, ×80. □O. Left MI, LO 5368:2, same sample as K, ×80. □P. Right MI, LO 5360:10, same sample as F.

Kettnerites (*K.*) *abraham abraham* Bergman 1987

Figs. 13N, O, 18F, 26, 27B, D, E, I, J

Synonymy. – *Kettnerites* (*K.*) *abraham abraham* n.subsp. – Bergman, pp. 47–49, Figs. 13N, O, 18, 26, 27B, D, E, I, J.

Derivation of name. – In honour of my son, Carl Abraham.

Holotype. – LO 5364:3, right MI, Fig 26G.

Type locality. – Buske 1.

Type stratum. – Lower Visby Beds, unit e.

Material. – Fig. 2; more than 179 right MI, 146 left MI, 202 right MII, and 197 left MII.

Occurrence. – Figs. 4 and 5; Late Llandovery to Early Wenlock. Lower Visby Beds, unit b, to Högklint Beds unit b. Buske 1, Gnisvärd 1 and 2, Gustavsvik 1, 2, and 3, Häftingsklint 1 and 4, Halls Huk 1, Ireviken 1, 2, and 3, Korpklint 1, Lickers 1, Lickershamn 2, Nygårdsbäckprofilen 1, Nyhamn 1, 2, and 4, Rönnklint 1, Saxriv 1, Själsö 1, Snäckgärdsbaden 1, Vattenfallsprofilen 1, Ygne 2.

Diagnosis. – Right MI: Large, slightly slanting denticles of almost equal size, covering inner margin. Shank ending with sharp tip; subrounded large thin-walled flange, pronounced spur. In dorsal view the myocoele opening is tubular.

Left MI: Denticles large, decreasing in size toward posterior end. Undenticulated ridge tapering posteriorly, pronounced spur.

Right MII: Two large, pre-cuspidal denticles of equal size, followed by a large cusp and a smaller denticle. Shank almost parallel, ramus short and wide.

Left MII: Double cusp, parts arranged laterally almost side by side, the outer, slightly more anterior cusp is somewhat smaller. Intermediate dentary pronounced.

Description. – Right MI, dorsal side: Length 0.34–1.51 mm, width about 1/4 of length. The inner and outer margins are almost parallel, in some cases slightly tapering. The anterior end forms a fairly large fang. The coarse denticulation extends along the convex inner margin for about 0.7 of the jaw length. About 20 denticles, slanting except for the anteriormost ones, and almost equal in size, end as one or a few small knobs immediately anterior to the inner wing. The transition between the inner margin and the undenticulated ridge is at a lateral, outward bend of the ridge. The undenticulated ridge is fairly high, narrow, forming the posteriormost, sharp tip of the shank. The inner wing is of normal size, representing about a quarter of the jaw length. The inner wing is deeply downfolded, almost triangular; its anterior part wide and rounded. The basal furrow is wide and long.

The basal plate is not recorded.

The flange is thin-walled, large, subrounded, slightly downbent along its outer margin and its anterior part elevated. The flange ends in a small concavity, anterior of which is a spur, representing the outermost part of the ligament rim. The basal portion is short and in dorsal view seems to end like a tube. The basal angle is usually about

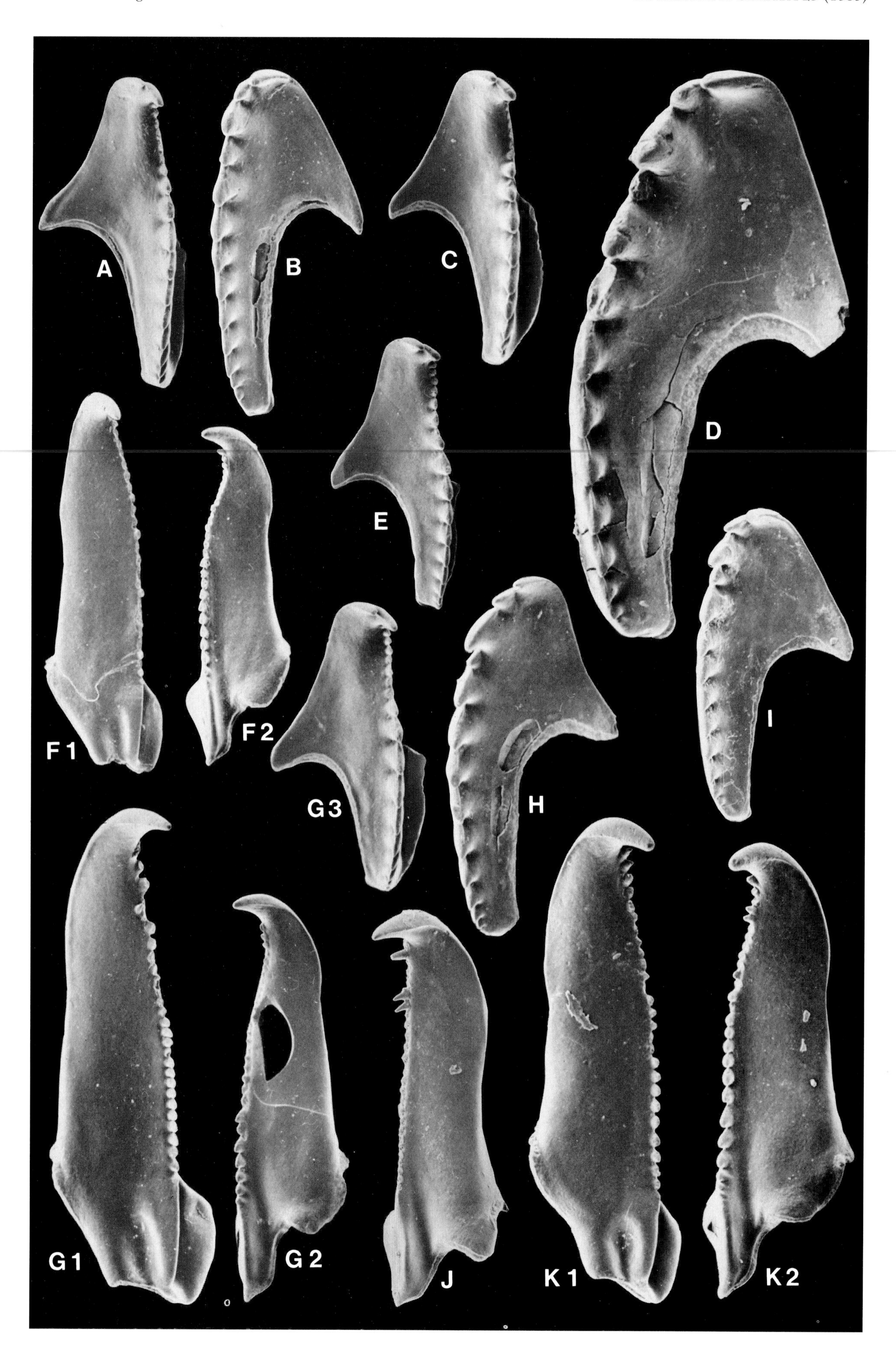
A
B
C
D
E
F 1
F 2
G 3
H
I
G 1
G 2
J
K 1
K 2

25–30°. The outer margin anterior to the basal portion has a slight concavity in the middle part.

Ventral side: The strongly enclosed myocoele opening represents about ⅓ of the jaw length. A shallow furrow surrounds the flat myocoele rim along its anterior part. The rim is elevated, forming two high, short, narrow and sharp triangular ridges on the anterior inner and outer margin. The spur is distinct. A sharp ridge continues along the inner margin of the shank.

Left MI, dorsal side: Length 0.34–1.80 mm, width less than ¼ of length. The inner and outer margins are almost parallel, ending anteriorly in a normally developed fang. The inner margin is slightly convex to almost straight with about 20 large (the anteriormost ones of almost equal size), slightly spaced denticles ending just anterior to the inner wing. The transition from the inner margin to the undenticulated ridge is smooth. The undenticulated ridge is low and becomes narrower toward the posterior end, where it almost reaches the posterior margin. The inner wing is of normal size, slightly more than 0.2 of the jaw length, almost rectangular and slightly downfolded. The basal furrow is of normal length, deep and parallel to the outer margin of the basal portion, which has a downfolded posterior part. On the anteriormost outer part of the basal portion there is a concavity and a spur similar to the one on the right MI.

Ventral side: The strongly enclosed myocoele opening represents about one third of the jaw length. The ligament rim is narrow and somewhat depressed in relation to the ventral surface. Along the inner margin of the opening and the inner wing, a narrow, fairly low ridge is present. On the outer posterior part there is a high, narrow, sharp ridge forming a spur. The spur dominates the small low ridge which continues along the whole of the outer margin of the basal portion.

Right MII, dorsal side: Length 0.26–1.35 mm, width about half the length. Two large pre-cuspidal denticles of equal sized, with the anteriormost one positioned to the right of the posterior one. The cusp is large, slanting towards the posterior, followed by one minor denticle. The denticulated ridge is laterally convex, composed of 9–10 large denticles, which decrease in size and slant towards the posterior. The shank, occupying about half the jaw length, tapers slightly to the posterior, with a rounded posterior extremity. The posterior outer margin is almost straight to slightly concave. The bight is shallow; the bight angle varies from about 55° to 75°. The ramus is short and wide. The anterior outer margin is usually nearly straight, though a small sinus may exist.

Ventral side: The slightly enclosed myocoele opening represents ⅔–¾ of the jaw length. The anterior and inner margins are slightly curved and surrounded by a fairly thin rim.

Left MII, dorsal side: Length 0.38 mm–1.48 mm, width about half the length or slightly more. The double cusp with its slightly smaller outer part positioned somewhat posteriorly, points almost at right angle to the extension of the denticulated ridge or has a slight posterior slant. The intermediate dentary is pronounced, composed of 4–7 denticles irrespective of the size of the jaw. The denticles are of equal size, except the posteriormost one, which is normally larger. The 8–10 fairly large denticles of the denticulated ridge slant posteriorly and are of almost equal size or decrease slightly towards the posterior. The shank occupies about half of the jaw length and tapers slightly to the posterior. The inner wing is about half of the jaw length and almost triangular, rounded. The bight is shallow; the bight angle varies from acute to right. The ramus is nearly triangular, fairly wide, with a convexity occupying the main part of the anterior margin.

Ventral side: The slightly enclosed myocoele opening represents ½–⅔ of the jaw length. The almost straight anterior and inner margins of the myocoele are surrounded by a rim of normal width.

Discussion. – Two different dentary forms among the MI's have been recorded, viz. the usual form with large denticles (e.g. Fig. 26O, P) and a less common form with fairly thin and more numerous denticles (e.g. Fig. 26D, E).

Comparison. – The right MI (738/51) reported as *Kettnerites polonensis* by Männil & Zaslavskaya (1985a) is very similar (conspecific?) to *K.* (*K.*) *abraham abraham* (see chapter 'Comparison with paulinitid faunas from other areas').

Fig. 27. A, C, F, G, H, K: *Kettnerites* (*K.*) *abraham isaac*; B, D, E, I, J: *K.* (*K.*) *abraham* cf. *abraham*. All specimens in dorsal view. □A. Left MII, LO 5767:5, Vattenfallsprofilen 1, Högklint Beds, unit b, 70-6LJ, ×120. □B. Right MII, LO 5764:6, Vattenfallsprofilen 1, Högklint Beds, unit a, 70-20LJ, ×120. □C. Left MII, LO 5772:5, Stave 1, Slite Beds, Slite Marl, central part, 75-11CB, ×120. □D. Right MII, LO 5368:3, Vattenfallsprofilen 1, Upper Visby Beds, 76-8LJ, ×80. □E. Left MII, LO 5765:1, Lickershamn 2, Högklint Beds, unit a, 73-57LJ, ×120. □F. Vattenfallsprofilen 1, Högklint Beds, unit d, *Valdaria testudo* level, sample RM; F1 left MI, AN 2713, ×80; F2 right MI, AN 2714, ×80. □G. Vattenfallsprofilen 1, Högklint Beds, unit b, 70-6LJ; G1, holotype, left MI, LO 5767:7, ×80; G2 right MI, LO 5767:6, ×80. □H. Right MII, LO 5763:4, Ansarve 1, Högklint Beds, SW facies, upper part, 79-46LJ, ×80. □I. Right MII, LO 5760:2, Vattenfallsprofilen 1, Upper Visby Beds, 70-14LJ, ×120. □J. Right MI, LO 5764:7, same sample as B, ×120. □K. Same sample as F; K1 left MI, AN 2715, ×80; K2 right MI, AN 2716, ×80.

Kettnerites (*K.*) *abraham isaac* Bergman 1987

Figs. 13P, 18F, 27A, C, F–H, K

Synonymy. – □1987 *Kettnerites* (*K.*) *abraham isaac* n.subsp. – Bergman, pp. 49–51, Figs. 13P, 18, 27A, C, F–H, K.

Derivation of name. – Isaac, on the Abraham 'lineage'.

Holotype. – LO 5767:7, left MI, Fig. 27G:1.

Type locality. – Vattenfallsprofilen 1.

Type stratum. – Högklint Beds, unit d, *Pterygotus* Beds.

Material. – Fig. 2; more than 62 right MI, 59 left MI, 107 right MII, and 98 left MII.

Occurrence. – Figs. 4 and 5; Early Wenlock. Högklint Beds, unit b, c, d, and the SW facies. Ansarve 1, Ar 1, Lauter 1, Langhammarsviken 2, Lauterhornsvik 3, Lickershamn 2,

Nors Stenbrott 1, Saxriv 1, Stutsviken 1, Svarven 1, Vattenfallsprofilen 1.

Diagnosis. – Right MI: Denticles increase in size from normal on falcal arch to large on posterior part of the dentary along the reverse s-shaped inner margin. Large, thin-walled flange, long, s-shaped outer margin of basal portion, ending in a spur-like structure. Shank ends sharply.

Left MI: Anterior taper, inner margin convex, covered by fairly large denticles. Outer margin of basal portion relatively long.

Right MII: Two large pre-cuspidal denticles of equal sized, followed by a large cusp and a smaller denticle. Shank subparallel; ramus short and wide.

Left MII: Double cusp, parts arranged almost side by side. The outer, slightly more anterior cusp, is somewhat smaller. Intermediate dentary pronounced.

Description. – Right MI, dorsal side: Length 0.34–1.51 mm, width ¼ of length. The jaw tapers anteriorly from the fairly wide basal portion, ending in a normally developed fang, bent slightly upwards in relation to the dorsal surface. The convex inner margin has 25–30 fairly large to large denticles from the falcal arch posteriorly, ending abruptly just before the inner wing. They slant and increase in size posteriorly, with the exception of the somewhat larger anteriormost one and the slightly smaller posteriormost ones. The undenticulated ridge continues as a smooth ridge, slightly bent towards the inner margin, forming the posteriormost tip of the sharply pointed shank. The inner wing is of normal size, downfolded, almost triangular, with a rounded prominent anterior 'corner'. The outer margin of the shank is sigmoidal; the basal furrow is wide and long.

The basal plate is not recorded.

The flange is large, thin-walled, elevated, the margin s-shaped and rounded, ending in a minor concavity and a spur-like structure (Figs. 27G:2, 27K:2) at the anterior part of the basal portion, where the ligament rim meets the flange. The posterior, outermost part of the flange is moderately downfolded. The basal portion is wide and fairly long. The basal angle is about 35°. The outer margin, anterior to the basal portion, is concave, ending in the sickle-shaped falx.

Ventral side: The strongly enclosed myocoele opening represents about one third of the jaw length. The anterior part of the opening is surrounded by a flat ridge which in turn is surrounded by a narrow groove. The rim increases in size toward the posterior inner and outer margins. It forms a ridge in the posteriormost inner margin of the shank.

Left MI, dorsal side: Length 0.33–1.37 mm, width slightly less than ¼ of length. The jaw tapers from basal portion to the pointed fang, which is moderately bent upwards in relation to the dorsal surface. The inner margin is convex, denticulated along its entire length, representing about 0.7 of the jaw length, extending from the falcal arch posteriorly and reaching the inner wing. The denticles are fairly large, of equal size, slanting, and slightly spaced. The anteriormost 3–4 denticles are often the largest and point forward. The undenticulated ridge forms a convex continuation of the denticulated inner margin, starting beside the anteriormost part of the almost rectangular, normal-sized inner wing. The inner margin of the wing is nearly parallel to the undenticulated ridge. The basal furrow is short and shallow, oriented intermediately between the undenticulated ridge and the outer margin of the basal portion. The basal angle varies from 25° to 40°. The outer margin of the basal portion is long, downfolded along the posterior half, forming a small outer wing on the anterior part and ending in a concavity. Posterior to the concavity is a spur-like extension of the ligament rim. The outer margin is concave with the sickle-shaped falx occupying the anterior half.

Ventral side: The strongly enclosed myocoele opening represents about ⅓ of the jaw length. It is elongated, surrounded by the ligament rim, the surface of which is flat to concave on its anterior part but forms a high, narrow ridge along the inner margin. On the anterior, outer margin of the basal portion, a small ridge with a furrow along its outer margin grows into a narrow, relatively high ridge on the posteriormost part.

Right MII: Very close resemblance to the right MII of *K.* (*K.*) *abraham abraham*. Among some *K.* (*K.*) *abraham isaac* specimens, the sinus in the anterior outer margin is more pronounced, and the shank is slightly less tapering. Length 0.26–1.41 mm.

Ventral side: The slightly enclosed myocoele opening is very similar to the nominal species, but the anteriormost inner part of the rim has a protruding corner.

Left MII: Very close resemblance to the left MII of *K.* (*K.*) *abraham*. The pre-cuspidal denticle seems to be more needle-shaped among the *K.* (*K.*) *abraham isaac* than in the left MII of *K.* (*K.*) *abraham abraham*. Length 0.33–1.00 mm.

Ventral side: Very similar to the nominal species.

Remarks. – At present the MII is not so well preserved, and the material is fairly small at each locality. The variability of the morphological characteristics of the MII of the two subspecies overlaps. Although some differences noted in the description are at hand, the morphological characters overlap and are thus not useful for identification. It is not improbable that with a very large, well preserved material, some other characteristics for distinguishing the MII of the two subspecies will be found. More probably, however, additional material will increase the amount of overlap and thus continue to make the identification of the MII on the subspecific level further impossible.

Discussion. – The transition from *Kettnerites* (*K.*) *abraham abraham* to *K.* (*K.*) *abraham isaac* is observable in the Högklint Beds, unit b, e.g. Vattenfallsprofilen 1 (70-6LJ) and in Lickershamn 2 (73-57LJ, 73-65LJ, 75-105CB samples not included in the locality list).

Kettnerites (*K.*) *jacobi* Bergman 1987

Figs. 13Q, 18F, 28A, B

Synonymy. – □1987 *Kettnerites* (*K.*) *jacobi* n.sp. – Bergman, pp. 51–53, Figs. 13Q, 18, 28A, B.

Derivation of name. – In honour of my son Jakob.

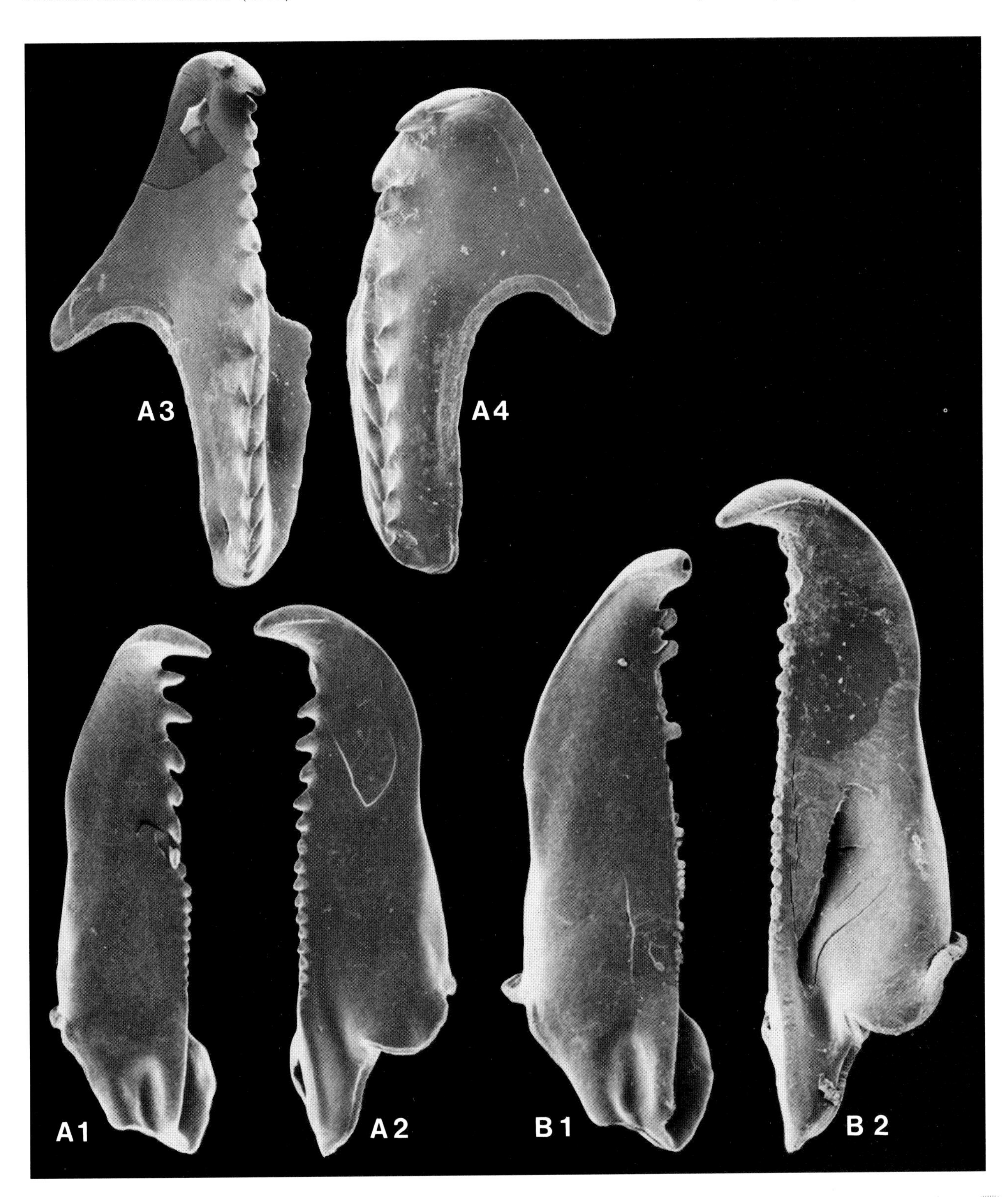

Fig. 28. Kettnerites (K.) jacobi. All specimens in dorsal view. □A. Slitebrottet 2, Slite Beds, Slite Marl, 83-31LJ; A1 left MI, LO 5773:5, ×80; A2, holotype, right MI, LO 5773:6, ×120; A3 left MII, LO 5805:2, ×120; A4, right MII, LO 5805:1, ×120. □B. Stutsviken 1, Högklint Beds, unit c, 77-7CB; B1 left MI, LO 5769:1, ×80; B2 right MI, LO 5769:2, ×80.

Holotype. – LO 5773:6, right MI, Fig. 28A:2.

Type locality. – Slitebrottet 2.

Type stratum. – Slite Beds, unit g.

Material. – Fig. 2; More than 30 right MI, 27 left MI, 39 right MII, and 37 left MII.

Occurrence. – Figs. 4 and 5; Wenlock. Slite Beds, unit f, Slite Marl, unit g, unit Lerberget Marl – *Pentamerus gothlandicus*; Broa 2, Haganäs 1, Myrsne 1, Slitebrottet 1 and 2, Stave 1, Stora Mafrids 2, Tjeldersholm 1, Valbytte 1, Vallstena 2.

Diagnosis. – Right MI: Large jaws with 15–20 large, anteriorly widely spaced denticles, decreasing in size and becoming more densely spaced toward the posterior end. Shank sharp-ended; flange thin-walled, rounded, with spur-like structure.

Left MI: Almost a mirror image of the right MI except for the basal portion. Inner wing anteriorly rounded, spur-like structure distinct.

Right MII: Two large pre-cuspidal denticles; cusp of moderate size; large wide shank, ending fairly pointed.

Left MII: Thin needle-shaped pre-cuspidal denticle; large cusp; pronounced, slightly spaced intermediate dentary; subtriangular ramus.

Description. – Right MI, dorsal side: Length 0.55–1.68 mm, width about ⅓ of length. The surface is flat anteriorly and convex posteriorly. The jaw tapers slightly anteriorly, ending in a well developed fang, bent some 10° upwards in relation to the surface. Slightly convex to almost straight inner margin. The large, 15–20, normally 16–18 denticles, anteriorly slightly spaced, decrease in size toward the posterior end and are slightly slanting or perpendicular towards the extension of the inner margin, covering about 0.6 of the total jaw length. The denticulated ridge ends in small denticles and continues as the undenticulated ridge well before reaching the same level as the anteriormost part of the inner wing, or on the outer side, before reaching the bight. The undenticulated ridge is smooth, high and long, representing 0.28–0.35 of the jaw length. It is almost straight, sometimes slightly bent towards the outer margin, forming the posteriormost part of the jaw and ending in the sharp tip of the shank. The inner wing is of normal size, about ¼ or less of the jaw length, almost triangular, with rounded corners, the widest part on the anterior half, steeply downfolded.

The basal plate is not recorded.

The basal furrow is short, squeezed between the undenticulated ridge and the large, rounded, elevated, thin-walled flange. The posterior, outermost part of the flange is downfolded. Anterior to the flange there is a small concavity in connection with a vague rim and a spur-like extension (Fig. 28A:2). The basal portion is of normal size, almost triangular with a basal angle of about 30–40°. The outer margin anterior to the basal portion is rounded with a shallow concavity in the middle, ending anteriorly in a sickle-shaped falx.

Ventral side: The strongly enclosed myocoele opening represents about ⅓ of the total jaw length. The ligament rim surrounds the opening along the anterior and inner margins and forms a small spur at the anterior, outer position.

Left MI, dorsal side: Length 0.51–2.40 mm, width ¼ of length. The dorsal surface is slightly concave. The jaw tapers gradually towards the anterior, ending in a pointed fang, moderately bent upwards in relation to the surface. The inner margin is faintly to moderately convex, covered for about 0.6 of the jaw length by 15–20 (normally 16–18) denticles, of which the anteriormost 5–6 are very large and widely spaced; the denticulation ends well before reaching the level of the inner wing. The denticles slant slightly, decreasing in size posteriorly, ending as small knobs. The anteriormost denticle is sometimes slightly smaller than the second and often directed forward. This is occasionally also the case with the anterior second and third denticles. There is a smooth transition to the long, straight, low and narrow undenticulated ridge. The inner wing is of normal size, about ¼ of the jaw length, rounded, moderately downfolded, forming the posteriormost part of the shank. The basal furrow is short, narrow and oriented slightly toward the outer margin. The posterior half of the outer margin of the basal portion is somewhat downfolded; the anterior half has a small concavity with a spur-like rim in front of it. The basal portion is of normal size, skew, with a basal angle of about 30–40°. In the middle of the outer margin, at its transition into the fang, and anterior to the basal portion, there is a vague concavity.

Ventral side: The strongly enclosed myocoele opening, ¼–⅓ of the jaw length, is in its slightly curved anterior part surrounded by a low ligament rim. Along the outer margin the rim becomes more distinct with anterior spur-like extensions.

Right MII, dorsal side: Length 0.35–2.12 mm, width more than half the jaw length. The inner margin is slightly convex. Two large pre-cuspidal denticles form the rounded, almost pointed anterior margin. Together with the cusp they are bent strongly upwards and slant slightly towards the posterior. The cusp is of moderate size, followed by a minor intermediate denticle. The post-cuspidal dentary comprises 8–9 large denticles, decreasing in size and slanting slightly towards the posterior. The shank, more than half the jaw length, is wide with almost parallel sides. The bight is fairly deep, the bight angle about 50–60°. The ramus is large, almost triangular, with an extremity pointing slightly posteriorly. The anterior outer margin varies from slightly convex to concave.

Ventral side: The slightly enclosed myocoele opening represents ¾, or somewhat less, of the jaw length. The anterior and inner margins are almost straight and surrounded by a fairly thin rim with a corner slightly protruding anteriorly on the anterior inner margin.

Left MII, dorsal side: Length 0.42–2.00 mm, width about half the jaw length. The cusp is large, with a small needle-shaped pre-cuspidal denticle along the anterior sickle-shaped margin, oriented in the same direction as the cusp. The intermediate dentary has 6–8, slightly spaced denticles of normal size. There is an abrupt transition to the post-cuspidal dentary, which consists of 7–10 fairly large denticles, slanting slightly towards the posterior. The shank, about half the jaw length, tapers slightly towards the posterior. The inner wing is almost triangular, laterally slightly less extended than the shank, its anterior part being widest. The bight is normal; the bight angle about 70°; the ramus is fairly large, nearly triangular with its extremity pointed slightly posteriorly. The anterior outer margin is smoothly sigmoidal.

Ventral side: The slightly enclosed myocoele opening represents about half, or slightly more, of the jaw length. The opening is smoothly curved, surrounded by a rim of normal width.

Discussion. – The MI elements of *Kettnerites* (*K.*) *abraham abraham*, *K.* (*K.*) *abraham isaac* and *K.* (*K.*) *jacobi* are characterised by a large, rounded, thin-walled flange, a spur decreasing in size from late Visby time, and large denticles which increase in size from Visby time to Slite time. The denticle formula of the MII is persistent through time, though the size of the pre-cusps decreases to a needle-like pre-cusp in the Slite Beds. *K.* (*K.*) *abraham abraham* evolves gradually into the Högklint subspecies, i.e. *K.* (*K.*) *abraham isaac.* It is possible that *K.* (*K.*) *jacobi* has evolved from the Högklint type (Figs. 13N–Q). Supporting this theory are the equivalent shape and size of the flange and the denticulation of the left and right MI, and the similar denticle formula of the MII's. However, there is a lack of information from the lower Slite Beds.

Remarks. – In the more than 14 samples yielding *K.* (*K.*) *abraham isaac,* the MII is consistently recorded about twice as frequently as the MI. There are a number of conceivable explanations for this discrepancy. First, the MII, with its denticle formula, is easier to determine than the MI. Second, I might have included other MII's belonging to some unidentified polychaete taxa in the count. Third, the MI is more fragile than the MII. Fourth, although this is difficult to prove, the MII may be represented by a double number of jaws in each apparatus. Fifth, the recorded frequencies are caused by the random variation in the preservational processes. Most probably the skew proportion of the two jaws is a question of preservation and identification.

Kettnerites (*K.*) *bankvaetensis* Bergman 1987

Figs. 13I–M, 18E, 29–32

Synonymy. – □1987 *Kettnerites* (*K.*) *bankvaetensis* n.sp. – Bergman, pp. 53–56, Figs. 13I–M, 18, 29–32.

Derivation of name. – Latin *bankvaetensis,* from the type locality Bankvät 1, where this species is common and dominates the annelid fauna in some levels.

Holotype. – LO 5829:5, right MII, Fig. 31F.

Type locality. – Bankvät 1.

Type stratum. – Hamra Beds, unit b.

Material. – Figs. 2 and 3; more than 250 elements each of right and left MI's and MII's.

Occurrence. – Figs. 4 and 7; Early Wenlock to Late Ludlow, Högklint to Hamra Beds, unit c, Sundre Beds lower(?) part; Ängmans 2, Baju 1, Bankvät 1, Bofride 1, Bottarve 1 and 2, Bringes 3, Däpps 1, Faludden 2, Fie 3, Fjäle 2 and 3, Fjärdinge 1, Gane 2, Gannor 1, Garnudden 3 and 4, Gisle 1, Gothemshammar 2 and 6, Gutenviks 1, Haganäs 1, Hide 1, Juves 4, Kättelviken 5, Kauparve 1, Krakfot 1, Kroken 2, Kullands 2, Möllbos 1, Nabban 2, Nyan 1, Nygårds 2, Ollajvs 1, Slitebrottet 1, Stora Myre 1, Storugns 1B, Strands 1, Suderbys 3, Valby Bodar 1, Valleviken 1, Vallstena 2, Västerbjärs 1, Vike 2 and 3, Yxne 1.

Diagnosis. – Right MI: Inner margin almost straight with closely spaced denticles of normal size. Shank sharply ended, flange angular, highly elevated posteriorly.

Left MI: Inner margin almost straight, denticles decrease in size posteriorly, ending without reaching the practically rhombic inner wing. Ligament rim forms a short ridge on posteriormost outer part of basal portion.

Right MII: Two fairly large pre-cuspidal denticles, the anterior one slightly larger; large cusp followed by one or two fairly small intermediate denticles.

Left MII: Double cusp with cusps of equal size, the anterior one positioned slightly more exteriorly. Shank tapers to a fairly pointed extremity.

Description. – Right MI, dorsal side: Length 0.35–1.63 mm, width about 1/4 of length. The inner and outer margins taper slightly toward the anterior end: a normally developed fang, moderately bent upward in relation to the dorsal surface. The inner margin is slightly convex among Wenlock jaws while it is straight among the Ludlow ones. The inner margin is covered by 24–44 normal-sized, close-spaced denticles, which are more or less perpendicular to the extension of the inner margin. The denticles occupy about 0.6–0.7 of the jaw length, and the number of denticles is to some degree coupled to jaw length. They extend from the anteriormost part of the falcal arch down to the inner wing, except in specimens from the Wenlock, where the denticulation ends before reaching the inner wing. The denticles are almost equal in size except for the posteriormost ones which decrease in size, ending as crenulations. The transition between the denticulated inner margin and the undenticulated ridge is characterized by crenulations on a small, winding, narrow ridge on top of the main ridge. The undenticulated ridge is fairly narrow, high, and its anterior part bends toward the outer margin. It continues down to the posteriormost part of the jaw, forming the sharp tip of the shank. The inner wing is of normal size, about a quarter of the jaw length, rounded and deeply downfolded, with its widest part at the anterior end.

The basal plate is not recorded.

The basal furrow is fairly long and deep posteriorly. The flange is angular, the posterior part highly elevated, with the posteriormost part somewhat downfolded, and the anterior part abutting the anterior end of the jaw. The liga-

Fig. 29 (p. 58). *Kettnerites* (*K.*) *bankvaetensis.* All specimens in dorsal view, ×120. □A. Right MI, LO 5836:7, Gandarve 1, Halla Beds, 71-81LJ. □B. Gothemshammar 6, Halla Beds, unit c, 73-24LJ; B1 left MI, LO 5795:1; B2 right MI, LO 5795:2; B3 left MII, LO 5795:3; B4 right MII, LO 5795:4. □ C. Vike 2, Slite Beds, *Pentamerus gothlandicus* Beds, 83-4CB; C1 left MI, LO 5782:1; C2 right MI, LO 5782:2; C3 left MII, LO 5782:3; C4 right MII, LO 5782:4. □D. Left MI, LO 5836:6, Gandarve 1, Halla Beds, 71-81LJ. □ E. Gothemshammar 2, Halla Beds, unit c, 75-35CB; E1 left MI, LO 5791:1; E2 right MI, LO 5791:2; E3 left MII, LO 5791:3; E4 right MII, LO 5791:4.

Fig. 30 (p. 59). *Kettnerites* (*K.*) *bankvaetensis.* All specimens in dorsal view. □ A. Möllbos 1, Halla Beds, unit b, 77-28LJ; A1 left MI, LO 5792:1, ×60; A2 right MI, LO 5792:2, ×60; A3 left MII, LO 5792:3, ×60; A4 right MII, LO 5792:4, ×60. □B. Gothemshammar 6, Klinteberg Beds, unit a, 73-30LJ; B1 left MI, LO 5799:1, ×80; B2 right MI, LO 5799:2, ×80; B3 left MII, LO 5799:4, ×80; B4 right MII, LO 5799:5, ×120; B5 left MII, LO 5799:6, ×120; B6 left MII, LO 5799:3, ×80. □C. Fjärdinge 1, Klinteberg Beds, unit b, 77-5CB; C1 left MI, LO 5796:1, ×80; C2 right MI, LO 5796:2, ×80; C3 right MII, LO 5796:3, ×80.

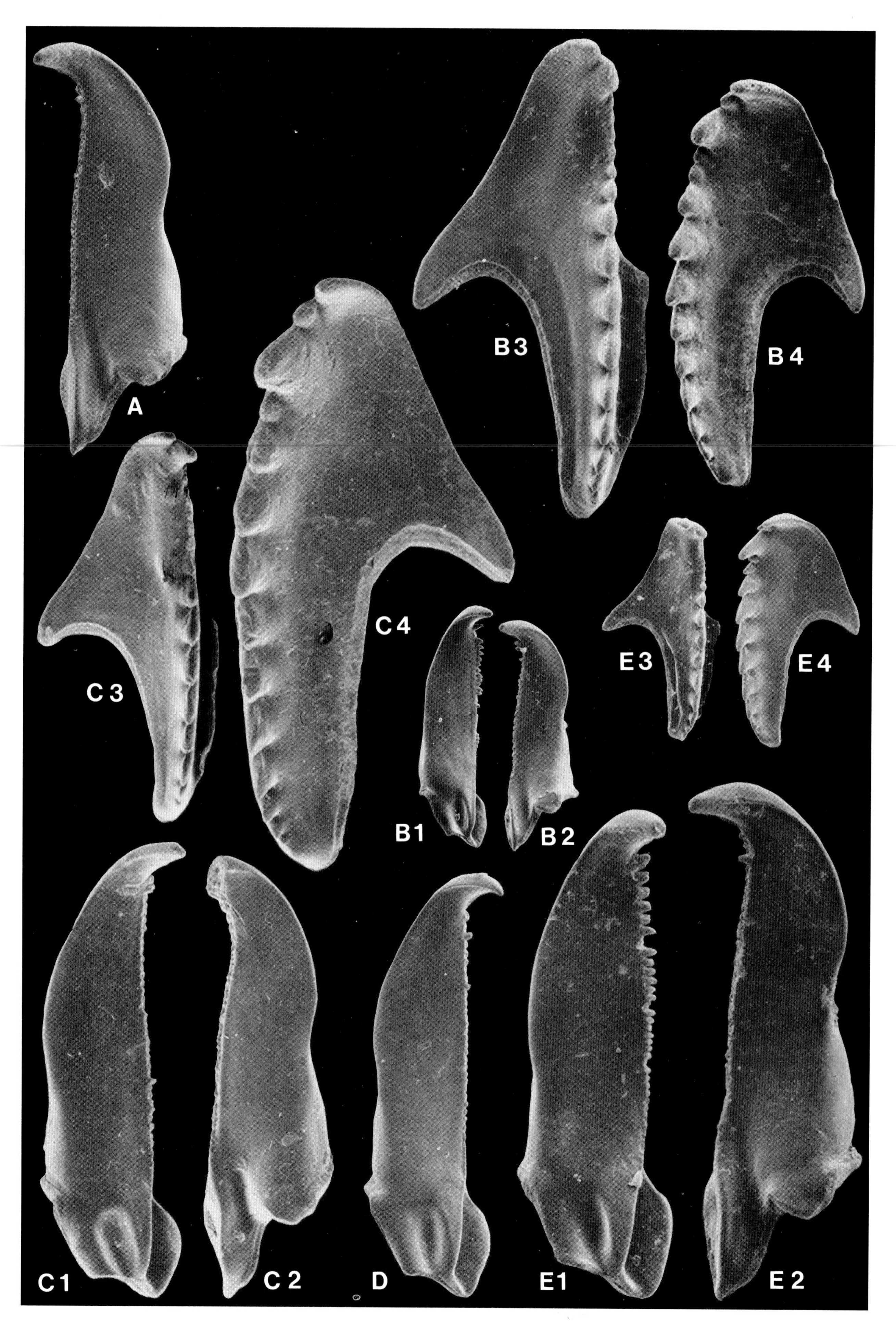

Fig. 29 (caption on p. 57).

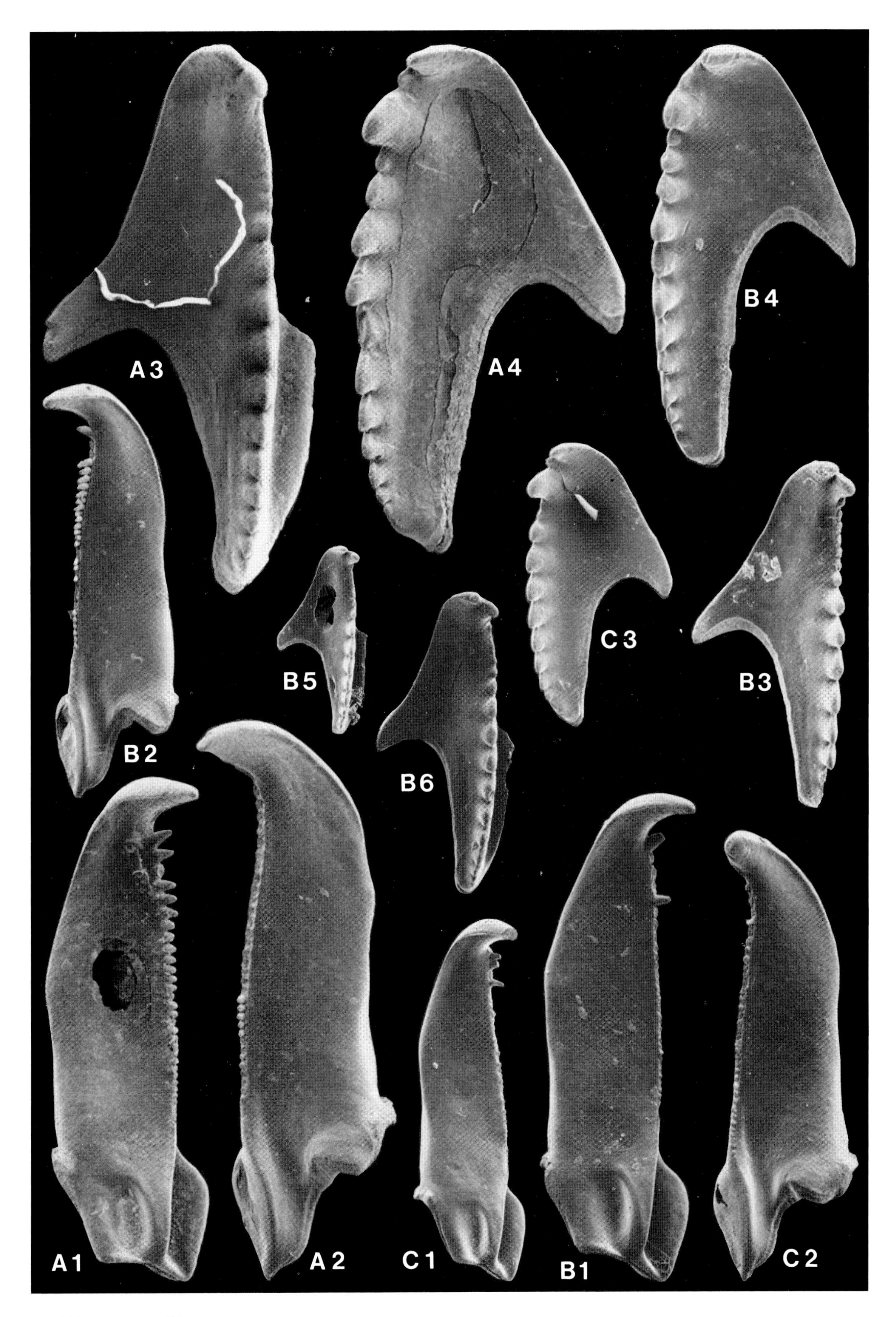

Fig. 30 (caption on p. 57).

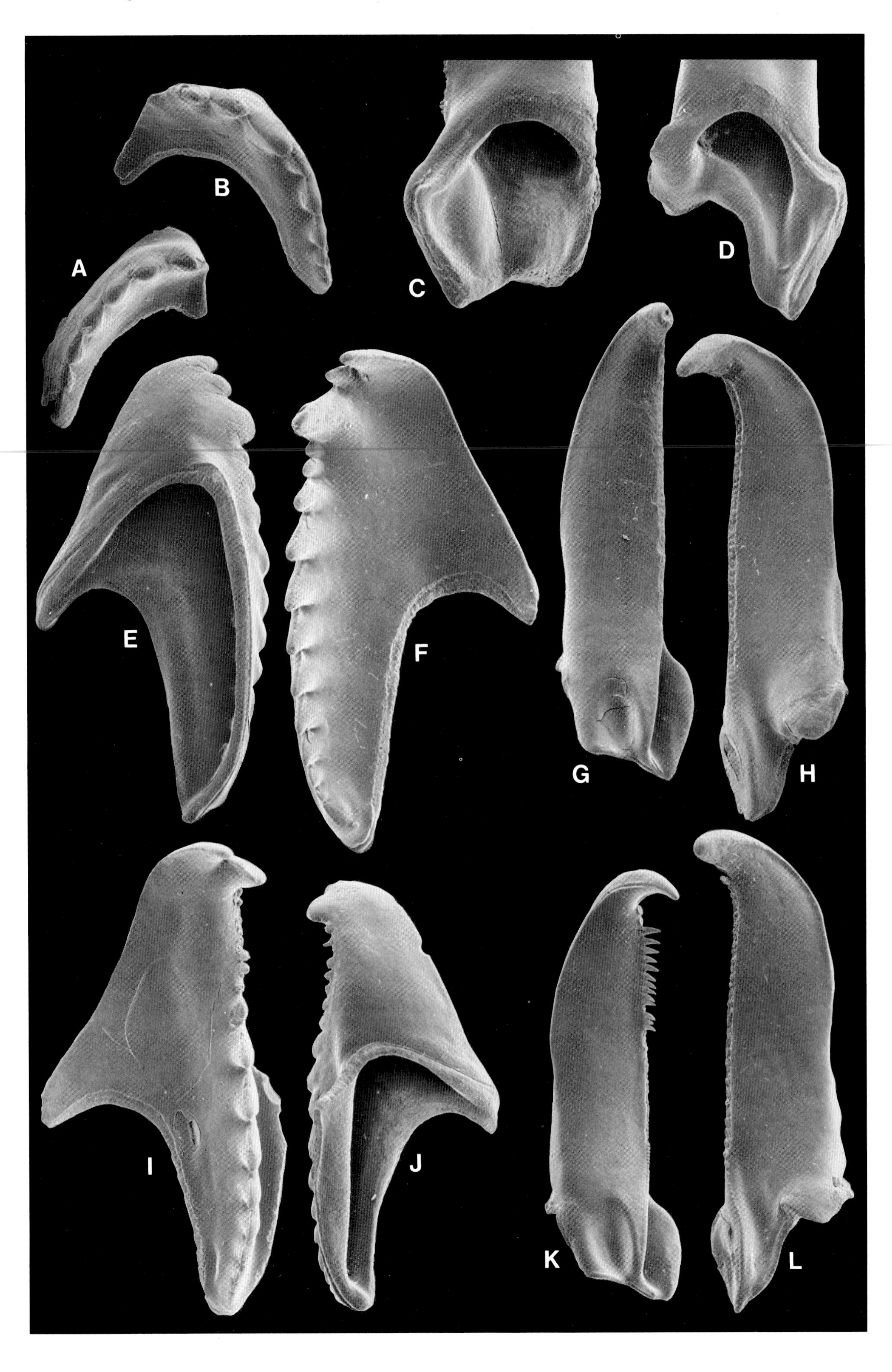
A
B
C
D
E
F
G
H
I
J
K
L

ment rim forms a short ridge with a spur-like structure along the anteriormost outer margin of the basal portion (Figs. 30A:2, B:2, C:2, 31D, H). The basal angle is between 28° and 50°.

Ventral side: The strongly enclosed, rounded myocoele opening, represents about 1/4 of the jaw length. The curved opening is surrounded by a broad and flat ligament rim, which rises to a fairly small and narrow ridge on the anterior outer part of the inner wing.

Left MI, dorsal side: Length 0.46–1.75 mm, width about 1/4–1/5 of length. The inner and outer margins are almost parallel, ending anteriorly in a normally developed fang which is slightly bent upward compared with the relatively flat dorsal surface. The inner margin is slightly convex among the Wenlock jaws but almost straight among the Ludlow ones. The inner margin is covered by 30–48 closely spaced denticles of normal size, extending posteriorly from the falcal arch to well in front of the inner wing. The denticles decrease in size toward the posterior, where they end as small knobs. There is a smooth transition to the undenticulated ridge, which continues as a low, narrow and straight ridge down to the posterior margin. The inner wing is of normal size, almost rhombic, somewhat rounded, with its inner margin parallel to the undenticulated ridge. The basal furrow, to the left of the ridge, is relatively short, deep and oriented parallel to the outer margin of the basal portion. The basal angle is between 20° and 35°, and the outer margin of the outer face of the basal portion is downfolded, its anterior part forming a small flange or a narrow fold, the ligament rim. The outer margin, anterior to the basal portion, is slightly concave, followed anteriorly by a slightly angular falx.

Ventral side: The myocoele opening is strongly enclosed, slightly crescent-shaped, and about one quarter as long as the jaw. It is anteriorly delimited by a broad, concave ligament rim, which forms a narrow and high ridge along the inner side of the innermost part of the inner wing.

Right MII, dorsal side: Length 0.38–1.30 mm, width slightly more than half the length. There are two fairly large pre-cuspidal denticles, the anterior one somewhat larger and situated slightly to the right of the second one. In three specimens the pre-cuspidal dentary is composed

Fig. 31 (opposite page). *Kettnerites* (*K.*) *bankvaetensis.* Bankvät 1, Hamra Beds, unit b, 82-30CB, except G and H. All specimen are in dorsal view except C, D, E and J (ventral). □A. Left MIV, LO 5829:10, ×120. □B. MIII, LO 5829:9, ×120. □C. Basal part left MI, LO 5829:8, ×120. □D. Basal part of right MI, LO 5829:7, ×120. □E. Right MII, LO 5829:6, ×120. □F. Holotype, right MII, LO 5829:5, ×120. □G. Left MI, LO 5841:1, Glasskär 3, Burgsvik Beds, lowermost part, 82-18CB, ×80. □H. Right MI, LO 5841:2, same sample as G, ×80. □I. Left MII, LO 5829:3, ×120. □J. Left MII, LO 5829:4, ×120. □K. Left MI, LO 5829:1, ×80. □L. Right MI, LO 5829:2, ×80.

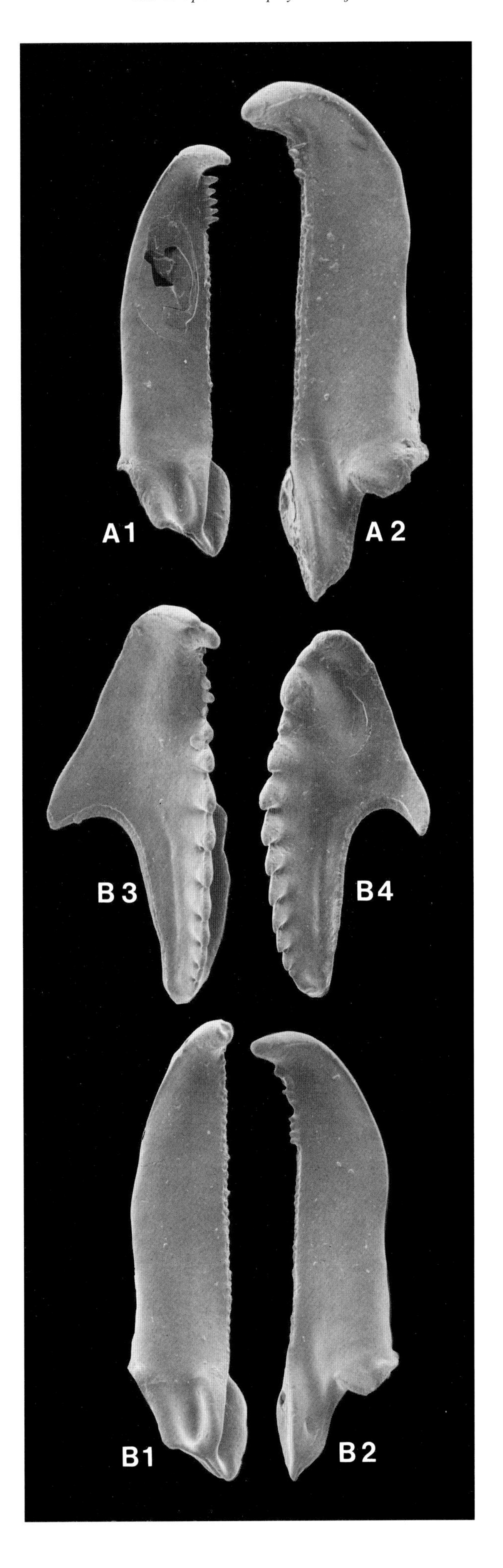

Fig. 32 (right). *Kettnerites* (*K.*) *bankvaetensis.* All specimens in dorsal view. □A. Västerbackar 1, Sundre Beds, middle–upper part, 75-2LJ; A1 left MI, LO 5838:4, ×80; A2 right MI, LO 5838:5, ×80. □B. Kauparve 1, Hamra Beds, lower–middle part, 76-13CB; B1 left MI, LO 5844:1, ×60; B2 right MI, LO 5844:2, ×60; B3 left MII, LO 5844:4, ×80; B4 right MII, LO 5844:5, ×60.

of one larger denticle followed posteriorly by two small ones. The cusp is large, its continuation is seen as a minor swelling of the jaw. It is followed by one or two small denticles, the intermediate dentary. The transition to the dentary of the denticulated ridge is marked by 2–3 denticles which increase in size. The 6–9 large denticles of the ridge decrease in size posteriorly and slant in the posterior direction. The shank, occupying slightly more than half the jaw length, has a wide anterior part. The inner margin of the shank is convex, the outer straight, forming a fairly pointed extremity. The bight is fairly deep with an acute bight angle from about 50° to 75°. The ramus is of moderate to fairly large size, and a vague sinus can be seen along the anterior outer margin of some jaws.

Ventral side: The slightly enclosed myocoele opening represents about ⅔, or slightly more, of the jaw length. A fairly wide rim forms the almost straight anterior margin of the myocoele opening, largest on the inner side.

Left MII, dorsal side: Length 0.25–2.08 mm, width half the length or slightly more, irrespective of the size of the jaw. The double cusp has parts of equal and moderate size, the two parts being almost side by side, the anterior one to the left. The intermediate dentary is distinct with 5–9 small to fairly small denticles, increasing in size to the posterior. The number of intermediate denticles and denticles on the ridge has no relation to jaw size. The transition to the almost straight, denticulated ridge is gradual, with the normally 7–9 denticles increasing in size. The shank usually occupies about half the jaw length; its anterior part is wide, it taper posteriorly, and its extremity is fairly pointed, slightly bent outwards. The inner wing is prominent, length about half the jaw length, its almost triangular anterior part being widest and smoothly rounded. The bight is of moderate size; the bight angle usually between 70° and 80°. The ramus is fairly large and the outer anterior sinus of varying size.

Ventral side: The slightly enclosed myocoele opening represents half, or slightly more, of the jaw length. A fairly wide ligament rim forms the anterior margin of the opening, nearly straight and almost perpendicular to the extension of the jaw.

Comparison. – It is almost impossible to distinguish between the MI's of *K.* (*K.*) *bankvaetensis* and the corresponding element of *K.* (*K.*) *martinssonii* from the middle Wenlock (e.g. Vike 2, Slite Beds, unit *Pentamerus gothlandicus* or slightly older). From this stratigraphical level there is no clear character which can be used for the differentiation of the elements. The shape of the elements evolves, and those from only slightly younger levels (e.g. Nygårds 2, Halla Beds, unit b) show at least small differences: both left and right MI of the *K.* (*K.*) *bankvaetensis* have an almost straight inner margin, and on the MI's the inner wing is more nearly rectangular. The inner wing of the right MI's of *K.* (*K.*) *bankvaetensis* has a straight inner margin and lacks the posteriormost tip of the shank characteristic of *K.* (*K.*) *martinssonii.* The change of morphology continues through time, and there is no particular similarity between the two species in the Ludlow strata. The basal portions of the MI's are less angular than the almost quadratic basal portions of *K.* (*K.*) *polonensis* cf. *polonensis* and the denticles are smaller than in that taxon.

The MII's have the same denticle formula as several of the other paulinitid species on Gotland (e.g., *Kettnerites* (*K.*) *polonensis, K.* (*K.*) *burgensis, K.* (*K.*) *abraham, K.* (*K.*) *jacobi, K.* (*K.*) *huberti,* including their varieties and the valle variety of *K.* (*A.*) *sisyphi*). However, the pre-cuspidal denticles of *K.* (*K.*) *bankvaetensis* with the anteriormost denticle slightly larger than the following one, and with the tapering and fairly sharply ending shank, makes the element distinguishable. The left MII is less easy to distinguish; the anteriormost cusp is situated slightly external to the second cusp. The shank tapers towards the posterior and is fairly pointed at the end.

Kettnerites (*K.*) *burgensis* (Martinsson 1960)

Figs. 13H, 18B, 33G, H, I, J, K, L

Synonymy. – □1960 *Paulinites burgensis* n.sp. – Martinsson, pp. 3–5, Fig. 1, 1–5. □1987 *Kettnerites* (*K.*) *burgensis* (Martinsson 1960) – Bergman, pp. 56–59, Figs. 13H, 18, 33G–L.

Holotype. – 'Paulinites' burgensis Martinsson 1960 pp. 3–5, Pl. 1:1–5, a jaw assemblage of left and right MI and MII, and carriers. The carriers are not sufficiently preserved to allow any description.

Type stratum. – Hemse Beds, Hemse Marl SE part.

Type locality. – Västlaus 1.

Material. – Fig. 3; more than 60 right MI, 72 left MI, 83 right MII, 59 left MII.

Occurrence. – Figs. 4 and 5; Middle Ludlow, Hemse Beds, Hemse Marl NW, unit c, d, Hemse Marl SE, and Marl uppermost part. Bodudd 3, Botvide 1, Fie 3, Gerumskanalen 1, Gogs 1, Gutenviks 1, Kärne 3, Klasård 1, Kullands 2, Linviken 2, Vaktård 2, 3, 4 and 5, Västlaus 1.

Diagnosis (emended). – Right MI: Slender jaw, coarsely denticulated, anteriorly with paucidentate or with slightly spaced dentary. Pointed shank, angular, fairly thick flange, distinct spur.

Left MI: Slender jaw, coarsely denticulated like right MI. Basal portion angular. Inner wing small, rounded; distinct spur.

Right MII: Two large pre-cuspidal denticles, cusp of moderate size or one large pre-cusp and a cusp of moderate size. Post-cuspidal dentary ending well before reaching the posterior margin. Slender, pointed ramus.

Left MII: Large cusp, with or without a very large denticle immediately posterior to it. Widely spaced intermediate dentary with fairly large denticles. Slender shank. Almost triangular, slender, pointed ramus.

Description. – Right MI, dorsal side: Length 0.18 to more than 2.61 mm, width about ¼ of the length, more among the small specimens and less among the large ones. The jaw tapers very slightly towards the anterior, ending in a large, sickle-shaped fang which is only very slightly bent upwards in relation to the almost flat dorsal surface. The inner margin is slightly convex to almost straight. Its denticula-

Fig. 33. Kettnerites (*K.*) *burgensis*. All specimens in dorsal view. All specimens ×120 except I, K and L. A–F: *K.* (*K.*) cf. *burgensis*, Vaktård 4, Hemse Beds, Hemse Marl, SE part, 81-35LJ). G–L: *K.* (*K.*) *burgensis*, Västlaus 1, Hemse Beds, Hemse Marl, SE part, 82-31LJ. □A. Left MII, LO 5822:5. □B. Right MII, LO 5822:6. □C. Left MII, LO 5820:3, ×170. □D. Left MI, LO 5822:7. □E. Right MI, LO 5822:8. □F. Right MI, LO 5822:9. □G. Right MII, LO 5806:8. □H. Left MII, 5806:7. □I. Left MI, 5806:5, ×80. □J. Left MI, 5846:2. □K. Right MI, 5846:1, ×60. □L. Left MI, 5806:6, ×80.

tion extends about ⅔ of the jaw length, from the anterior part of the falcal arch to near the inner wing. The denticles are large anteriorly, decrease in size towards the posterior, ending as small knobs. There are two main types of denticulation as well as intermediate varieties: the extremely paucidentate form and the form with only widely spaced denticles. Paucidentation seems to be slightly more common among the larger specimens and is almost lacking among the smallest. Paucidentate forms have more triangular denticles, laterally elongated along the length of the jaw. The anterior denticles are usually perpendicular to the extension of dentary, while the posterior denticles may or may not slant slightly towards the posterior. The number of denticles varies between 14 and 18, with an average of 15, and a slight positive correlation to jaw size. The undenticulated ridge is fairly low, short, about 0.2 of the jaw length, almost straight, its posterior part bent slightly to the outside. The transition to the undenticulated ridge is very smooth. The inner wing is almost triangular, small and deeply downfolded, widest anteriorly with its corner varying from angled to rounded. The posteriormost part of the inner wing forms, together with the undenticulated ridge, the sharp-ended posterior part of the fairly small, almost triangular shank. The basal furrow is fairly short and shallow.

The basal plate is not recorded.

The flange is moderately thick-walled, its posterior part highly elevated and only slightly downfolded along the outer posterior margin. A conspicuous spur marks the anterior margin of the basal portion. The basal angle is about 20–25°. The outer margin of the basal portion varies, but there is always a concavity posterior to the large conspicuous spur. The outer margin, anterior to the basal portion, is almost straight with a wide minor concavity on the posterior part.

Ventral side: The strongly enclosed myocoele opening represents ⅓–¼ of the jaw length, smaller jaws having proportionally larger openings. In front of the opening a narrow groove surrounds the somewhat sunk, smoothly rounded ligament rim. The outer anteriormost margin of the rim is transformed into a prominent, pointed, high and narrow spur. The groove and spur are also seen in dorsal view. A high, sharp ridge is present along the inner wing, highest on its anterior part and decreasing evenly in size posteriorly.

Left MI, dorsal side: Length 0.19–2.67 mm, width slightly less than ¼ of length. The large, sickle-shaped fang is slightly bent upwards in relation to the dorsal surface. The jaw tapers continuously anteriorly from the posterior, middle part of the jaw. The inner margin is slightly convex, covered by moderately large to large denticles over slightly less than ⅔ of the jaw length. The anteriormost two denticles are fairly small, the third usually the largest, the following 10–17 decreasing in size to the posterior, the dentition ending as small knobs well before reaching the inner wing. The dentary varies on the anterior part from paucidentate to slightly spaced. The denticles are oriented more or less perpendicular to extension of the jaw, though the posterior dentary usually slants slightly towards the posterior. The denticulated inner margin continues smoothly onto the straight, undenticulated ridge without any notable point transition. The inner wing is almost rectangular, fairly small, about ⅕ of the jaw length. The anterior outer end is rounded, the inner margin parallel to the undenticulated ridge. The basal furrow is very short, parallel to the outer margin of the basal portion. The basal portion is fairly wide and short, somewhat angular. The basal angle is about 30–40°. Along the anterior half of the outer margin of the basal portion runs the ligament rim, with a conspicuous spur. The outer margin, anterior to the basal portion, is almost straight or with a wide and shallow concavity in its posterior part.

Ventral side: The strongly enclosed myocoele opening represents about ⅓ of the jaw length among the smaller jaws, to ¼ among the larger ones. The smoothly rounded, narrow ligament rim is delimited anteriorly by a narrow groove. Near its outer part the rim is transformed into a narrow, short ridge. The spur is similar to the one on the right jaw.

Right MII, dorsal side: Length 0.24–2.27 mm, width slightly more than half the length. Two types of denticulation exist with only some relation to jaw size. In the first type, two large pre-cuspidal denticles are followed by a cusp of moderate size, only slightly larger than the pre-cusps, and posterior to the cusp there are one or two smaller intermediate denticles. The second type has a double cusp with parts of almost equal size, followed by one or two smaller intermediate denticles. The number of post-cuspidal denticles of both types are 8–10, the number having no relation to jaw size or anterior dentary. The denticles are fairly large, slanting slightly towards the posterior, the dentition ending before reaching the fairly pointed posterior extremity. The shank varies from wide to fairly slender, slightly tapering, occupying about half of the jaw length. The inner margin is slightly convex. The deep bight has an acute angle usually between 50° and 60°. The slender ramus is almost triangular and has a pointed extremity. The anterior outer margin varies from nearly straight to slightly concave.

Ventral side: The slightly enclosed myocoele opening represents about ⅔ of the jaw length. The ligament rim is fairly narrow, the margin of the opening curves smoothly, with an inner–anterior bulge.

Left MII, dorsal side: Length 0.32–1.90 mm, width slightly less than half the length. The cusp is large and single, or slightly smaller and double, almost equal-sized, forming a sickle-shaped anterior margin. The intermediate dentary is represented by 4–6 fairly large, widely spaced denticles. The 8–10 fairly large, slightly slanting post-cuspidal denticles end at the posteriormost part of the fairly slender shank, which occupies about half of the jaw length. The inner wing is almost triangular, with its widest part anteriorly. The bight angle varies from 60° to almost 90°. The ramus is fairly slender and pointed, its posterior part often bent to the posterior. The anterior outer margin has a sinus.

Ventral side: The slightly enclosed myocoele opening represents about ⅔ of the jaw length. The myocoele opening is crescent-shaped, its rim fairly narrow.

Remarks. – The dentition of the MI and MII varies from paucidentate in some MI's to only widely spaced denticulation. The right MII has one or two pre-cuspidal denticles, and the left MI one cusp followed by a denticle of normal size, or a cusp followed by a large denticle. The differences are usually consistent within and between samples and are not regarded as being of taxonomic importance above the population level. From some samples only one of the forms has been encountered. This might be an artifact due to the small samples. Some specimens do not fit exactly into the description of this species and so have been determined as *K.* (*K.*) cf. *burgensis* (Figs. 33A–F).

Comparison.- The basal portion of the MI is similar to the corresponding part of *K.* (*K.*) *polonensis* cf. *polonensis* but differs in having a more pronounced spur and a more highly elevated flange. However, the size of the spur is not a particularly good character because it is difficult to estimate when seen from above. The dentition is a more conspicuous character, though in 'wide-spaced' forms the similarity to *K.* (*K.*) *polonensis* cf. *polonensis* is close. The left MII is the best element by which to differentiate the two species (Figs. 33A, H, 43A:3, 44A:3, B:3, C:3).

Kettnerites (*K.*) *huberti* Bergman 1987

Figs. 13R, S, T, U, 18D, 34, 35, 36

Synonymy. – □1970 *Paulinites polonensis* Kielan–Jaworowska 1966 – Szaniawski, pp. 465–466, Pl. I:5A–D. □1987 *Kettnerites* (*K.*) *huberti* n.sp. – Bergman, pp. 59–62, Figs. 13R–U, 18, 34–36.

Derivation of name. – Named in honour of Dr Hubert Szaniawski, Warsaw, who was the first to describe specimens of the species.

Holotype. – LO 5833:7, right MI, Fig. 36H.

Type locality. – Bankvät 1.

Type stratum. – Halla Beds, unit b.

Material. – Fig. 3; more than 250 right MI, 250 left MI, 250 right MII, 250 left MII.

Occurrence. – Figs. 4 and 6; Wenlock to Late Ludlow. Slite Beds, Slite Marl, undifferentiated, to Hamra Beds, unit c. Amlings 1, Ängmans 2, Autsarve 1, Bandlunde 1, Bankvät 1, Barkarveård 1, Bodudd 1, 2, and 3, Botvide 1, Djaupviksudden 4, Faludden 2, Fie 3, Gannor 1 and 3, Garnudden 3 and 4, Gogs 1, Grogarnshuvud 1, Grundård 2, Gyle 2, Hallsarve 1, Hide 1, Hoburgen 2, Hummelbosholm 1, Juves 3, Kapelludden 1, Kärne 3, Kättelviken 5, Klasård 1, Kroken 2, Lambskvie 1, Linviken 2, Myrsne 1, När 2, Nisse 1, Nyan 2, Öndarve 1, Petsarve 2, Petsarve 15, Ronnings 1, Snauvalds 1, Stora Kruse 1, Strands 1, Tjängdarve 1, Tomtbodarne 1, Träske 1, Vaktård 4 and 5, Västlaus 1, Vidfälle 1.

Diagnosis. – Right MI: Inner margin normally strongly convex with small denticles in falcal arch and large to very large denticles on posterior part of the denticulated, arched inner margin. Shank blunt-ended, flange fairly thick-walled, basal portion wide and angular.

Left MI: Inner margin convex, with somewhat smaller denticles than right MI. Almost rectangular inner wing. Basal portion wide and angular.

Right MII: Two large pre-cuspidal denticles, large cusp, large denticles on the shank, nearly parallel, slightly convexo-concave to almost straight sides of the shank, large, wide, approximately triangular ramus.

Left MII: Keel along the anterior outer margin ending in small but distinct pre-cusp. Large, approximately triangular ramus.

Description. – Right MI, dorsal side: Length 0.32–1.40 mm, width ¼ of length or slightly less. The inner and outer margins are almost parallel or taper in the anterior direction, ending in a fairly large fang, moderately bent upwards in relation to the dorsal surface. The inner margin is usually convex but varies from almost straight to strongly convex, sometimes within one sample (e.g. the Ludlow Bankvät 1). The inner margin is denticulated along 0.6–0.7 of the jaw length, from the anteriormost part of the falcal arch almost to the anteriormost part of the inner wing. The denticulation is very varied. The anterior 8–14 denticles are fairly small, slightly spaced, with the anteriormost denticles directed forward. Succeeding denticles gradually increase in size toward the central part of the dentary and then decrease from the lower middle part of the posterior dentary. This is more pronounced in the larger jaws. The denticulation ends beside or slightly posterior to the anterior part of the inner wing, normally very abruptly, without any decline in size of the posteriormost denticles (except in, e.g., Hallsarve 1, Gannor 3, and Bankvät 1). The almost triangular inner wing is deeply downfolded, anteriorly rounded, widest part anteriorly, with a sometimes concave inner margin. The undenticulated ridge is fairly large, often bent slightly towards the outer margin, sometimes with a smaller narrow costa parallel to and on top of the anteriormost part of the ridge. The ridge, the extremity of which is bent towards the inner margin, forms the posterior part of the jaw, the blunt-ended shank. The basal furrow is of normal length and width. The flange is fairly thick-walled, very varied in shape from angular with the outer margin bent down (Fig. 36J, Bankvät 1, Hamra Beds) to rounded (Fig. 34E, Gannor 3, Hemse Beds), with a furrow along its outer margin of the basal portion and its outermost margin folded upward. Much, but not all, of the variation depends on the preservation, i.e. compaction of the jaws increases the size of the more or less pronounced folds and furrows. The basal portion is wide and distinct, but there is a large variation in shape. The basal angle is low to very low, almost zero. The outer margin is normally concave, but variation occurs from straight to strongly concave, especially in the Ludlovian material. The outer margin ends in a rounded falx.

Ventral side: The strongly enclosed myocoele opening occupies about ⅓ of the jaw length. The anterior margin is

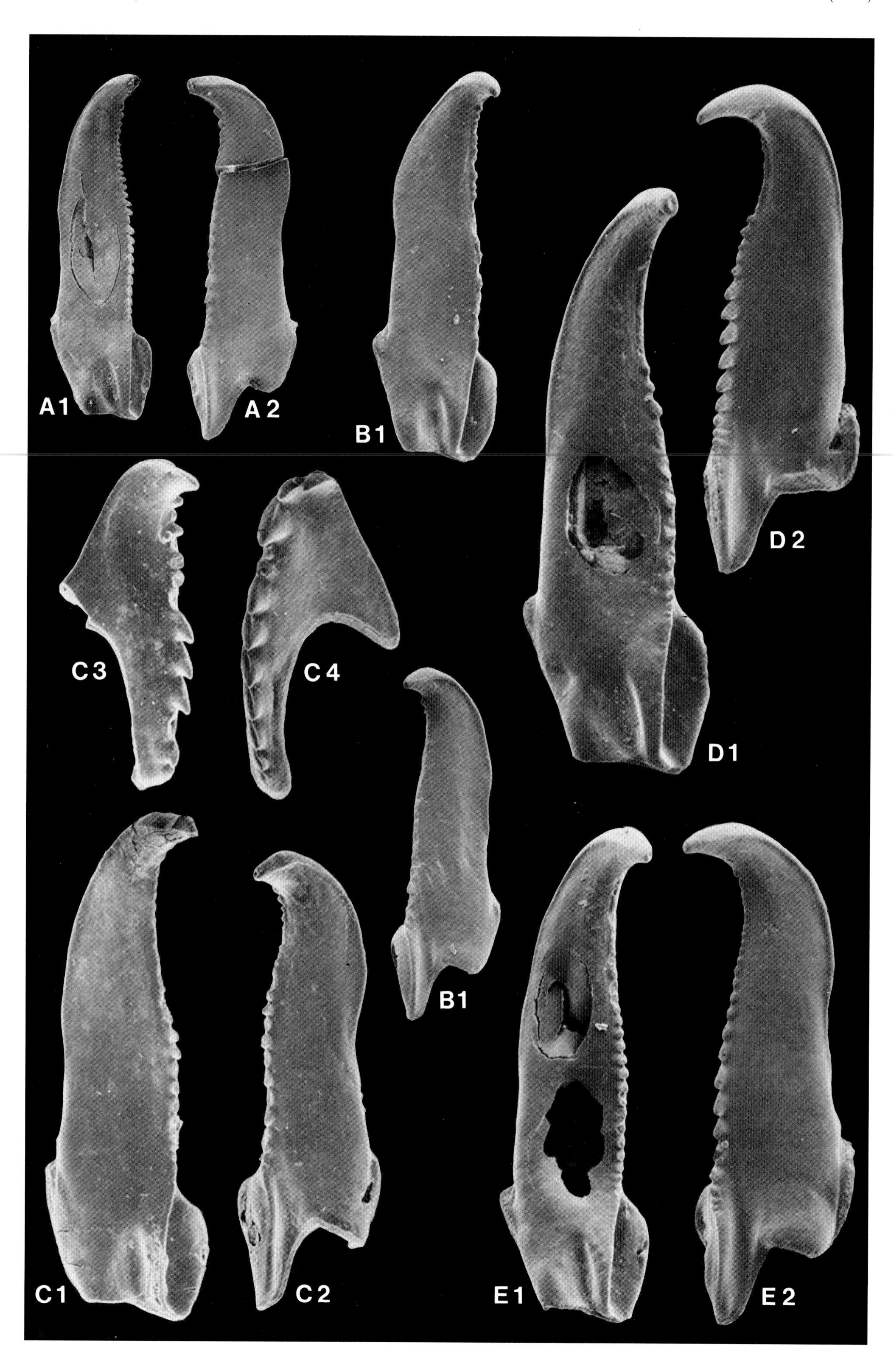
A1
A2
B1
D2
C3
C4
D1
B1
C1
C2
E1
E2

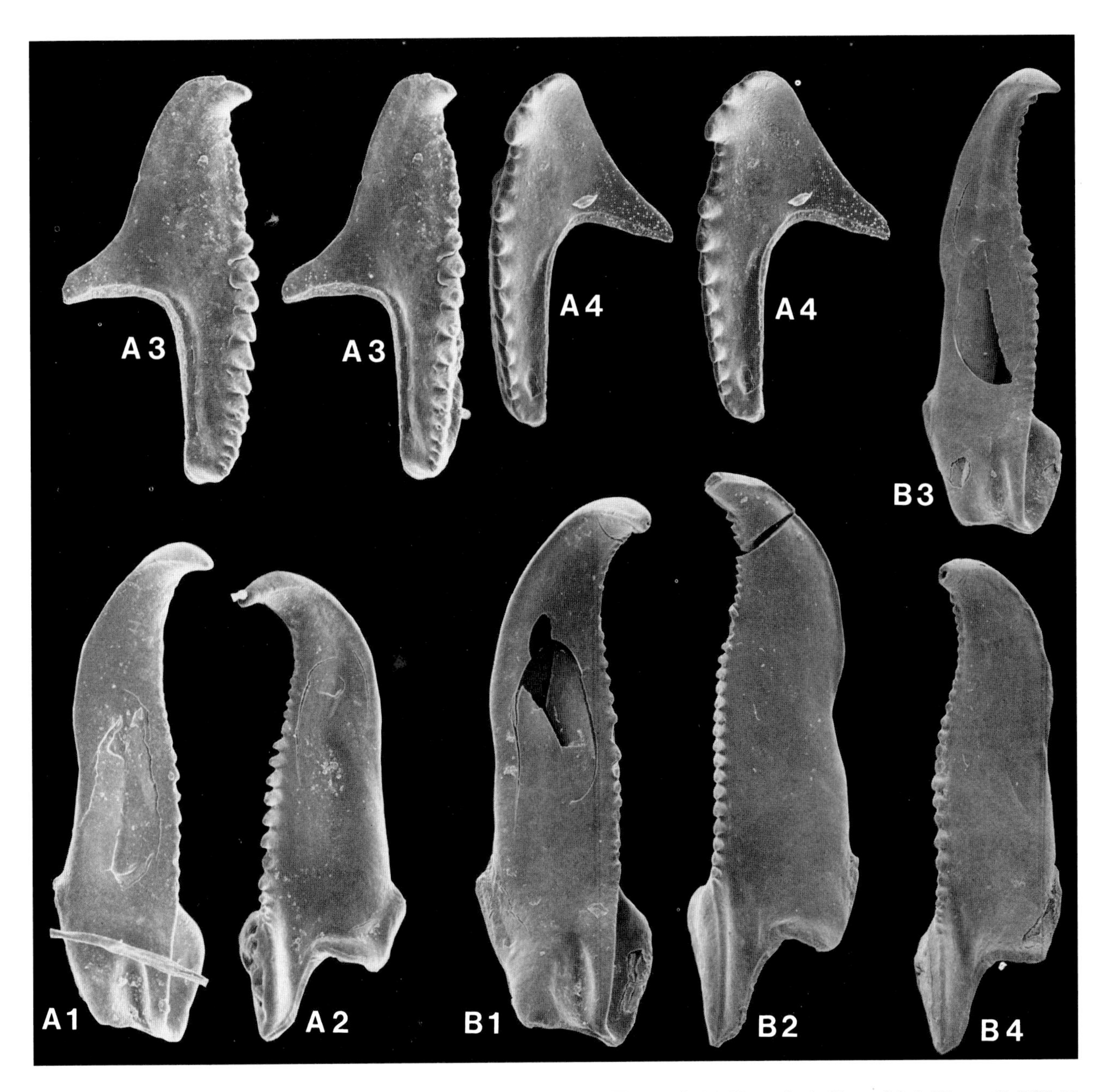

Fig. 35. Kettnerites (*K.*) *huberti.* All specimens in dorsal view, ×120, except B4. □A. Vaktård 4, Hemse Beds, Hemse Marl, SE part, 81-35LJ; A1 left MI, LO 5822:10; A2 right MI, LO 5822:11; A3 left MII, stereopair, LO 5822:12; A4 right MII, stereopair, LO 5822:13. □B. Strands 1, Hamra Beds, unit b, 75-14LJ; B1 left MI, LO 5840:1; B2 right MI, LO 5840:2; B3 left MI, LO 5840:3; B4 right MI, LO 5840:4, ×80.

crescent-shaped and surrounded by the ligament rim, a roll-shaped ridge which in turn is separated from the anterior of the jaw by a narrow groove. In the anteriormost part of the inner wing the ridge forms a narrow, high (in lateral view, triangular) process.

Left MI, dorsal side: Length 0.30–1.52 mm, width less than ¼ of length. The jaw tapers anteriorly, ending in a fairly large fang, bent moderately upwards relative to dorsal surface. In a few jaws, the inner and outer margins are almost straight and parallel, though the inner margin is normally distinctly convex. Denticulation covers 0.5–0.7 of the jaw length, from the anteriormost part of the falcal arch to near the anteriormost part of the inner wing. The denticles in the falcal arch range in number and size from about 5 moderately large, widely spaced denticles to about 15 small ones, the latter type being more common. In some jaws the denticles are more or less widely spaced crenulations on a low ridge. Posterior to the falcal arch follow larger slanting denticles, though the difference in size between the anterior and posterior ones is less than in the right MI. The denticulation ends abruptly or with some small denticles. The undenticulated ridge is low and fairly narrow, forming a smooth, straight continuation of the denticulated inner margin. The somewhat downfolded

Fig. 34. Kettnerites (*K.*) *huberti.* All specimens in dorsal view, ×120. □A. Kärne 3, Eke Beds, lowermost part, 71-198LJ; A1 left MI, LO 5843:4; A2 right MI, LO 5843:5. □B. Gannor 1, Eke Beds, lowermost part, 71-123LJ; B1 left MI, LO 5811:5; B2 right MI, LO 5811:6. □C. Hide 1, Slite Beds, Slite Marl, 73-2LJ; C1 left MI, LO 5784:1; C2 right MI, LO 5784:2; C3 left MII, LO 5784:3; C4 right MII, LO 5784:4. □D. Hallsarve 1, Hemse Beds, Hemse Marl, uppermost part, 69-28LJ; D1 left MI, LO 5803:1; D2 right MI, LO 5803:2. □E. Gannor 3, Hemse Beds, Hemse Marl, SE part, 71-125LJ; E1 left MI, LO 5814:1; E2 right MI, LO 5814:2.

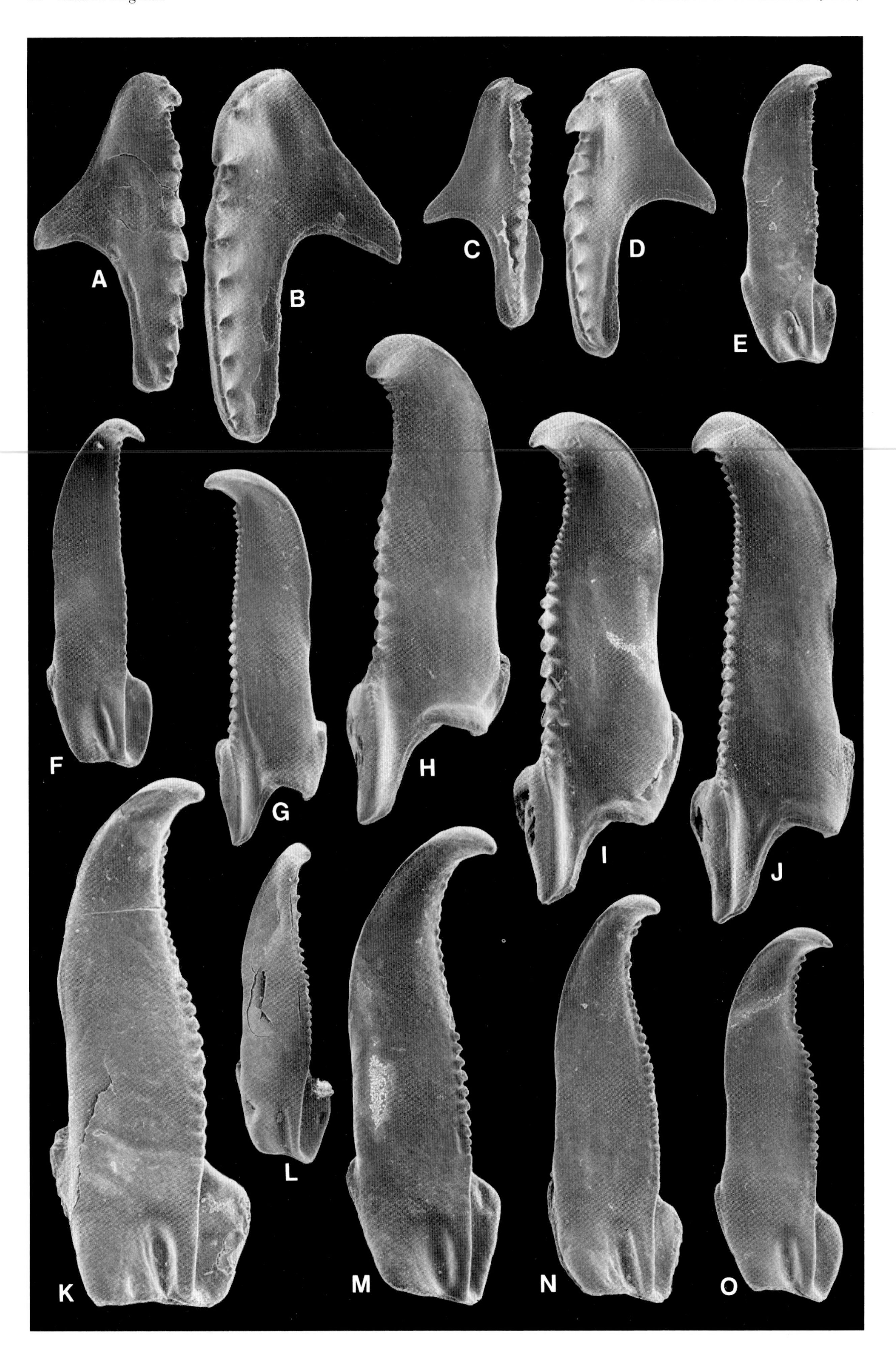
A
B
C
D
E
F
G
H
I
J
K
L
M
N
O

inner wing is fairly large, 0.25–0.30 of the jaw length, almost rectangular with its inner margin straight to slightly convex, and oriented parallel to the denticulated ridge or with the anterior corner pointing away from the latter. The basal furrow is fairly long and narrow. The posterior margin is long and almost straight. The outer margin of the basal portion represents almost 0.3 of the jaw length and is convex to straight with a smaller 'outer wing' on its anteriormost part. The basal portion is wide, angular and prominent. The basal angle varies but is usually low. The outer margin in front of the basal portion is more or less concave, ending in a large, sickle-shaped falx.

Ventral side: The strongly enclosed myocoele opening, occupies about ⅓ of the jaw length. The anterior margin of the opening is crescent-shaped, drawn slightly forwards along the inner margin of the jaw. A fairly narrow, rounded ridge is surrounded by a groove. The ridge forms a small process on the anterior part of the inner wing and a sharp edge along the ventral inner margin of the inner wing. Along the outer margin it is unfolded.

Right MII, dorsal side: Length 0.22–1.76 mm, width about half the length or slightly less. Two large pre-cuspidal denticles, with distinct cutting edges, form a large protruding anterolateral part of the jaw, with an acute to right-angled margin. The cusp is fairly large and posteriorly directed; it is followed by 2–3 intermediate denticles increasing in size towards the posterior, and normally 6–8 fairly large succeeding denticles. All denticles along the slightly convex to almost straight inner margin slant. The inner wing is very narrow. The shank is long, normally slightly more than half the jaw length, with convexo-concave to almost parallel sides. The posterior extremity of the Wenlock specimens is directed outwards (Fig. 34C:4), while the termination on the Ludlow specimens is more or less straight and blunt. The shank is also slender among the Wenlock specimens. The bight angle varies among different populations, from acute (Fig. 34C:4) to almost right angled (Figs. 35A:4, 36D). The ramus is fairly short and broad, the anterior outer margin has a more or less pronounced sinus.

Ventral side: The slightly enclosed myocoele opening represents ⅔–¾ of the jaw length. The ligament rim is narrow, slightly curved, the opening extending forwards along the inner margin.

Left MII, dorsal side: Length 0.24–2.03 mm, width about half the length. A small but distinct pre-cuspidal denticle forms the anteriormost part of the jaw; its continuation is represented by a characteristic smooth keel running along the anterior outer margin and disappearing in the sinus at the base of the ramus. The cusp is of normal size, usually followed by 4–6, sometimes 8–10, intermediate denticles of equal size. The size of these denticles varies among the different populations. The inner wing is fairly short and rounded. The inner and outer margins of the shank are almost parallel and the shank represents somewhat less than half of the jaw length. Normally the posteriormost part of the shank projects posteriorly, though in the Wenlock material, a slightly outward direction is normal. The bight angle is more or less 90°. The ramus is fairly short with a distinct sinus in the outer margin at the basal part.

Ventral side: The slightly enclosed myocoele opening represents about half of the jaw length. The slightly crescent-shaped opening is surrounded by a narrow ligament rim which gets wider at the outer side.

Remarks. – Wenlock and Lower Ludlow populations seem to be less varied than those of the Middle and Upper Ludlow. This could be an effect of the very limited material from the older strata. The population from the Middle Wenlock locality, Hide 1, is very similar to the normal-shaped specimens from, e.g., the Upper Ludlow locality Bankvät 1. However, the high number of intermediate and post-intermediate denticles found among some left MII from the Ludlow, distinguish them from the normal form. I suspect that the left MII with the larger number of denticles (Fig. 36C) should be grouped with the slender MI (Figs. 35A:1, B:3). Thus, the species seems to show a phylogenetic trend towards at least three different morphs. The normal morph has an MI with a pronounced convex inner margin carrying large, spaced denticles (e.g. Figs. 34C:1, C:2, E:1, E:2, 36H, I, K, M). The second morph has a less wide and almost straight MI, and slightly smaller denticles (Figs. 36F, G, J). The third, less common, morph has a slender s-shaped left MI and a right MI with reverse s-shape and smaller denticles. As late as in the Late Ludlow, the morphs can be found in the same samples. This implies that they are still to be regarded as one species.

Discussion. – This species seems to have been widely spread in the shallow sea of Baltica. The Upper Ludlow species from the Mielnik borehole in Poland (identified by Szaniawski 1970 as *Paulinites polonensis* Kielan-Jaworowska), known from three apparatuses and isolated elements, is very similar to the Gotland species and is probably conspecific with the latter. The type specimen of *Oenonites aspersus* (Hinde 1880, p. 373, Pl. 14:7) is compressed and partly buried in the rock; only a very slight resemblance to *K.* (*K.*) *huberti* may be noted. *Kettnerites* (*K.*) *huberti* is also present in the Ludlow of the Welsh borderland (courtesy Alison Brooks).

Comparison. – Both the left and right MI of the normal morph (i.e. having a convex inner margin with large denticles) are characteristic and easy to identify. The second morph is less heavily denticulated and resembles the corresponding elements of *K.* (*A.*) *sisyphi klasaardensis*, but differs in having small denticles in the anterior part and its largest denticles in the middle part of the denticulated inner margin. The basal portion is also slightly different;

Fig. 36. Kettnerites (*K.*) *huberti.* Bankvät 1, Hamra Beds, unit b. C–G from sample 82-34CB, the remaining ones from 81-39LJ. All specimens are in dorsal view, A–G ×120 and H–O ×80. □ A. Left MII, LO 5833:9. □B. Right MII, LO 5833:10. □C. Left MII, LO 5834:3. □D. Right MII, LO 5834:4. □E. Left MI, LO 5834:5. □F. Left MI, LO 5834:1. □G. Right MI, LO 5834:2. □H. Holotype, right MI, LO 5833:7. □I. Right MI, LO 5833:8. □J. Right MI, LO 5833:2. □K. Left MI, LO 5833:3. □L. Left MI, LO 5833:6. □M. Left MI, LO 5833:1. □N. Left MI, LO 5833:5. □O. Left MI, LO 5833:4.

usually the basal plate is lacking on the right MI of *K.* (*K.*) *huberti*.

The right MII has two pre-cuspidal denticles, which is fairly common within *Kettnerites*, e.g., *K.* (*K.*) *polonensis*, *K.* (*K.*) *bankvaetensis*, and some of *K.* (*K.*) *burgensis*. The pre-cuspidal denticles of *K.* (*K.*) *huberti* are of equal size, in contrast to those of *K.* (*K.*) *bankvaetensis*. Further, the ramus of the MII of *K.* (*K.*) *huberti* is large and almost triangular, which distinguishes the species from *K.* (*K.*) *polonensis* and *K.* (*K.*) *burgensis*, these having a more or less needle-shaped ramus (Figs. 18A, B, D, E).

Kettnerites (*K.*) *martinssonii* Bergman 1987

Figs. 12A–G, 18C, 37–40

Synonymy. – □1882 *Eunicites cristatus* (Hinde 1879) – Hinde, p. 10, Pl. 1:6, right MII. □1882 *Arabellites anglicus* Hinde 1880 – Hinde, p. 18 (pars.), Pl. 2:50, left MII. □1882 *Oenonites aspersus* Hinde 1880 – Hinde, p. 13, Pl. 1:21, 22, left MI. □1960 *Paulinites* sp. – Martinsson, pp. 5–6, Fig. 1:6, a jaw apparatus. □1979 *Oenonites aspersus* Hinde – Bergman, p. 99 (pars.), Fig. 28:4B, E, F. □1987 *Kettnerites* (*K.*) *martinssonii* n.sp. – Bergman, pp. 62–66, Figs. 12A–G, 18, 37–40.

Derivation of name. – In honour of the late Professor Anders Martinsson, who increased our knowledge of the geology and biostratigraphy on Gotland considerably and was the initiator and coordinator of the Project Ecostratigraphy. He was also the first to describe polychaete apparatuses from Gotland. He found two apparatuses and named one but left this species under open nomenclature. This apparatus belongs to the very common species now named *Kettnerites* (*K.*) *martinssonii*.

Holotype. – LO 5764:1, right MII, Fig. 37G.

Type locality. – Vattenfallsprofilen 1.

Type stratum. – Högklint Beds, unit b.

Varieties. – *K.* (*K.*) *martinssonii* var. mulde.

Material. – Figs. 2 and 3; more than 250 elements each of right MI, left MI, right MII, left MII.

Occurrence. – Figs. 4 and 9; Latest Llandovery to Late Ludlow, Lower Visby Beds to Hamra Beds. Ajmunde 1, Amlings 1, Ansarve 1, Ar 1, Aursviken 1, Autsarve 1, Baju 1, Bandlunde 1, Blåhäll 1, Bodudd 1 and 3, Bofride 1, Bottarve 2, Buske 1, Däpps 1 and 2, Djupvik 1, 2, 3, and 4, Fardume 1, Fie 3, Fjäle 3, Fjärdinge 1, Follingbo 2 and 12, Gamla Hamn 1, Gandarve 1 and 2, Gannor 1 and 3, Garnudden 3 and 4, Gerete 1, Gerumskanalen 1, Glasskär 1 and 3, Gläves 1, Gnisvärd 1 and 2, Gogs 1, Gothemshammar 1, 2, 6 and 7, Grogarnshuvud 1, Grymlings 1, Gustavsvik 2, Gutenviks 1, Gyle 1 and 2, Häftingsklint 1 and 4, Haganäs 1, Hägur 1, Hällagrund 1, Halls Huk 1, Hide 1, Hide Fiskeläge 1, Hoburgen 2, Hörsne 3 and 5, Hummelbosholm 1, Ireviken 2 and 3, Kappelshamn 1, Kättelviken 5, Kauparve 1, Klinteberget 1, Kluvstajn 2, Korpklint 1, Kroken 2, Kullands 1 and 2, Langhammarsviken 2, Lauter 1, Lauterhornsvik 2, Lickershamn 2, Lilla Hallvards 1 and 4, Loggarve 2, Möllbos 1, Mölner 1, Mulde 2, Mulde Tegelbruk 1, Nabban 2, När 2, Närshamn 2 and 3, Nygårds 1 and 2, Nygårdsbäckprofilen 1, Nyhamn 4, Öndarve 1, Paviken 1, Petsarve 2 and 15, Rågåkre 1, Rangsarve 1, Ronnings 1, Rönnklint 1, Saxriv 1, Sigvalde 2, Slitebrottet 1 and 2, Snäckgärdsbaden 1, Snoder 2 and 3, Sproge 4, Stave 1, Stora Myre 1, Storugns 1B, Strandakersviken 1, Stutsviken 1, Svarvare 1, Svarven 1, Talings 1, Tjeldersholm 1, Tomtbodarne 1, Träske 1, Valbybodar 1, Valle 2, Valleviken 1, Vallmyr 1, Vallstena 2, Valve 3, Värsände 1, Västlaus 1, Vattenfallsprofilen 1, Vidfälle 1, Vike 2 and 3.

Diagnosis. – Right MI: Subtriangular, sharply ended shank, thin-walled, narrow, highly elevated flange. Almost triangular, small basal portion.

Left MI: Denticles of normal size to fairly large, slightly to widely spaced in anterior half, somewhat posteriorly directed, almost rectangular inner wing.

Right MII: One large pre-cuspidal denticle, cusp of moderate size. Ramus fairly large, almost triangular, wide in basal part, anterior outer margin nearly straight. Shank fairly wide, tapers to posterior.

Left MII: Single cusp, followed by 5–8 intermediate denticles of equal size. Ramus fairly large, almost triangular, wide in basal part.

Fig. 37 (opposite page). *Kettnerites* (*K.*) *martinssonii*. All specimens in dorsal view, ×120, except B2, E and I1. □A. Buske 1, Lower Visby Beds unit e, 79-40LJ; A1 left MI, LO 5362:1; A2 right MI, LO 5362:2. □B. Follingbo 2, Slite Beds, Slite Marl NW part, 75-10CB; B1 left MI, LO 5783:1; B2 right MI, LO 5783:2, ×80. □C. Right MII, LO 5767:1, Vattenfallsprofilen 1, Högklint Beds, unit b, 70-6LJ. □D. Left MII, LO 5763:1, Ansarve 1, Högklint Beds, SW facies, upper part, 79-46LJ. □E. Right MII, LO 5767:2, Vattenfallsprofilen 1, Högklint Beds, unit b, 70-6LJ, ×80. □F. Left MII, LO 5771:1, Valle 2, Slite Beds, *Pentamerus gothlandicus* Beds, 66-145SL. □G. Holotype, right MII, LO 5764:1, Vattenfallsprofilen 1, Högklint Beds, unit a, 70-20LJ. □H. Vattenfallsprofilen 1, Högklint Beds, unit b, 70-6LJ; H1 left MI, LO 5767:3; H2 right MI, LO 5767:4. □ I. Vattenfallsprofilen 1, Högklint Beds, unit a, 70-20LJ; I1 left MI, LO 5764:3, ×80; I2 right MI, LO 5764:2, ×80. □J. Slitebrottet 2, Slite Beds Slite Marl, 83-31LJ; J1 left MI, LO 5773:1; J2 right MI, LO 5773:2.

Fig. 38 (p. 72). *Kettnerites* (*K.*) *martinssonii* var. mulde. All specimens in dorsal view. □A. Mulde Tegelbruk 1, Mulde Beds, 82-7CB; A1 right MII, LO 5789:1, ×120; A2 left MII, LO 5789:2, ×120. □B. Däpps 1, Mulde Beds, upper part, 81-56LJ; B1 left MI, LO 5787:1, ×80; B2 left MI, LO 5787:2, ×80; B3 right MI, LO 5787:3, ×80; B4 right MI 5787:4, ×80; B5 left MII, LO 5787:5, ×60; B6 right MII, LO 5787:6, ×60.

Fig. 39 (p. 73). *Kettnerites* (*K.*) *martinssonii*. D and E are *K.* (*K.*) *martinssonii* var. mulde. A, C, E: Snoder 2, Hemse Beds, Hemse Marl, NW part, 82-14CB. B, D: Kullands 2, Hemse Beds, Hemse Marl, NW part, 84-312DF. All specimens in dorsal view. All ×120, except E. □A1, left MII, LO 5808:8; A2 right MII, LO 5808:7. □B1 right MI, LO 5832:2; B2 left MI, LO 5832:1. □C1 left MI, LO 5808:3; C2 right MI, LO 5808:4. □D1 left MI, LO 5832:3; D2 right MI, LO 5832:4. □E1 left MI, LO 5808:1, ×80; E2 right MI, 5808:2, ×80; E3 left MII, LO 5808:5, ×80; E4 right MII, LO 5808:6, ×60.

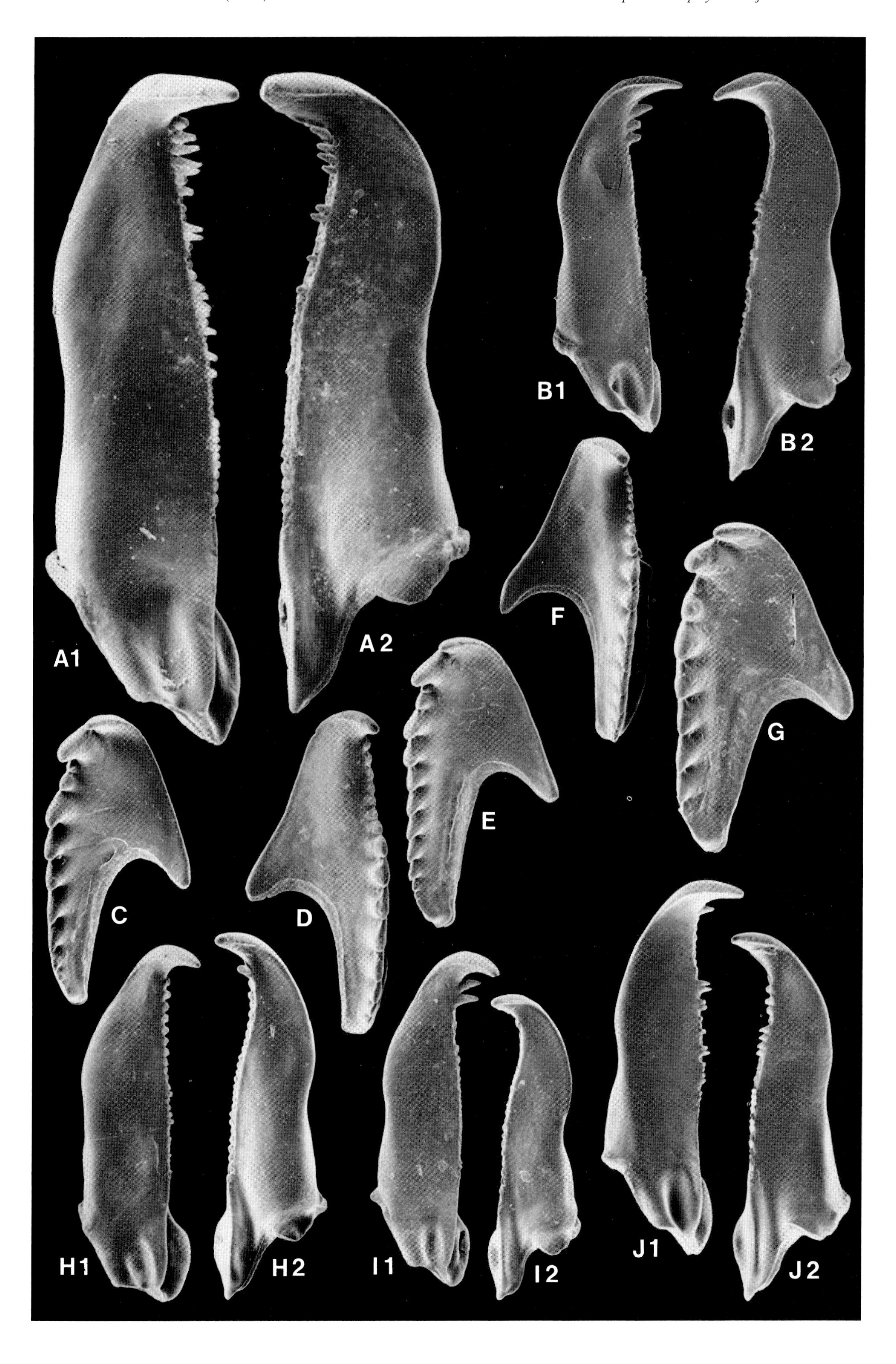
A1
A2
B1
B2
F
G
E
C
D
H1
H2
I1
I2
J1
J2

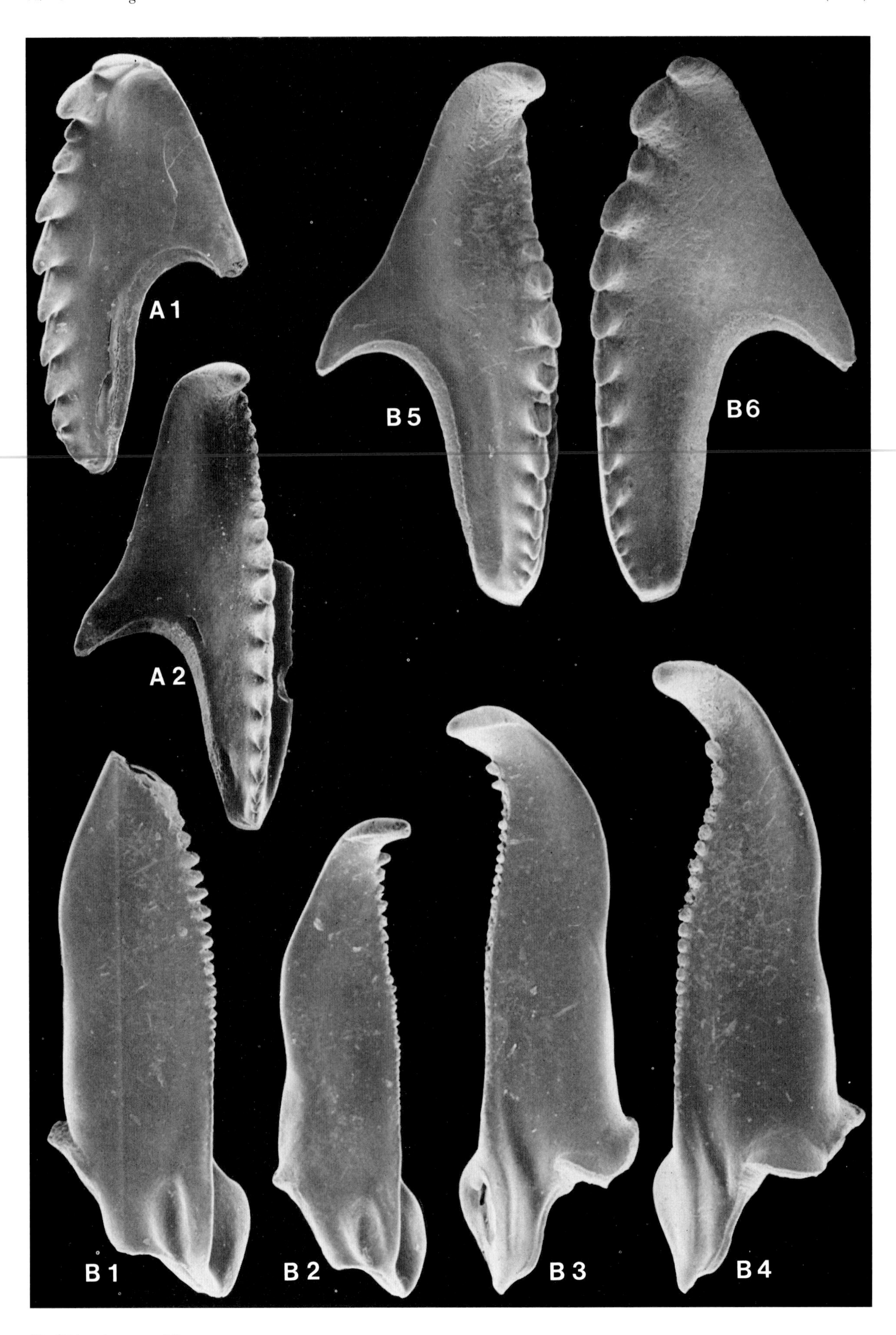

Fig. 38 (caption on p. 70).

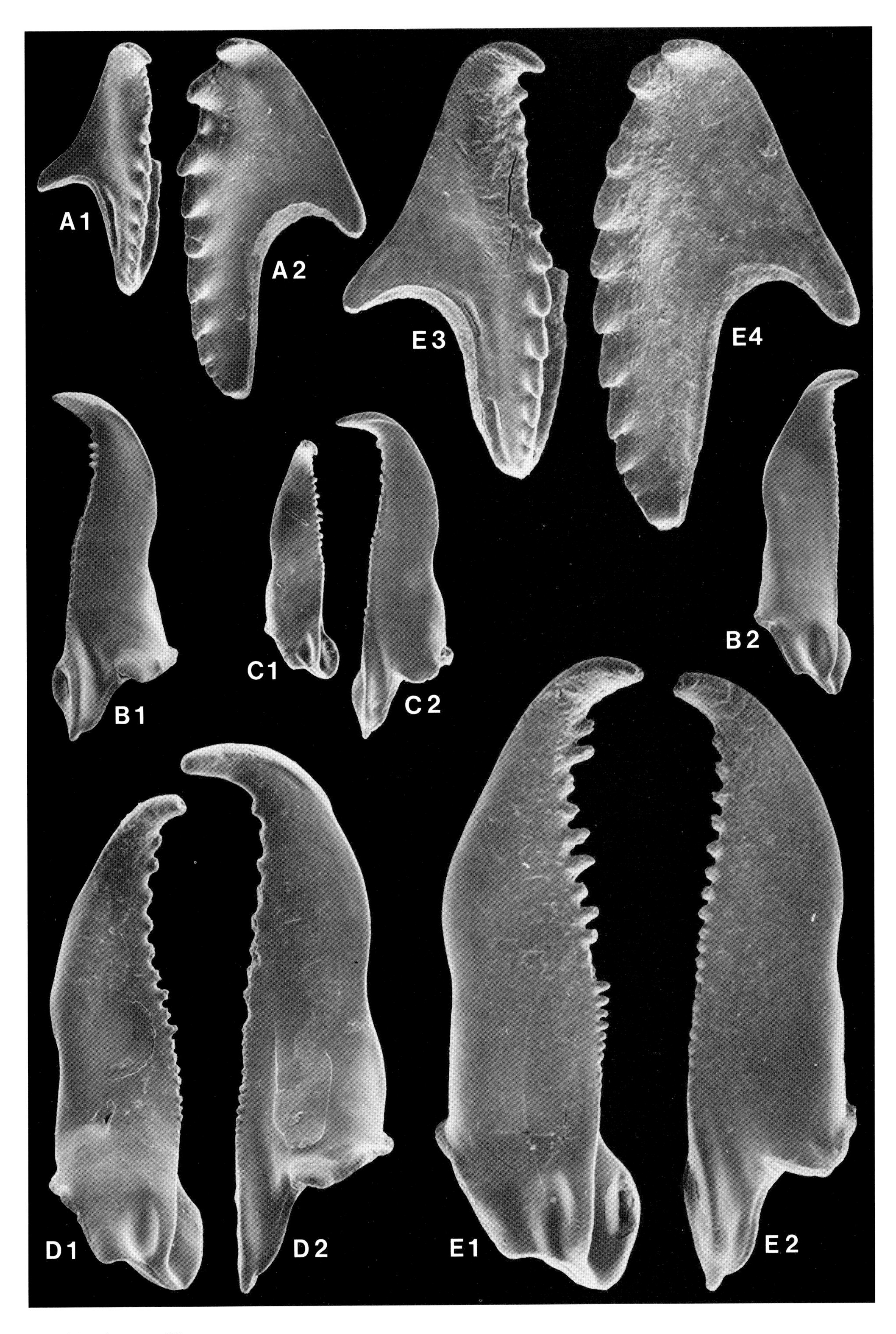

Fig. 39 (caption on p. 70).

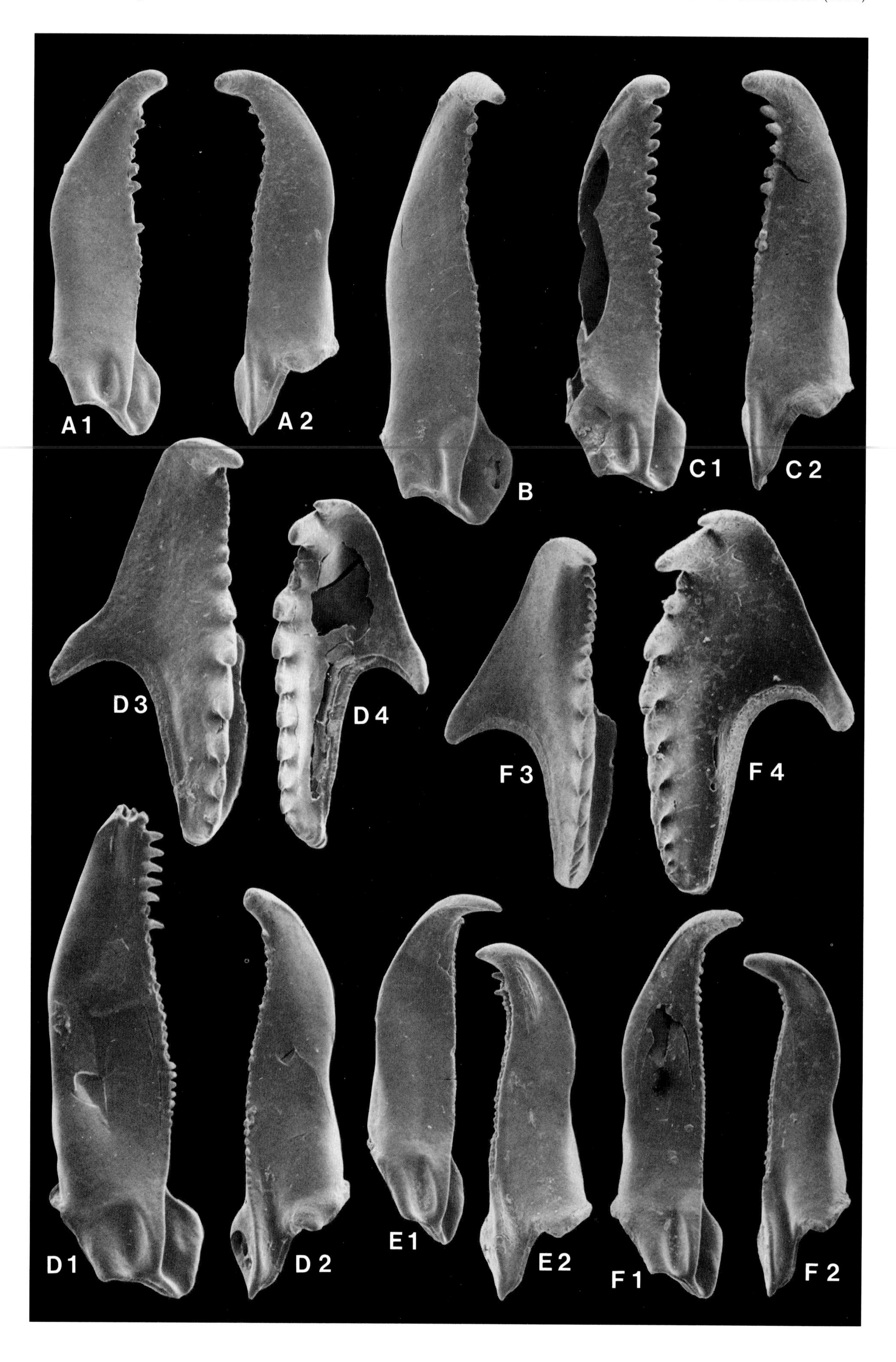
A 1
A 2
B
C 1
C 2
D 3
D 4
F 3
F 4
D 1
D 2
E 1
E 2
F 1
F 2

Description. – Right MI, dorsal side: Length 0.43–1.23 mm (mulde variety 0.65–2.28 mm), width ⅓–¼ of length. The jaw tapers anteriorly, ending in a normal to fairly large fang bearing a fairly distinct cutting edge. The inner margin is convex, denticulated with about 30 denticles of almost equal size in Early Wenlock specimens. During the Middle Wenlock to Late Ludlow, the denticles increase in size, particularly the anterior ones, and the number decreases to about 20. The dentition covers the inner margin from the falcal arch, ending, with slightly spaced denticles, before reaching the inner wing. The large jaws from the Late Wenlock strata often have large to very large, slightly spaced denticles. The continuation of the inner margin onto the undenticulated ridge is smooth. The undenticulated ridge is long, high and narrow, its middle part slightly bent outward, ending as a prominent tip. The tip forms the sharp-ended shank together with the posterior part of the inner wing. The inner wing is of normal size, deeply downbent, wide, its anterior half rounded. The basal furrow is deep and long. The flange is thin-walled, highly elevated, narrow, ending postero-laterally as a slightly downbent tip or blunt end. The flange passes over anteriorly into the fairly large and rounded ligament rim. This rim has a rough surface and forms a more or less pronounced ridge, sometimes with a groove on its inner side. A conspicuous, wide, outer border is present in the mulde variety from the late Middle Wenlock (see Remarks).

The basal plate is not recorded.

The basal portion is wide and narrow, almost triangular, with a basal angle of about 35–50°. The posterior part of the jaw is widest, with a concavity in its middle part whence it extends forming the sickle-shaped falx.

Ventral side: The strongly enclosed, crescent-shaped myocoele opening represents one third of the jaw length. It is surrounded by a fairly narrow ligament rim.

Left MI, dorsal side: Length 0.40–1.54 mm (mulde variety 0.73–2.18 mm), width ⅓–¼ of length. The inner margin is almost straight (Figs. 37H:1, 39B:2) to convex with fairly large, slightly spaced denticles on the anterior part (Figs. 39D:1, E:1, 40A:1, B). The anteriormost denticles may be oriented anteriorly. Denticles, about 30 in number in Early Wenlock specimens, decrease in number but increase in size with stratigraphic time. In Ludlow specimens the anterior denticles are large to very large and the posterior ones are small knobs. Their number varies greatly but is normally about 20. The denticulation ends as crenulations well before reaching the inner wing. The latter is slightly downfolded, often posteriorly projecting, almost rectangular, nearly angular, but highly variable in shape. The denticulated inner margin has a smooth and straight continuation onto the low and narrow undenticulated ridge, which ends before reaching the posterior margin, with a low ridge oriented towards the posterior, inner corner of the inner wing. The basal furrow is of normal size. The posterior margin is short and concave. The posterior part of the outer margin of the basal portion is downfolded and has a fairly narrow flange on its anterior part. The basal portion is fairly small, the basal angle usually 30–40°. The outer margin between the basal portion and the large flat to rounded falx is concave. The cutting edge of the fang is fairly distinct.

Ventral side: The strongly enclosed, crescent-shaped myocoele opening represents ¼–⅓ of the jaw length. The opening is very similar to the one on the right MI.

Right MII, dorsal side: Length 0.35–1.13 mm (mulde variety 0.38–1.80 mm), width more than half the length. A large, pre-cuspidal denticle with a distinct cutting-edge, forms the anteriormost, fairly acute margin of the jaw. It is followed by a somewhat larger, moderate-sized cusp, directed slightly to the posterior or almost perpendicular to the denticulated ridge. Posterior to the cusp there are one or two minor denticles of which the anteriormost is the smaller, followed by 6–9 large to very large denticles, slanting towards the posterior direction and decreasing slightly in size towards the posterior; the posteriormost ones are in some cases very small. The number of denticles is independent of jaw length. The anterior part of the shank is wide, tapering slightly towards the posterior, and occupies slightly more than half of the jaw length. The ramus is fairly large, usually without any sinus in the outer margin, almost triangular, wide in the basal part. The bight angle is 55–75°.

Ventral side: The slightly enclosed, crescent-shaped myocoele opening is ½–⅔ of the jaw length. The opening is extended slightly forwards along the inner margin. The ligament rim is narrow.

Left MII, dorsal side: Length 0.33–1.22 mm (mulde variety 0.38–1.73 mm), width about half the length or slightly less. The single cusp with its distinct cutting edge forms the anteriormost part of the jaw. The intermediate dentary has 5–8 denticles of equal and moderate size, followed by 8–10 larger denticles, reaching maximum size in the central part of the dentary. The number of denticles is unrelated to the size of the jaw. The inner wing is pronounced, representing about half of the jaw length. The inner margin is almost parallel with the denticulated ridge. The shank occupies about half or slightly less of the jaw length, its anterior part being very wide and tapering towards the posterior, ending in a fairly sharp tip. The ramus is fairly large, almost triangular, wide in the basal part, with a sinus in the anterior outer part. The bight angle is 60–85°.

Ventral side: The slightly enclosed myocoele opening represents slightly more than half the jaw length. The anterior margin is somewhat curved with its main orienta-

---

Fig. 40. Kettnerites (*K.*) *martinssonii*. D are *K.* (*K.*) *martinssonii* var. mulde. All specimens in dorsal view. □A. Glasskär 3, Burgsvik Beds, 82-18CB; A1 left MI, LO 5828:1, ×80; A2 right MI, LO 5828:2, ×80. □B. Left MI, LO 5844:3, Kauparve 1, Hamra Beds, lower–middle part, 76-13CB, ×80. □C. Gannor 1, Eke Beds, 71-123LJ; C1 left MI, LO 5811:1, ×120; C2 right MI, LO 5811:2, ×120. □D. Värsände 1, Mulde Beds, 75-22CB; D1 left MI, LO 5793:1, ×120; D2 right MI, LO 5793:2, ×80; D3 left MII, LO 5793:3, ×80; D4 right MII, LO 5793:4, ×80. □E. Sigvalde 2, Hemse Beds, lower middle part, 71-115LJ; E1 left MI, LO 5819:1, ×120; E2 right MI, LO 5819:2, ×120. □F. Träske 1, Hemse Beds, unit b, 71-103LJ; F1 left MI, LO 5818:1, ×60; F2 right MI, LO 5818:2, ×60; F3 left MII, LO 5818:3, ×120; F4 right MII, LO 5818:4, ×60.

tion oblique to the extension of the jaw. The ligament rim is narrow.

Remarks. – During the latest Llandovery and Early to Middle Wenlock the MI jaws of this species are usually 0.5–1 mm in size. The shape of the jaws varies slightly from population to population. The inner margin is usually almost straight in the left MI. During the later part of the Wenlock and in the Early Ludlow, i.e. Halla, Mulde, Klinteberg, and Hemse Marl, NW part, however, the maximum size of the jaws increased and a new shape of the larger MI evolved (Fig. 12A–G). The inner margin became pronouncedly convex, the flange grew larger, the outer anterior part of the flange became conspicuously wider, forming a border (e.g. Fig. 38B:1–4). I have named this variety 'mulde' after the Mulde Beds where these characteristics become most conspicuous.

During the Middle and Later Ludlovian the maximum size seems to have decreased slightly and the jaws became more curved and somewhat thinner, with comparably larger denticles. The posterior, outer margin of the basal portion of the left MI is normally considerably shorter and slightly concave (Fig. 40A, B, and C).

The dentition formula and the morphology of the MII are virtually unchanged from the latest Llandovery to Late Ludlow. The size changes follow those of the MI.

Comparison. – The right MII with its large pre-cuspidal denticle, followed by a fairly large cusp is the best character for identification. Large MI's younger than the middle Wenlock, with their large dentary, convex inner margins and conspicuous basal portion, are also easy to identify. The Left MII, however, is considerably more difficult to identify. The corresponding element of some of the forms of *Kettnerites* (*A.*) *sisyphi* is virtually impossible to distinguish from *K.* (*K.*) *martinssonii.* There is no safe character to rely on, though the cusp of the left MII of *K.* (*A.*) *sisyphi* is normally slightly wider and somewhat flattened obliquely to the extension of the jaw. The *K.* (*K.*) *martinssonii* left MII jaw might give a slightly coarser impression and is often larger than in *K.* (*A.*) *sisyphi.* Comparisons with *K.* (*K.*) *bankvaetensis* were given above.

Kettnerites kosoviensis Zebera from the Silurian of Bohemia is probably closely related to *K.* (*K.*) *martinssonii.* They are probably not conspecific. Zebera's species (1935) was revised by Snajder (1951) who collected new material from the same and other localities used by Zebera, and described the polychaete fauna there according to a biological species concept. The badly preserved elements from the Bohemian differ from the Gotland ones in that the number of denticles of the MII is somewhat higher and the denticles of the MI seem to be thinner. However, the jaws from Bohemia are very large, flattened and crushed, making a comparison very difficult. New material from Bohemia, showing three-dimensional morphology and variability, is needed to resolve this question.

Kettnerites (*K.*) *polonensis* (Kielan-Jaworowska 1966)

Figs. 13A–G, 18A, 41, 42B, E, 43, 44

Synonymy. – □1980 *Paulinites polonensis* Kielan-Jaworowska 1966 – Wolf, pp. 86–87, Pl. 12:102. □1987 *Kettnerites* (*K.*) *polonensis* (Kielan-Jaworowska 1966) – Bergman, pp. 66–69, Figs. 13A–G, 18, 41, 42B, E, 43, 44.

Holotype. – *'Paulinites' polonensis* Kielan-Jaworowska 1966, pp. 126–129, Pl. 29:2.

Type locality. – Erratic boulder.

Type stratum. – Not known, see discussion below.

Varieties. – *K.* (*K.*) *polonensis* var. gandarve, *K.* (*K.*) *polonensis* var. sjaustre.

Material. – Fig. 3; more than 250 right MI, 250 left MI, 250 right MII, 250 left MII.

Occurrence. – Figs. 4 and 10; Early Wenlockian to Late Ludlovian, Högklint Beds to Sundre Beds. Amlings 1, Baju 1, Bankvät 1, Barkarveård 1, Botvide 1, Djaupviksudden 4, Fågelhammar 3, Fakle 1, Faludden 2, Fjärdinge 1, Gannes 2, Gannor 1 and 3, Garnudden 1, 3, and 4, Gerumskanalen 1, Glasskär 1, 2, and 3, Gläves 1, Godrings 1 and 2, Gothemshammar 1, 2, 6, 7 and 8, Grogarnshuvud 1, Gröndalen 1, Gyle 1 and 2, Hallsarve 1, Herrvik 2, Hide 1, Hoburgen 2, Holmhällar 1, Kapelludden 1, Kättelviken 5, Klinteberget 1, Kroken 2, Kullands 2, Lambskvie 1, Loggarve 1 and 2, Möllbos 1 and 2, Mulde 2, Mulde tegelbruk 1, När 2, Närshamn 2, Nyan 2, Nygårds 2, Rågåkre 1, Ronehamn 2, Rudvier 1, Sibbjäns 2, Sjaustrehammar 1, Slitebrottet 2, Sigvalde 2, Stora Vikare 2, Strandakersviken 1, Strands 1, Trädgården 1, Träske 1, Valby Bodar 1, and Västerbackar 1.

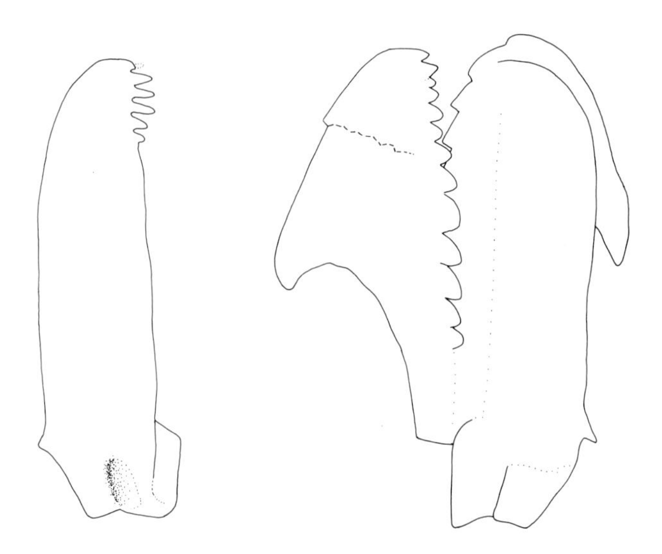

Fig. 41. Kettnerites polonensis Kielan-Jaworowska 1966, camera-lucida drawing of the type specimen apparatus consisting of the right MI, left and right MII (Z. Pal. No. O.439/3), ×235. The MIII element was not fused to the apparatus in 1985. Note the following characteristics: the right and left MI have almost rectangular inner wings and prominent spurs; the right MII lacks the pronounced cusp and left MII has a double cusp.

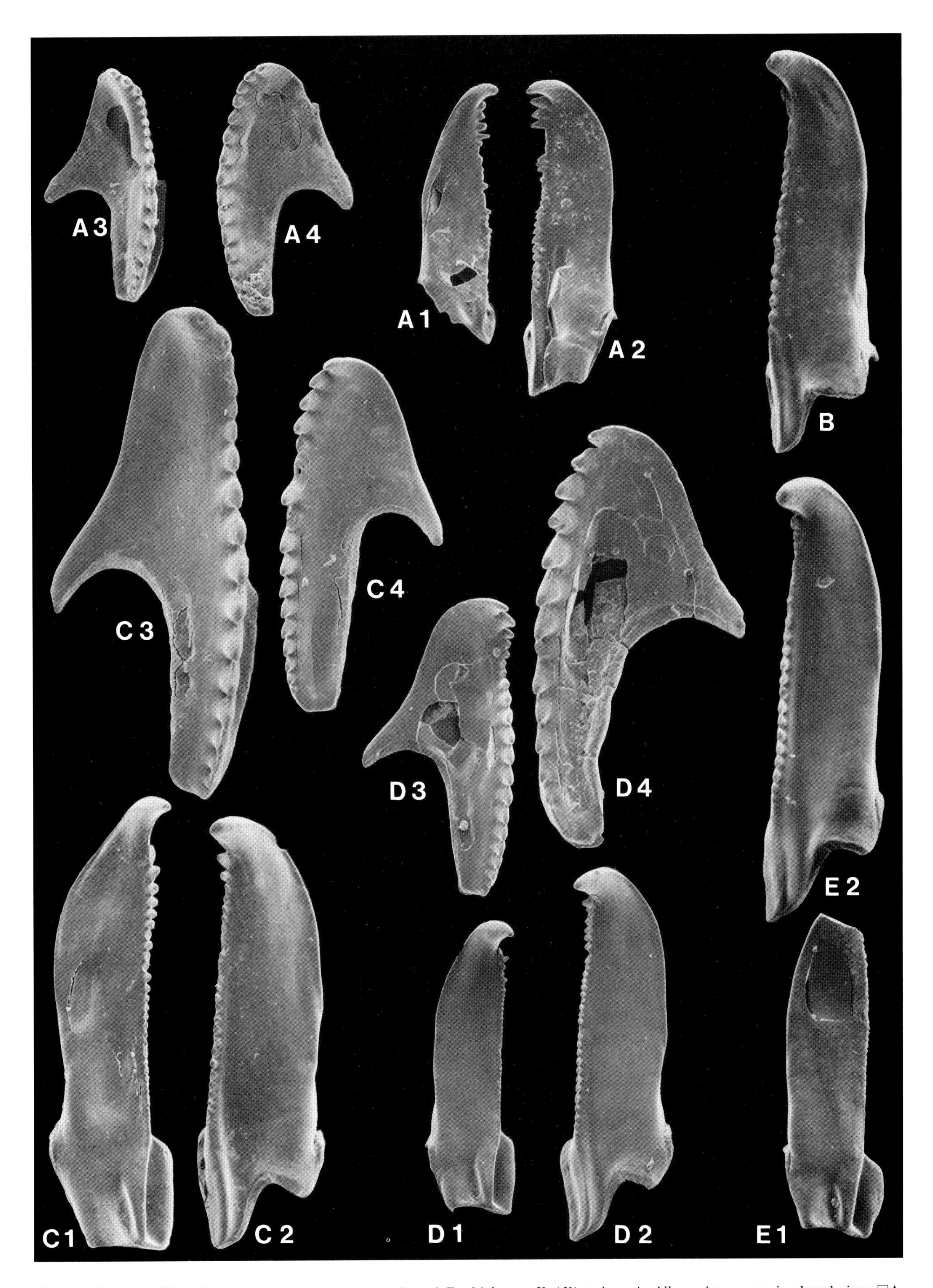

Fig. 42. Kettnerites (*K.*) *polonensis* var. gandarve, except B and E which are *K.* (*K.*) *polonensis.* All specimens are in dorsal view. □A. Sjaustrehammar 1, Hemse Beds, unit d, 82-19LJ; A1 left MI, LO 5816:1, ×170; A2 right MI, LO 5816:2, ×170; A3 left MII, LO 5816:3, ×120; A4 right MII, LO 5816:4, ×120. □B. Right MI, LO 5813:3, Gannor 3, Hemse Beds, Hemse Marl SE part, 71-126LJ, ×120. □C. Gandarve 1, Halla Beds, 71-81LJ; C1 left MI, LO 5836:4, ×100; C2 right MI, LO 5836:5, ×100; C3 left MII, LO 5839:2, ×120; C4 right MII, LO 5839:3, ×100. □D. Gothemshammar 7, Halla Beds, unit c, 77-45LJ; D1 left MI, LO 5837:1, ×80; D2 right MI, LO 5837:2, ×80; D3 left MII, LO 5837:3, ×80; D4 right MII, LO 5837:4, ×80. □E. Gannor 1, Eke Beds lower part, 71-124LJ; E1 left MI, LO 5812:1, ×80; E2 right MI, LO 5812:2, ×80.

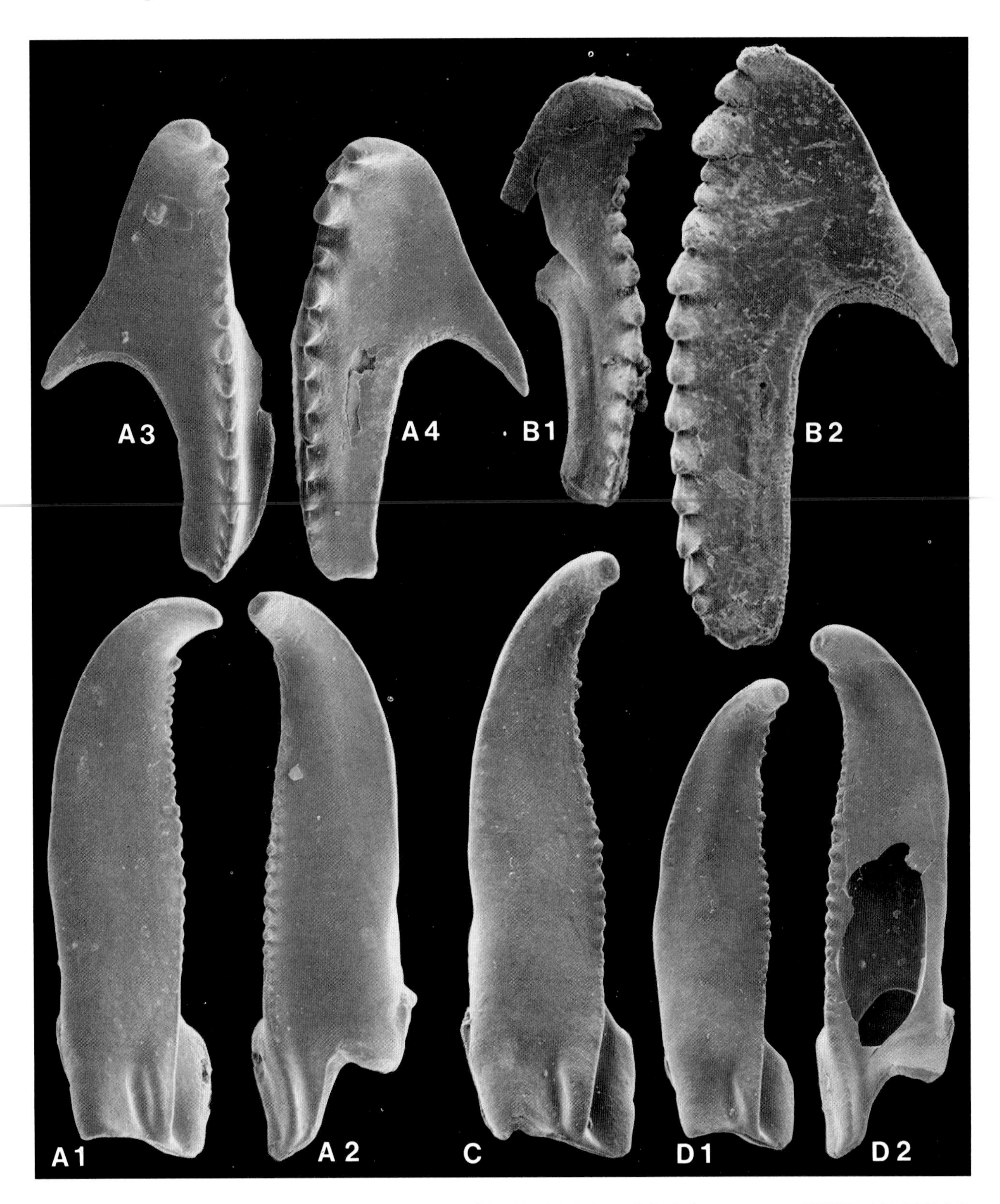

Fig. 43. Kettnerites (K.) polonensis. All specimens are in dorsal view. □A. Glasskär 3, Burgsvik Beds, lowermost part, 82-18CB; A1 left MI, LO 5827:3, ×60; A2 right MI, LO 5827:4, ×60; A3 left MII, LO 5827:5, ×60; A4 right MII, LO 5827:6, ×60. □B. Lambskvie 1, Hemse Beds, unit c, 75-45CB; B1 left MII, LO 5817:1, ×60; B2 right MII, LO 5817:2, ×60. □ C. Left MI, LO 5842:1, Glasskär 1, Burgsvik Beds, lowermost part, 82-15CB, ×80. □D. Same sample as C; D1 left MI, LO 5842:2, ×80; D2 right MI, ×60.

Fig. 44 (opposite page). *Kettnerites (K.) polonensis.* All specimens in dorsal view. □A. Faludden 2, Hamra Beds, unit c, 76-16CB; A1 left MI, LO 5830:1, ×120; A2 right MI, LO 5830:2, ×120; A3 left MII, LO 5830:3, ×210; A4 right MII, LO 5830:4, ×210. □B. Sibbjäns 2, Hamra Beds, unit b, 82-32LJ; B1 left MI, LO 5845:1, ×120; B2 right MI, LO 5845:2, ×120; B3 left MII, LO 5845:3, ×170, B4 right MII, LO 5845:4, ×170. □C. Västerbackar 1, Sundre Beds, middle–upper, 75-2LJ; C1 left MI, LO 5838:1, ×80; C2 right MI, LO 5838:1, ×60; C3 left MII, LO 5838:2, ×80; C4 right MII, LO 5838:3, ×210. □D. Left MI, LO 5845:5, Sibbjäns 2, Hamra Beds, unit b, 82-32LJ, ×60.

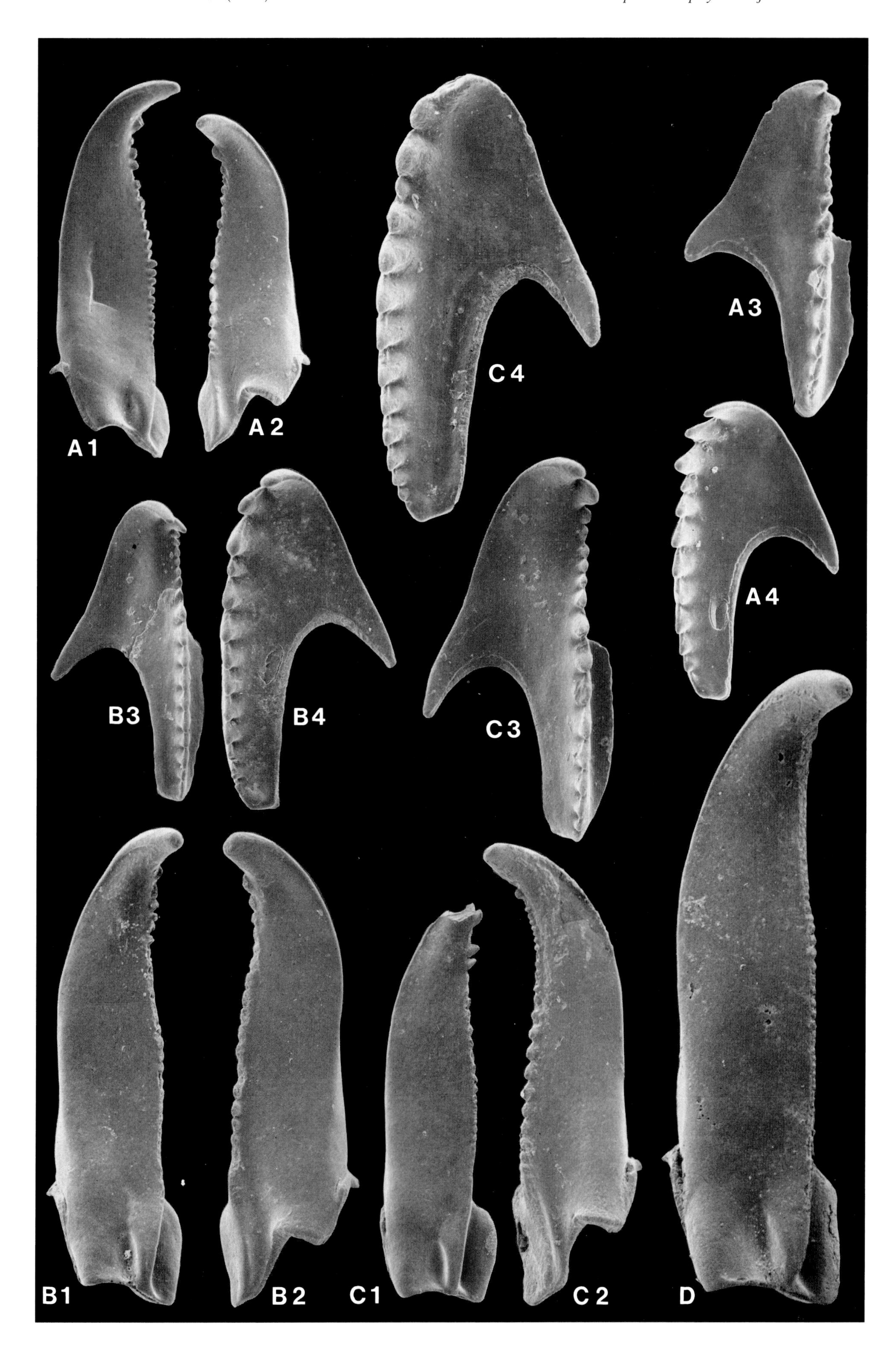
A1
A2
C4
A3
A4
B3
B4
C3
B1
B2
C1
C2
D

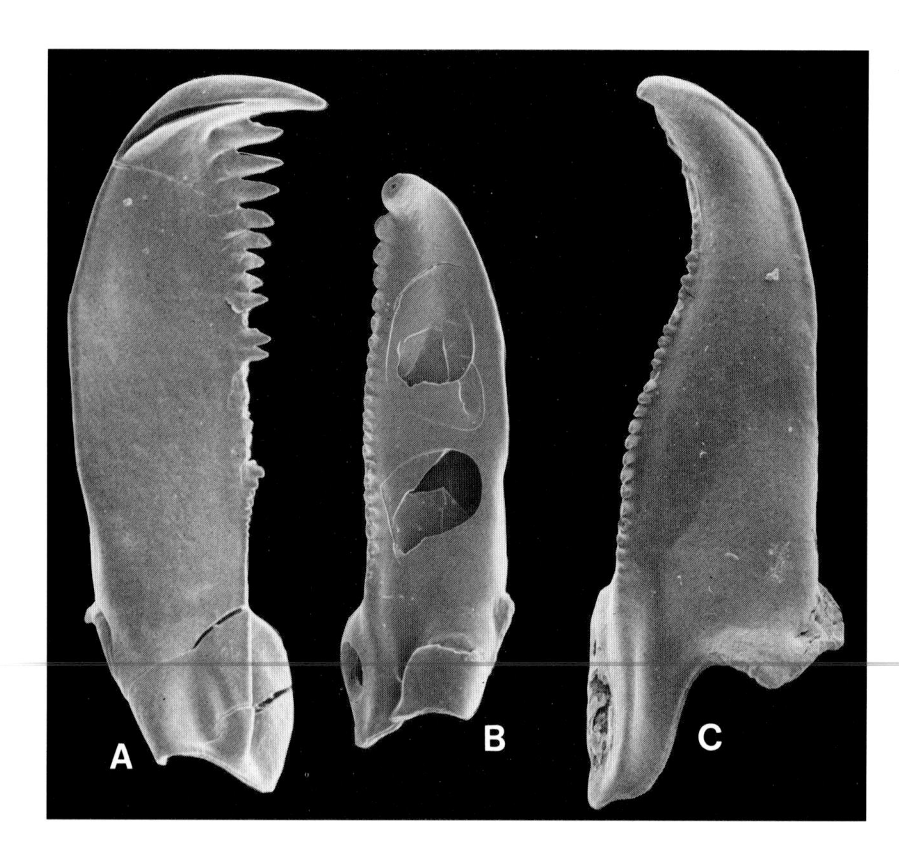

Fig. 45. Kettnerites (*K.*) cf. *polonensis*. All specimens are in dorsal view. □A. Left MI, LO 5827:1, Glasskär 3, Burgsvik Beds, 82-18CB, ×80. □B. Right MI, LO 5827:2, same sample as A, ×80. □C. Right MI, LO 5842:3, Glasskär 1, Burgsvik Beds, 82-15CB, ×80.

K. (*K.*) *polonensis* var. gandarve (Fig. 42A, C, D) is recorded from the following Middle to Late Wenlock localities (not included in the *K.* (*K.*) *polonensis* list): Fardume 1, Gandarve 1, and Gothemshammar 7.

Diagnosis (emended). – Right MI: Large fang, inner margin slightly convex, covered by denticles of varying size. Basal portion angular, shank ending bluntly, flange angular and thick-walled with a groove and a spur on the anterior outer part.

Left MI: Convex inner margin with large denticles anteriorly, decreasing in size posteriorly. Basal portion angularly square with a groove along its outer margin. Prominent rectangular inner wing.

Right MII: Two large pre-cuspidal denticles of equal size; shank long, straight and blunt-ended. Ramus needle-like, long, straight, narrow, with pointed extremity.

Left MII: Large double cusp of equal size; shank large and almost straight. Ramus needle-like, long, straight, narrow, with pointed extremity.

Description. – Right MI, dorsal side: Length 0.38–2.82 mm, width about ¼ of length. The jaw tapers, ending in a large fang, bent upwards in relation to jaw surface; the lower middle part of the inner margin is slightly convex. The denticulation covers 0.60–0.65 of the jaw length. The large anteriormost denticle is followed by about five denticles that decrease to fairly small size. Posterior to these are 5–8 denticles which slowly increase in size, followed by a series of about 10 large denticles. The posteriormost denticles are often small, reduced to indistinct knobs, probably due to wear. Rarely, jaws with denticles of more or less even size occur. The undenticulated ridge is large and slightly sigmoidal, ending in the posteriormost part of the bluntly ended shank. The deeply downfolded inner wing is of normal size, sigmoidal, and almost parallel to the undenticulated ridge. The basal furrow is fairly narrow; the thick-walled, rectangular flange is highly elevated along the posterior margin and downbent along the outer margin. The ligament rim on the posterior outer margin of the basal portion is about 0.15 of the jaw length and consists of a longitudinal, narrow groove with a small spur on its anteriormost part.

The basal plate is rarely recorded in place, except sometimes in the gandarve variety (Fig. 42A:2) and in *K.* (*K.*) cf. *polonensis* (Fig. 45B). It is somewhat triangular, longer than high, and with a high, convex inner margin.

The basal portion is almost triangular and the basal angle usually 13–30°. The outer margin, anterior to the basal portion, is almost straight, ending in the prominent sickle-shaped falx.

Ventral side: The strongly enclosed crescent-shaped myocoele opening is about ⅓ of the jaw length. In front of the opening a narrow groove surrounds the somewhat sunk, smoothly rounded ligament rim. The outer, anteriormost margin of the rim forms a prominent, short, pointed, high and narrow spur. The groove and spur are visible also in the dorsal view.

Left MI, dorsal view: Length 0.43–2.60 mm, width about ⅓ of length. The large fang is relatively strongly bent upwards in relation to the jaw surface. The jaw tapers anteriorly from the posterior middle part. The inner margin is slightly convex, covered by moderately large denticles for about 0.5–0.6 of the jaw length. Normally the size of the 20–25 denticles decreases posteriorly in a regular way. At some localities (e.g. Glasskär 1, Sibbjäns 2) the anteriormost 1–2 denticles are large and followed by some 10 small denticles, which in turn are followed by 8–12 large ones. The denticulated inner margin continues smoothly onto the straight undenticulated ridge which does not reach the posterior margin. The inner wing is rectangular, pronounced, fairly large, 0.25–0.37 of the jaw length, parallel with the undenticulated ridge. The basal furrow is short and fairly wide, parallel to the outer margin of the angular, nearly rhombic, basal portion. The basal angle is medially about 20° (varies between 15° and 30°). The ligament rim runs along the anterior half of the outer margin of the basal portion, forming a spur, followed laterally by a groove. The outer margin anterior to the basal portion has a very variable outline (cf. 42C:1 and 43A:1).

Ventral side: The strongly enclosed, crescent-shaped myocoele opening represents about ¼–⅓ of the jaw length. The smoothly rounded narrow ligament rim is surrounded anteriorly by a narrow groove. Toward its outer part the rim is transformed into a narrow ridge similar to the one on the right MI.

Right MII, dorsal side: Length 0.38–2.05 mm, width somewhat more than half the length. Two pre-cuspidal denticles of equal size, almost as large as the moderately sized cusp, form the sickle-shaped anterior part. The jaw is robust, with fairly coarse denticles. Posterior to the slightly slanting cusp, along the almost straight inner margin, there are two, or, in a few cases, one intermediate denticle, the anteriormost one relatively small, the second larger. The intermediate dentary is not particularly distinguished. All denticles on the denticulated ridge, including the intermediate ones, slant slightly towards the posterior. The number of denticles varies from 6 to 9, more or less depending on the size of the jaw. The shank is about 0.6 of the jaw length, coarse, almost straight with almost parallel inner and outer margins and with a blunt end. The bight is deep with an acute angle. The ramus is long, often straight, and fairly narrow, tapering distally and ending in a very pointed tip. Many specimens have a very vague sinus on the anterior outer margin at the base of the ramus.

Ventral side: The slightly enclosed, crescent-shaped myocoele opening represents ⅔–¾ of the jaw length. It extends to the anterior along the inner side and is surrounded by a normal rim slightly widened on its outermost part of the ramus.

Left MII, dorsal side: Length 0.31–1.90 mm, width about 0.6 of the length. A double cusp with parts of equal size, slanting slightly in the posterior direction, forms the anteriormost, sickle-shaped margin. It is followed by 6–7 intermediate denticles of moderate, almost equal size, and about 8–10 slightly slanting denticles on the straight denticulated ridge. The denticle size decreases towards the posterior end. The shank occupies about half of the jaw length. The inner margin of the prominent, laterally projecting, inner wing represents about half of the jaw length. The posterior part of the inner wing is smoothly rounded and its anterior ⅔ is parallel to the denticulated ridge. The inner and outer margins of the shank are almost parallel, and the shank ends fairly bluntly. The bight is deep with an acute bight angle. The ramus is long, tapering and pointed with a minor sinus in the anteriormost part of the outer margin.

Ventral side: The slightly enclosed myocoele opening is almost a mirror image of the right MII except that the opening is less extended anteriorly and thus represents ½–⅔ of the jaw length.

Remarks. – Normally this species is represented by large jaws. It is found from the Wenlock to Late Ludlow, particularly in the limestone beds on Gotland; it is fairly common in erratic boulders from the Baltic area (Kielan-Jaworowska 1966). The morphology becomes more and more conspicuous with time, with a progressively more developed, prominent, needle-like, pointed ramus of the MII and a very angular basal portion of the MI, often with a spur.

In some localities from the uppermost Slite Beds and the Halla Beds, an early *K.* (*K.*) *polonensis* variety, gandarve (after the locality Gandarve 1) is found. Typical specimens occur in the Middle to Late Wenlock strata; Fardume 1, Gandarve 1 (Fig. 42C), and Gothemshammar 7 (Fig. 42D). The variety is also found from the Hemse Beds (e.g. Sjaustrehammar 1), and related forms have been encountered from the Hamra Beds. The gandarve variety seems to occur partly in parallel with the *Kettnerites* (*K.*) *polonensis* lineage on Gotland. The most conspicuous character of the gandarve variety is the even-sized denticles on the right MII, where no cusp can be identified. The inner margin of the MI is almost straight and the spur is not pronounced. The basal portion is extremely angled, almost like a square.

Another variety, *K.* (*K.*) *polonensis* var. sjaustre, is described separately below.

Discussion. – The type specimen consists of an apparatus with the four main elements, including the MIII. It is a very small jaw apparatus, stored in glycerin, and is thus difficult to study. To the drawing of the main part of the apparatus (Fig. 41) may be added the following observations. The right MI has a distinct spur and a fused basal plate. A fused basal plate seems to be more common among the smaller right MI's than among larger ones within the species. Thus, the fused basal plate may simply reflect the small size of the specimen. The outer posterior part of the basal plate is strongly bent down, which changes the outline of the posterior part somewhat. The shank is slightly pointed and almost triangular, the inner wing is straight. The denticles are fairly large. The left MI has an angular basal portion, an almost rectangular inner wing, a small spur and large denticles. The right MII has no cusp, a needle-shaped ramus, and no distinct sinus. The right MII has a double cusp, five intermediate denticles and an almost straight anterior outer margin.

Kettnerites (*K.*) *polonensis* dominates the paulinitid fauna with large jaws (normal 1–1.5 mm in length) at Glasskär 1–3 and is characteristic in the lowermost Burgsvik Beds of Ludlow age.

Comparison. – The species is most easily identified by the needle-shaped ramus of the MII and the denticle formula. The formula with two pre-cuspidal denticles is not unusual in the right MII, but their large, equal size makes them more easy to distinguish. Of the MI, the almost quadratic basal portion with the spurs are the most conspicuous morphological details (see also *K.* (*K.*) *burgensis*), although the shape of the basal portions has a tendency to change somewhat with time. *K.* (*K.*) *bankvaetensis* has the same dentition formula but differs in shape and size, see comparison of *K.* (*K.*) *bankvaetensis*.

The left MII, no. 738/56, described by Männil & Zaslawskaya (1985a) from the Wenlock Srednij sequence is very similar to *K.* (*K.*) *polonensis* from Gotland.

Kettnerites (*K.*) *polonensis* var. sjaustre

Fig. 46A

Synonymy. – *Kettnerites* (*K.*) *polonensis* var. sjaustre – Bergman, pp. 71–73, Fig. 46A.

Material. – 1 right MI, 1 left MI, 2 right MII, 4 left MII.

Occurrence. – Fig. 4; Ludlow, Hemse Beds unit d, at present known only from the locality Sjaustrehammar 1.

Diagnosis. – Right MI: Inner margin slightly convex with about 22 fairly large denticles of almost equal size. Shank ends sharply. Flange thick-walled, angular. Basal portion wide, basal angle about 30°.

Left MI: About 28 denticles on the convex inner margin, denticles decrease in size posteriorly, ending as small knobs. Inner wing almost rhombic. Wide basal portion, basal angle about 30°.

Right MII: Two large pre-cuspidal denticles followed by a somewhat larger cusp. Two distinct intermediate denticles of unequal size. Large shank with almost parallel sides, ending bluntly. Ramus long, narrow with pointed extremity.

Left MII: Double cusp with the parts of equal size, distinct intermediate dentary, shank with almost parallel sides and ramus long and slender.

Description. – Right MI, dorsal side: Length 1.47 mm, width slightly more than a quarter of length. The jaw tapers from the basal part towards the anterior end. The inner margin is slightly convex, covered for just over half of the jaw length by 23 fairly large, slightly spaced denticles of equal size. The posterior part of the undenticulated ridge is slightly bent outward and is smooth, fairly thick, forming the very sharp posterior tip of the wide, almost triangular shank. The inner wing is strongly downfolded, of normal size, 0.24 of the jaw length, parallel to the undenticulated ridge. The basal furrow is fairly long, its posterior part narrow. The flange is very thick-walled, its posteriormost part elevated, the posterior and outer margins forming two nearly right angles. The anteriormost outer margin of the basal portion is wide, thick, with a deep, short groove separating it from the inner part of the jaw. The posterior outer part of the flange is slightly downfolded. The basal portion is wide; the basal angle about 30°. The outer margin, anterior to the basal portion, is almost straight and passes gradually into the large, sickle-shaped falx.

Ventral side: Not studied.

Left MI, dorsal side: Length 0.95 and 1.45 mm, width slightly more than a quarter of length. The jaw tapers from the anterior half to the large fang. The inner margin is convex from the undenticulated ridge to the anteriormost part of the falx and is covered by fairly large denticles decreasing in size towards the posterior end, ending as small crenulations close to the anterior margin of the inner wing. The denticulation covers 0.6 of the jaw length. The undenticulated ridge is wide and fairly low. The inner wing is of normal size, 0.26 of the jaw length, almost rhombic; the inner margin is parallel to the undenticulated ridge. Basal furrow short, deep with parallel sides. The posterior margin is wide, almost oblique to the extension of the inner margin. The outer margin of the basal portion is short, 0.24 of the jaw length, and is dominated by the narrow, convex, outer wing, formed by the ligament rim. The outer face of the basal portion is large and convex. The basal portion is almost rectangular, wide and short; the basal angle about 30°. The outer margin anterior to the basal portion is straight and passes gradually into the smooth, large, sickle-shaped falx.

Ventral side: not seen.

Right MII, dorsal side: Length 0.48–0.88 mm, width about half the length. Two large pre-cuspidal denticles of equal size form the anteriormost, acute-angled margin. The cusp is of moderate size, slanting slightly in the posterior direction. Two denticles of unequal size form the distinct intermediate dentary. It is followed by nine posteriorly slanting denticles decreasing in size towards the posterior on the almost straight denticulated ridge. The denticulation ends well before reaching the blunt end of the shank. The shank occupies about half of the jaw length; its almost straight sides are almost parallel. The bight is deep with an acute bight angle. The ramus is long and narrow with a pointed extremity. The anterior outer margin is straight.

Ventral side: Not studied.

Left MII, dorsal side: Length 0.31–1.19 mm, width about half the length. A double cusp with parts of equal size forms the anteriormost margin. A distinct intermediate dentary is composed of six relatively small denticles and a larger one at the transition to the denticulated ridge. Nine fairly large denticles, slanting posteriorly, decreasing slightly in size towards the posterior, form the almost straight, denticulated ridge. The shank, with almost parallel sides, occupies about half of the jaw length. The inner wing is almost triangular, representing about half of the jaw length, its anterior part widest. The bight is deep with an acute bight angle. The ramus is long, narrow, with a sinus in its anterior, outer margin.

Ventral side: The slightly enclosed, crescent-shaped myocoele opening represents about ½–⅔ of the jaw length. The opening is only somewhat extended to the anterior along the inner side of the jaw. The ligament rim is of normal size, slightly widened at the outer side, along the ramus.

Discussion. – Jaws very similar to *K.* (*K.*) *polonensis* var. sjaustre or even identical with this variety have been recorded from the Polish erratic boulder 410/3 (the boulder described by Kielan-Jaworowska 1966).

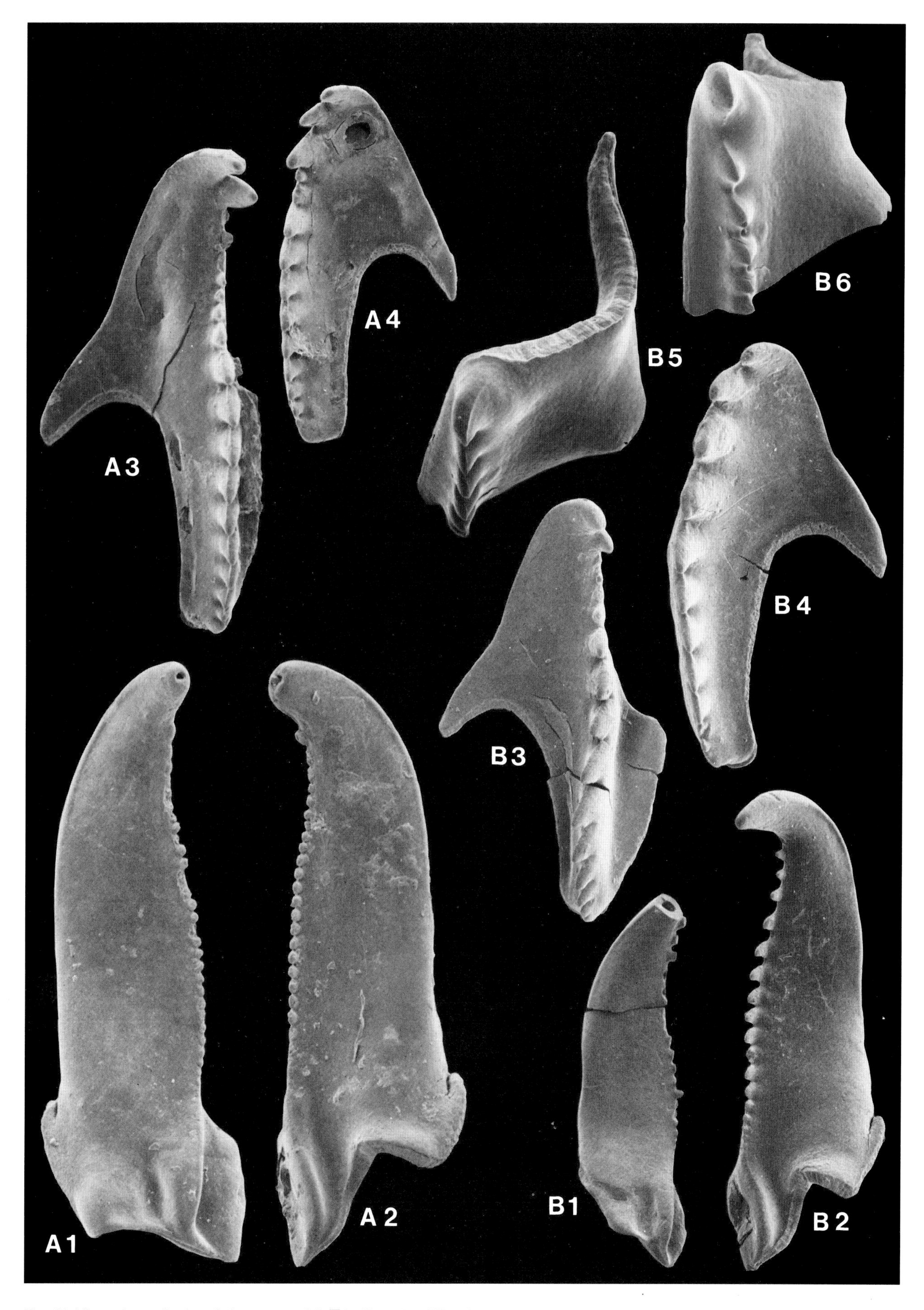

Fig. 46. All specimens in dorsal view, except B5. □A. *Kettnerites* (*K.*) *polonensis* var. sjaustre, Sjaustrehammar 1, Hemse Beds, unit d, 82-19LJ; A1 left MI, LO 5816:7, ×80; A2 right MI, LO 5816:8, ×80; A3 left MII, LO 5816:5, ×80; A4 right MII, LO 5816:6, ×80. □B. *K.* (*K.*) sp. A, Likmide 2, Hemse Beds, Hemse Marl SE, 82-28LJ; B1 left MI, LO 5800:1, ×60; B2 right MI, LO 5800:2, ×120; B3 left MII, LO 5800:3, ×60; B4 right MII, LO 5800:4, ×60; B5 basal plate, dorsal–lateral view, note the denticles on the inner side, LO 5800:5, ×210; B6 same specimen as B5, dorsal view, LO 5800:5, ×210.

Remarks. – The variety is named after the locality Sjaustrehammar 1 from the Hemse Beds, unit d, of Ludlow age. A characteristic element is the left MI (Fig. 46A:1, LO 5816:7).

K. (*K.*) *polonensis* var. sjaustre is a rare form on Gotland, found only at Sjaustrehammar 1. I have placed it as a variety within *K.* (*K.*) *polonensis* because of the fairly similar denticulation and ramus of the MII and the slight similarity of denticulation and habitus of the basal portion of the right MI.

Comparison. – The basal portion of the MI is wider than the corresponding part in *K.* (*K.*) *polonensis*.

Kettnerites (*K.*) sp. A

Fig. 46B.

Synonymy. – *Kettnerites* (*K.*) sp. A – Bergman, pp. 69–71, Fig. 46B.

Material. – One jaw of each, right MI, left MI, right MII, left MII.

Occurrence. – Fig. 4; Early Ludlow, known only from the locality Likmide 2. The sample may be the only one at present available from this horizon (personal communication, Lennart Jeppsson).

Diagnosis. – Right MI: Jaw tapers anteriorly, denticles fairly large, slanting, largest in the middle part of the convex, paucidentate ridge. Sharply pointed wide shank; thick-walled flange. Basal angle about 45°.

Left MI: Denticles along convex inner margin varying from large to small, decreasing in size posteriorly. Inner wing elongated, rounded, extended posteriorly. Basal angle about 45°.

Right MII: Two very large pre-cuspidal denticles, intermediate denticles lacking, denticulated ridge convex laterally, ramus slightly convexo-concave, slender and pointed.

Left MII: Double cusp with anteriormost cusp slightly larger, prominent, almost triangular inner wing, ramus long and needle-shaped.

Description. – Right MI, dorsal side: Length 1.52 mm, width about 1/4 of length, or slightly more. The jaw tapers anteriorly, ending in a large fang. The inner margin is slightly convex, covered by about 17 slanting denticles, widely spaced in the anterior part and largest on the middle part of the margin. The distance between denticles decreases posteriorly; on the posterior, middle part the denticles are closely spaced; the denticulation ends as a few knobs. The undenticulated ridge is fairly wide, rounded, sigmoidal, forming the posteriormost, sharp, pointed tip of the wide shank. The inner wing is downfolded, fairly short, and the inner margin is parallel to the undenticulated ridge. The basal furrow is wide and vaguely defined. The flange is thick-walled, elevated posteriorly. The outer side of the shank and the posterior side of the flange form a right angle, and so do the posterior side of the flange and the outer side of the flange. The anteriormost outer margin of the basal portion to the right of the deep and narrow groove is thick and wide.

The basal plate is not recorded in place – a basal plate possibly belonging to this variety is almost rectangular with a deep downfolded outer part. It is denticulated with a small cusp-like denticle followed by five denticles arranged parallel with the undenticulated ridge.

The basal portion is wide, the basal angle 40°. The outer margin in front of the basal portion is almost straight, with a minor concavity in its middle part, ending with a well-defined, sickle-shaped falx.

Ventral side: not studied.

Left MI, dorsal side: Length 1.06 mm, width about 1/3 of length. The jaw tapers anteriorly, ending in a prominent fang which is moderately bent upwards in relation to surface. The inner margin is slightly sigmoidal with large, widely spaced denticles in the anterior position. The denticulation, 0.6 of the jaw length, consists of about 17 denticles decreasing in size posteriorly and ending in a crenulation before reaching the inner wing. The transition into the undenticulated ridge is smooth; the ridge is slightly arched, fairly short and wide, parallel to the inner margin. The inner wing is elongatedly rounded, extending posteriorly, fairly short, about 0.22 of the jaw length, widest in the posterior middle half. The basal furrow is short, relatively wide and deep. The outer face and margin of the basal portion is short and narrow, the ligament rim is visible on anterior margin. The basal angle is about 40°. The lower middle part of the outer margin anterior to the basal portion has a small convexity. The falx is large and smooth rounded.

Ventral side: Not studied.

Right MII, dorsal side: Length 1.34 mm, width more than half the length. There are two very large pre-cuspidal denticles of equal size, the anterior one forming the almost pointed anterior end of the jaw. They are followed by nine very large denticles that slant slightly posteriorly; the anteriormost of these may be the cusp. Intermediate denticles are lacking. The posteriormost two denticles are very small. The inner margin is convex, the outer slightly concave, making the large shank taper slightly towards the blunt posterior end. The bight is deep with an acute bight angle. The ramus is large with a pointed extremity, the outer margin is convex, the inner concave. A small sinus is present at the anterior outer margin, at the base of the ramus.

Ventral side: Not studied.

Left MII, dorsal side: Length 1.33 mm, width more than half the length. The anteriormost part of the double cusp is slightly larger and forms the sickle-shaped anterior margin. The intermediate dentary with three small denticles of equal size passes into the denticulated ridge through two denticles of increasing size. These are followed by about eight denticles, decreasing in size to the posterior and slanting strongly towards the outer, posterior side. The almost straight shank occupies about half of the jaw length, with a blunt extremity. The nearly triangular, smoothly rounded, laterally extended inner wing is very prominent, occupying about half the jaw length, with its widest part anteriorly. The bight is fairly deep with an acute bight angle. The ramus is long and slender with a pointed extremity. The anterior outer margin of the jaw is almost straight with a minor sinus at the anteriormost part of the ramus.

Ventral side: Not studied.

Remarks. – The form is placed under open nomenclature due to its rareness. I believe that *K.* (*K.*) sp. A is closely related to *K.* (*K.*) *polonensis* because of the dentition and the form of the ramus of the MII, and the form of the basal portion of the MI, excluding the possibly correctly identified denticulated basal plate. The affinities of the taxon are not yet fully understood, as only one reconstructed apparatus, composed of five elements, is at hand.

A characteristic element is the right MI (Fig. 46B:2, LO 5800:2).

Kettnerites (*K.*) *versabilis* Bergman 1987

Figs. 14K–N, 18G, 47

Synonymy. – *Kettnerites* (*K.*) *versabilis* n.sp. – Bergman, pp. 73–76, Fig. 14K–N, 18, 47.

Derivation of name. – Latin *versabilis*, changeable, referring to the variable denticulation of the MI and MII.

Holotype. – LO 5361:4, right MI, Fig. 47E:2.

Type locality. – Buske 1.

Type stratum. – Lower Visby Beds, unit e.

Material. – Fig. 2; about 51 right MI, 52 left MI, 42 right MII, and 40 left MII.

Occurrence. – Figs. 4 and 6; Upper Llandovery. Lower Visby Beds, unit d and e, and Upper Visby Beds: Buske 1, Gustavsvik 1, Häftingsklint 1, Ireviken 1, Ireviken 3, Nygårdsbäckprofilen 1, Nyhamn 1 and 4, Lickershamn 2, Vattenfallsprofilen 1. The species has been encountered only from some samples from outside the Upper Visby Beds, unit e.

Diagnosis. – Right MI: Jaw tapering anteriorly, cross-section almost rounded, long slender pointed fang. Anterior denticles large; dentary paucidentate; wide, almost pentagonal basal portion with basal plate fused.

Left MI: Jaw tapering anteriorly, paucidentate dentary along the anterior part of the straight inner margin; wide almost pentagonal basal portion.

Right MII: Large single cusp bent strongly upwards in relation to dorsal surface, intermediate denticles may occur on the falcal arch.

Left MII: Large, slender, single cusp, wide-spaced intermediate denticles, fairly slender ramus with pronounced sinus on the anterior outer margin.

Description. – Right MI, dorsal side: Length 0.42–1.03 mm, width about 1/4 of length. The dorsal surface is convex, the jaw tapering slowly and unevenly towards the anterior. The inner margin is almost straight, denticulated along its entire length in larger jaws; in some jaws also along part of the falcal arch. In smaller jaws the dentary ends before reaching the undenticulated ridge. The anteriormost 1–3 denticles in the falcal arch are normally fairly small and followed by about 5–7 fairly large to very large denticles which slant slightly towards the posterior. In larger jaws these anterior denticles are more or less widely spaced to paucidentate. The denticles of the smaller jaws are only slightly spaced. The forms with these different types of dentary are fairly distinct and their sizes overlap. The denticulation ends at the posteriormost part of the denticulated ridge, posterior or close to the transition into the small undenticulated ridge at the same level as the inner wing. The undenticulated ridge ends before reaching the posterior margin. The inner wing is almost triangular, its anterior part widest. The basal portion is large and almost pentagonal. The basal furrow is fairly deep and of normal length. A fairly distinct furrow runs along the outer margin of the basal portion anterior to the basal plate, separating the characteristic wide outer margin from the inner part of the basal portion.

The basal plate varies from almost square in smaller specimens to rectangular in larger ones, with height about half the length of the plate and widely spaced denticles. The corners of the basal plate form almost right angles, and the outer margin is deeply downfolded. The basal angle is about 25–30°.

The outer margin anterior to the basal portion is rounded longitudinally with a more or less pronounced posterior concavity. The anterior outer margin along the fang is smoothly sickle-shaped and ends in a large fang with a prominent cutting-edge.

Ventral side: The strongly enclosed myocoele opening is about 1/3 of the jaw length. The crescent-shaped anterior margin is surrounded by a fairly narrow and somewhat sunken ligament rim in the anterior inner part. Along the anterior outer part, the rim is bent over towards the dorsal side and forms a sharp ridge along the inner margin of the inner wing.

Left MI, dorsal side: Length 0.43–1.17 mm, width slightly less than 1/4 of length. The fang is large and pointed, and the inner margin straight to almost straight. The dentary varies from normal-spaced in several of the smaller jaws to widely-spaced and paucidentate in some of the larger jaws. The small anteriormost denticle is followed by 1–2 slightly larger denticles, which in turn are followed by 12–16 denticles of uneven size, varying from large to very large. Usually the latter denticles slant slightly and decrease in size towards the posterior. The dentary ends as small knobs well before reaching the inner wing (Fig. 47E:1) except in some of the larger jaws with wide-spaced denticles, where it continues posteriorly to the undenticulated ridge (Fig. 47F:1). The transition to the undenticulated ridge is smooth, except in larger jaws with a wide-spaced dentary. Here, the denticulation ends at the anteriormost part of the undenticulated ridge, and the ridge is elevated and bent slightly to the outer side. The basal portion is almost pentagonal, wide, and has fairly straight margins. The fairly short and wide basal furrow is parallel to the outer margin of the basal portion. The inner margin of the inner wing is parallel to the undenticulated ridge. The inner wing is almost rectangular, about 1/4–1/5 of the jaw length. A more or less distinct furrow runs along the characteristic wide

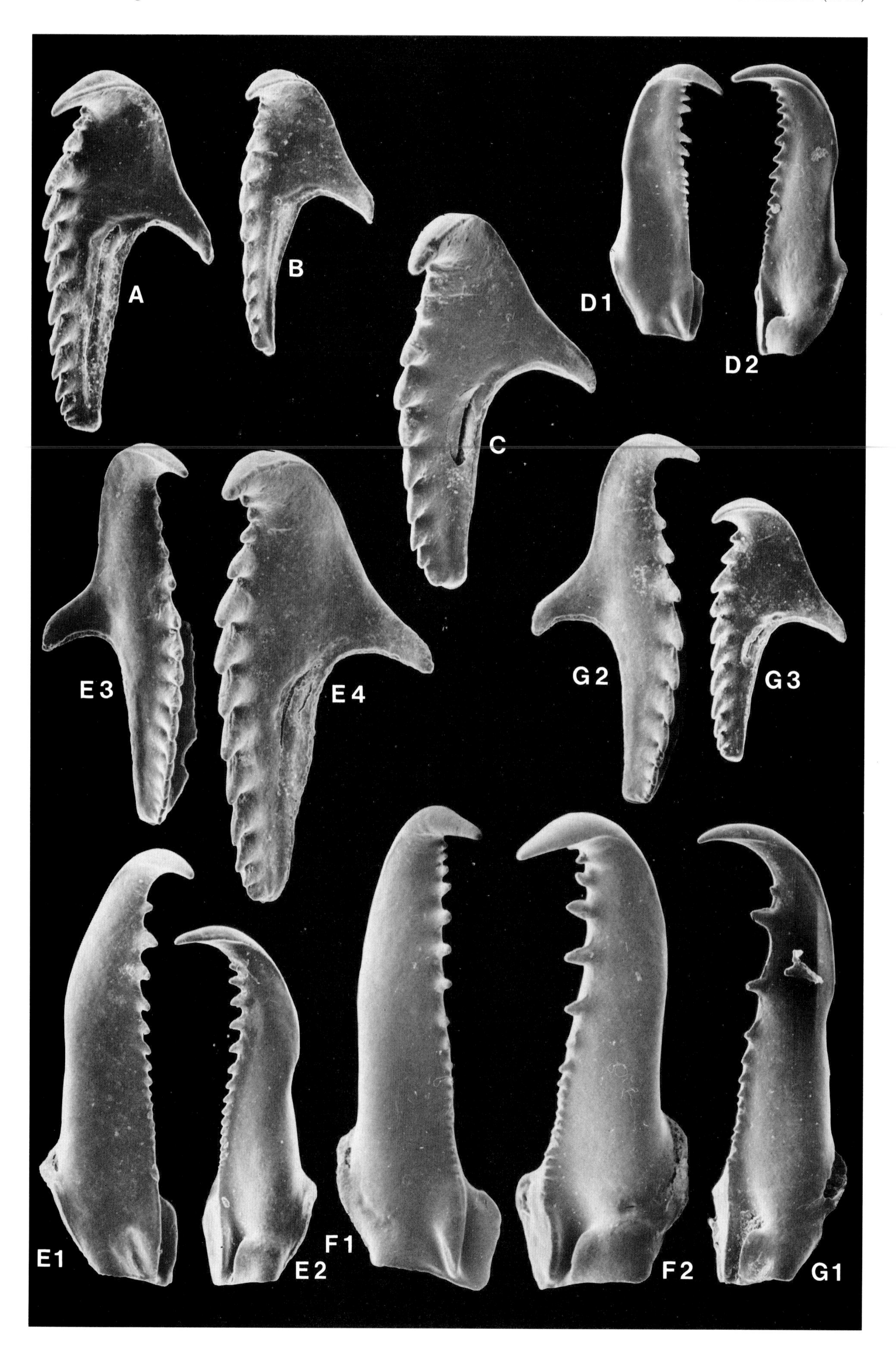
A
B
C
D1
D2
E3
E4
G2
G3
E1
E2
F1
F2
G1

outer margin of the basal portion, ending before reaching the posterior margin. The basal angle is 20–30°. The outer margin, anterior to the basal portion, has a slightly convex end. The smoothly sickle-shaped fang has a prominent cutting edge.

Ventral side: The strongly enclosed myocoele opening is almost a mirror image of that of the right MI, except that it represents about one quarter of the jaw length.

Right MII, dorsal side: Length 0.25–0.74 mm, width about half the length. The large, sickle-shaped, single cusp is bent strongly upwards in relation to the dorsal side. The 1–3 small denticles on the falcal arch represent the intermediate dentary on many of the jaws. The transition to the post-cuspidal dentary is gradual. The size of the denticles increases towards the posterior, reaching maximum size at the upper middle part of the dentary. The approximately 8 fairly large denticles slant and decrease in size towards the posterior, ending in a minor denticle. The shank represents about half the jaw length or slightly more. The outer margin of the shank is almost straight (Fig. 47A–C; the bight is of normal size, and the bight angle varies from 60° to almost 90°. The ramus is slender and slightly pointed, with a well developed sinus on the anterior outer margin.

Ventral side: The crescent-shaped, slightly enclosed myocoele opening, ⅔–¾ of the jaw length, has an anterior–inner bulge and is surrounded by a fairly wide ligament rim.

Left MII, dorsal side: Length 0.43–0.93 mm, width slightly less than half the length. The sickle-shaped, single cusp, is large, slender and strongly bent upwards in relation to the dorsal surface. The 1–3widely spaced intermediate denticles are followed by 9–11 slightly spaced post-cuspidal ones on the denticulated ridge. The cutting edge is very distinct. The shank occupies about half the jaw length and tapers to the posterior. The inner wing is fairly narrow and rounded, extending over slightly more than half the jaw length. The outer margin of the shank is almost straight. The bight is fairly deep; the bight angle varies from 60° to almost 90°. The ramus is slender, slightly pointed, with a pronounced smooth sinus in the anterior outer margin, forming a very prominent sigmoidal curvature.

Ventral side: The slightly enclosed myocoele opening is somewhat more than half the jaw length. In other respects it is almost a mirror image of the right MII.

Remarks. – It cannot be excluded that with more material it will be possible to further subdivide this variable taxon. Three groups based on the MI's can be distinguished:

(A) A long and slender form, of normal colour (dark-brown), with paucidentate, large denticles. The holotype (Fig. 47F:2) is included in this group.

(B) A slender form, of normal colour, with more or less widely spaced, fairly large denticles (Fig. 47E, D).

(C) A long slender form, black in colour, with a large paucidentate dentary (Fig. 47G:1). This form is more slender and has more pronounced paucidentate dentary than the holotype.

The jaws show several similar characters: The MI's of all three types taper evenly towards the anterior, and all the right MI's have a fused basal plate. The dorsal surface is fairly convex and the basal portion fairly large to large. However, the denticulation of MI's varies considerably, at least between forms A and B.

The two forms A and B might be different ontogenetic stages of the same species (e.g. form B juvenile and form A adult) or, less likely, be sexual dimorphs.

The size-ranges of the elements overlap to a great extent. The fairly rare C form is distinguishable by the very unusual opaque and black colour compared to the brownish colour that is normal in paulinitid jaws. Due to the limited material it cannot be resolved whether or not the C form represents another taxon, is a variety of the A form, or is an A form which has been altered diagenetically.

So far, I have not been able to make a reliable distinction between the MII elements associated with the different MI forms. If, indeed, particular MII elements are associated with each of the MI forms listed above, all these MII types might not even yet have been encountered in the very limited material available. However, differences in dentition formulas have been noticed among the MII elements (e.g. Fig. 47E:3 and 47G:2), but with the limited material it is not possible to appreciate the variation. The MII's which have been linked with the B form of MI (Fig. 47E:3, 4) could be associated with the C form of MI (Fig. 47G:2 and 3) and vice versa.

The A and B forms normally occur together and are also associated with the C form in rich localities. The C form (Fig. 47G:1) has so far been found only in Buske 1, Häftingsklint 1, and Nygårdsbäckprofilen 1, from which localities the largest collections of the species (also A and B forms) derive. About 8 right MI and 5 left MI have been encountered of form C.

Comparisons. – The basal portion of the MI's shows a slight similarity to that of *K.* (*K.*) *polonensis* var. sjaustre, but the MI of the latter has a wider outer margin (outer wing).

Fig. 47. Kettnerites (*K.*) *versabilis* including forms A, B, and C. Specimens F represent form A, those of D and E form B, and those of G form C. Specimens A, B, E and G are from Buske 1, Lower Visby Beds, unit e, 79-40LJ, the remaining specimens (C, D and F) are from Häftingsklint 1, Upper Visby Beds, 76-10CB. All specimens are in dorsal view, ×120, except F. □A. Right MII, LO 5361:1. □B. Right MII, LO 5361:2. □C. Right MII, LO 5762:1. □D. D1 left MI, LO 5762:2; D2 right MI, LO 5762:3. □E. E1 left MI, LO 5361:3; E2, holotype, right MI, LO 5361:4; E3 left MII, LO 5361:5; E4 right MII, LO 5361:6. □F. F1 left MI, LO 5762:4, ×80; F2 right MI, LO 5762:5, ×80. □G. G1 right MI, LO 5361:7; G2 left MII, LO 5361:8; G3 right MII, LO 5361:9.

Kettnerites (*Aeolus*) Bergman 1987

Synonymy. – □1987 *Kettnerites* (*Aeolus*) n. subgen. – Bergman, p. 77.

Derivation of name. – From the Greek mythology; Aeolus was the ruler of the winds, father of Sisyphus.

Type species. – *Kettnerites* (*Aeolus*) *sisyphi* Bergman 1987

Other species. – *K.* (*A.*) *microdentatus*, *K.* (*A.*) *fjaelensis*, *K.* (*A.*) *siaelsoeensis*.

Diagnosis. – Right MI: Fairly slender jaw, slender fang. Pronounced, almost rounded, often fairly wide basal portion.

Left MI: Slender jaw, basal portion with almost straight posterior and outer margins, inner wing almost rectangular.

Right MII: One or two very small to fairly small pre-cuspidal denticles. Cusp fairly large, its base swollen.

Left MII: Fairly large cusp without pre-cuspidal denticles, except for *K.* (*A.*) *sisyphi* var. valle which has one.

Remarks. – The denticulation of the left and right MI varies considerably, from fine, densely-spaced denticles of equal size to fairly large denticles of unequal size. The denticles of the left MI are slightly finer than those of the right MI. The dentition on the inner margin also varies in extent between the different subspecies of the corresponding element type. The dentition of the right MI is usually more extended than on the left MI.

The number, arrangement and shape of the denticles (especially the pre-cusps and the cusp) of the right MII and, to a lesser degree, of the left MII, varies between the subspecies. This variation also occurs between populations of *K.* (*A.*) *sisyphi sisyphi* from the Middle-Late Wenlock.

There is no particular variation of the dentition (form, extension), between the same element type (left and right of MI and MII) within the populations in a locality (sample).

Kettnerites (*Aeolus*) *sisyphi* Bergman 1987

Figs. 12K–O, 18H, 20, 48–53

Synonymy. – □1985a *Kettnerites aspersus* (Hinde 1880) – Männil & Zaslawskaya, pp. 117–118, Pl. 17:4e, right MI (738/55). □1987 *Kettnerites* (*Aeolus*) *sisyphi* n. sp. – Bergman, p. 77–87, Figs. 12K–O, 18, 20, 48–53.

Derivation of name. – As for the nominal subspecies.

Holotype. – As for the nominal subspecies.

Type locality. – As for the nominal subspecies.

Type stratum. – As for the nominal subspecies.

Subspecies included. – *K.* (*A.*) *sisyphi sisyphi*, *K.* (*A.*) *sisyphi klasaardensis*, *K.* (*A.*) *sisyphi* var. valle.

Material. – More than 250 elements of each of the four main elements.

Occurrence. – Late Llandovery to Late Ludlow. Lower Visby Beds to Hamra Beds.

Diagnosis. – Right MI: Jaw with almost parallel inner and outer margins and pronounced basal portion with large, deep, downfolded flange. Denticles of varying size, smallest in central part and largest in the posterior or anterior part, ending abruptly at the undenticulated ridge.

Left MI: Slender jaw, tapering anteriorly. Denticles fairly small, ending as small crenulations well before reaching the undenticulated ridge. Almost rectangular inner wing with inner margin straight and parallel to the undenticulated ridge.

Right MII: Two (except in *K.* (*K.*) *sisyphi klasaardensis* which has one) very small to fairly small, pre-cuspidal denticles. Cusp fairly large, more or less swollen in its basal part. Shank fairly pointed, bight angle almost 90°.

Left MII: Fairly large, single cusp, except in *K.* (*A.*) *sisyphi* var. valle which has one pre-cuspidal denticle. Four to six intermediate denticles. Shank fairly pointed; bight angle almost 90°.

Description. – Right MI, dorsal side: Length 0.21–2.1 mm, width about 1/4 of length. The slender jaw, narrowing at the middle, ends anteriorly in a pronounced sickle-shaped falx, and has a distinct cutting edge along the anterior outer margin. The inner margin is straight, denticulated from the falcal arch to the undenticulated ridge. The dentition is fairly dense, except in *K.* (*A.*) *sisyphi klasaardensis*, where the anterior denticles are slightly spaced. The size of the denticles varies considerably in some populations, but also within a single jaw. Usually they are fairly small at the middle of the dentary and larger in the posterior or anterior part. The anterior denticles are often oriented slightly forward or obliquely to the length axis of the jaw, and the posterior ones slant towards the posterior. The denticulated ridge ends abruptly with distinct denticles at the transition to the undenticulated ridge which is fairly high, narrow and often slightly bent outwards at the middle, ending in the posteriormost part of the shank.

A square basal plate with rounded corners is recorded from *K.* (*A.*) *sisyphi klasaardensis* and *K.* (*A.*) *sisyphi* var. valle.

The basal portion is fairly wide, the basal angle about 10–30°. The inner wing is deeply downfolded, fairly narrow at the posterior, widening anteriorly and ending in a fairly large, rounded anterior corner. The bight angle is almost 90°. The flange is normal to slightly thickened, wide and deeply downfolded along the posterior outer margin. A more or less conspicuous, laterally extended border runs along the anterior outer margin of the basal portion. The outer margin anterior to the basal portion and posterior to the falx has a more or less pronounced concavity.

Left MI, dorsal side: Length 0.35–3.0 mm, width about 1/4 of length. The jaw is slender, normally with a pronounced falx and a prominent cutting edge along the outer margin. Denticle size decreases towards the posterior. The denticulation ends as small knobs or crenulations in the middle, or slightly posterior to the middle, on the denticulated ridge, well before reaching the inner wing. The transition to the low, undenticulated ridge is smooth and gradual. The basal portion is fairly wide and conspicuously skewed rectangular to almost pentagonal. The inner wing is almost rectangular, angular or rounded, the basal furrow of normal size. The posterior outer margin of the basal portion is almost straight, the basal angle about 30°. The outer margin anterior to the basal portion and posterior to the falx is slightly concave.

Right MII, dorsal side: Length 0.25–1.38 mm, width about half the length. There are two very small to fairly small pre-cuspidal denticles of unequal size on the anteriormost part of the large cusp, except in *K.* (*A.*) *sisyphi klasaardensis* which has only one. The almost straight, denticulated ridge is composed of fairly large to large denticles, slanting posteriorly, with the largest ones in the middle. The shank occupies slightly more than half the jaw length; the almost straight inner and outer margins taper to the posterior. The bight is fairly deep, the bight angle varies between acute and 90°. The ramus is fairly narrow, the anterior outer margin, almost straight.

Left MII, dorsal side: Length 0.16–0.83 mm, width about half the length. There is a large, sickle-shaped cusp with a distinct cutting edge, and no pre-cuspidal dentary, except in *K.* (*A.*) *sisyphi* var. valle. The cusp is followed by 4–6 fairly large intermediate denticles. All the post-cuspidal denticles vary in number and size. The shank is about half the jaw length and tapers to the posterior. The inner wing is of normal size. The bight is deep, the bight angle acute to almost straight. The ramus is almost triangular to long and narrow, posteriorly somewhat pointed. The sigmoidal shape of the anterior outer margin is due to a pronounced concavity at the base of the ramus.

Discussion. – The difference between the subspecies is most clearly expressed in the pre-cuspidal dentary of the right MII. The nominal subspecies has two very small to small pre-cusps while those of *K.* (*A.*) *sisyphi* var. valle are of normal size. The subspecies *K.* (*A.*) *sisyphi klasaardensis* has only one minor pre-cusp.

Comparison. – The left and right MII of *K.* (*A.*) *sisyphi sisyphi* are very similar to the corresponding elements of *K.* (*A.*) *microdentatus*.

Kettnerites (*Aeolus*) *sisyphi sisyphi* Bergman 1987

Figs. 12L, M, O, 18H, 20, 48, 49

Synonymy. – □1987 *Kettnerites* (*Aeolus*) *sisyphi sisyphi* n. ssp. – Bergman, p. 79–81, Figs. 12L, M, O, 18, 20, 48, 49.

Derivation of name. – Named in commemoration of Sisyphus who also had a great problem to solve.

Holotype. – LO 5836:2, right MI, Fig. 48D:2.

Type locality. – Gandarve 1.

Type stratum. – Halla Beds.

Material. – Figs. 2 and 3; more than 250 right MI, 250 left MI, 250 right MII, 250 left MII.

Occurrence. – Figs. 4 and 7; Late Llandovery to Ludlow. Lower Visby Beds to Hemse Beds. Ajmunde 1, Ajstudden 1, Alby 1, Amlings 1, Ansarve 1, Autsarve 1, Bjärges 1 and 2, Blåhäll 1, Broa 2, Däpps 1 and 2, Djupvik 1, 2, and 3, Fardume 1, Fårö Skola 1, Follingbo 3, Gandarve 1 and 2, Gannor 3, Gardsby 1, Gerete 1, Gnisvärd 1, Godrings 1, Gothemshammar 6 and 7, Gröndalen 1, Grundård 2, Grymlings 1, Häftingsklint 1, Haganäs 1, Hägur 1, Halls Huk 1, Hide 1, Hörsne 5, Källdar 1 and 2, Kullands 2, Lau Backar 1, Lauter 1, Lerberget 1, Lickershamn 2, Lilla Hallvards 1, Loggarve 1 and 2, Lukse 1, Mölner 1, Munkebos 1, Mulde 2, Mulde Tegelbruk 1, Myrsne 1, Nygårds 1 and 2, Rangsarve 1, Robbjäns Kvarn 3, Rönnklint 1, Saxriv 1, Sigvalde 2, Slitebrottet 1 and 2, Snäckgärdsbaden 1, Snauvalds 1, Snoder 2 and 3, Sojvide 1, Sproge 4, Stora Myre 1, Stora Vikare 2, Svarven 1, Talings 1, Tjeldersholm 1, Trädgården 1, Valby Bodar 1, Valleviken 1, Vallstena 2, Valve 3, Värsände 1, Västerbjärs 1, Vattenfallsprofilen 1, Vidfälle 1.

Diagnosis. – Right MI: Fairly slender jaw, pronounced, rounded basal portion. Denticles of normal size, ending abruptly at the undenticulated ridge. Smallest denticles in the central part of the dentary and largest at the posterior end.

Left MI: Slender jaw, tapers anteriorly, ends in a fairly small fang. Denticles fairly small, covering about ½–⅔ of the margin, ending well before reaching the smooth transition to the undenticulated ridge.

Right MII: Two very small to small pre-cuspidal denticles. Cusp fairly large, slightly swollen at the base. Shank fairly pointed, bight angle almost right.

Left MII: Single, fairly large cusp. Four to six intermediate denticles. Shank slender, fairly pointed; bight angle almost right.

Description. – Right MI, dorsal side: Length 0.33–0.85 mm, width about ¼ of length. The slender jaw, narrowing at the middle, ends anteriorly in a pronounced sickle-shaped fang with a distinct cutting edge along the anterior outer margin. The inner margin is straight, denticulated from the falcal arch to the undenticulated ridge. The dentition varies considerably from specimens with fairly small denticles in the Early Wenlock and the Early Ludlow to specimens with larger denticles in the Middle Late Wenlock, particularly those from the Halla and Mulde Beds. The dentition of specimens from this stratigraphic interval is also more diversified, with fairly small denticles at the middle of the dentary and the largest at the posterior end. The anterior denticles are oriented slightly forward or obliquely to the length axis of the jaw, and the posterior ones slant towards the posterior. The denticulated ridge ends abruptly with distinct denticles at the transition to the undenticulated ridge. The undenticulated ridge is fairly high, narrow and often slightly bent outwards at the middle, ending in the posteriormost part of the shank.

The basal plate is not recorded.

The basal portion is fairly wide with rounded corners. The basal angle is about 10–30°. The inner wing is about ¼ of the jaw length. Its posterior part is fairly narrow, widening in the anterior and ending in a fairly large and conspicuous, rounded anterior corner. The bight angle is almost 90°. The flange is normal to slightly thickened, wide, deeply downfolded along the outer side and the posterior outer corner, and with a ligament rim along the outer margin of the basal portion. The outer margin, anterior to the basal portion and posterior to the falx, has a more or less pronounced concavity.

Ventral side: The crescent-shaped, strongly enclosed myocoele opening, about ⅓ of the jaw length, is bounded to the anterior and along the inner margin by a somewhat sunk ligament rim, forming a narrow ridge along the inner wing. On the outer side the rim is bent over and is visible from the dorsal side.

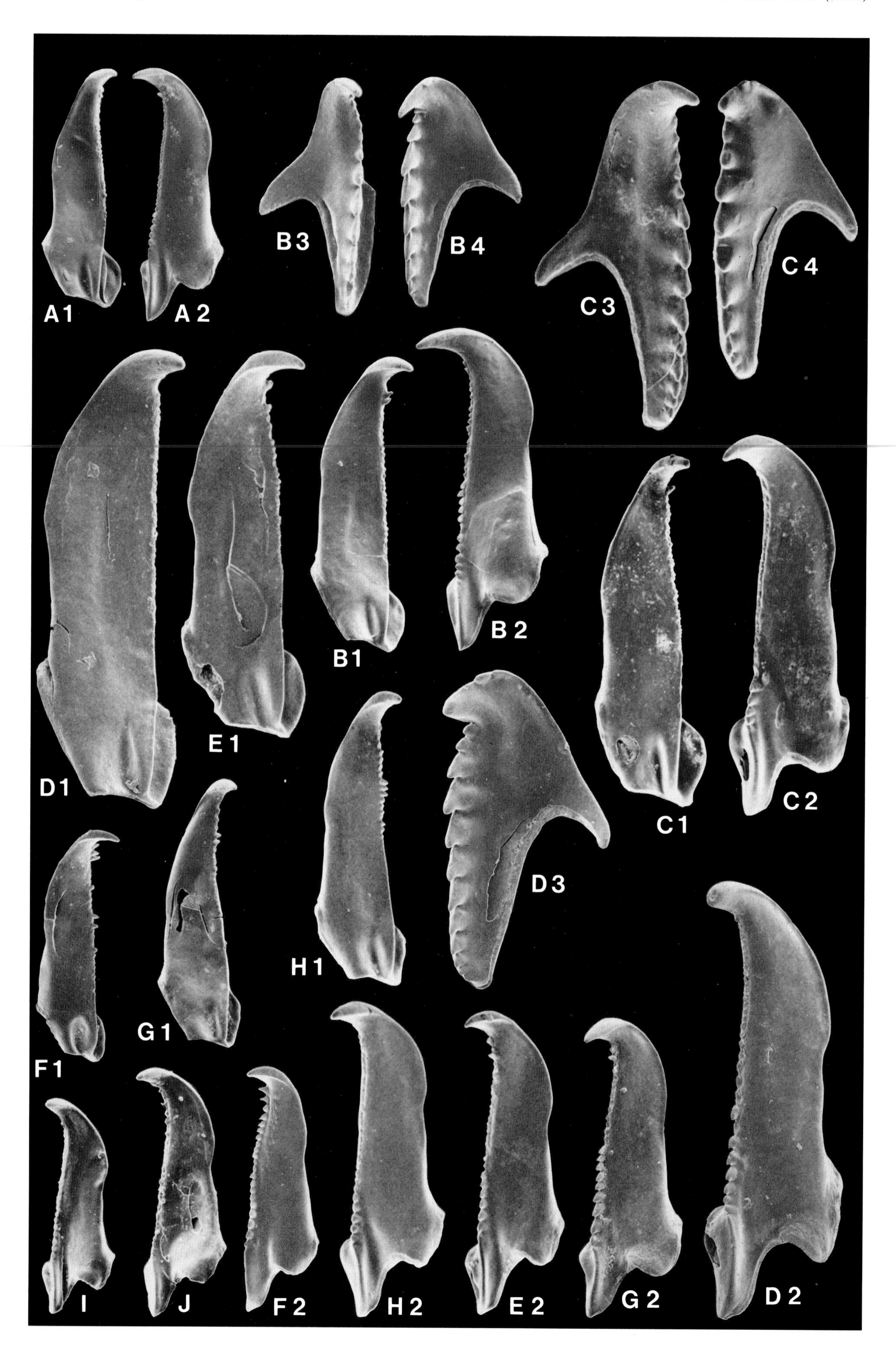
A1
A2
B3
B4
C3
C4
D1
E1
B1
B2
C1
C2
H1
D3
F1
G1
I
J
F2
H2
E2
G2
D2

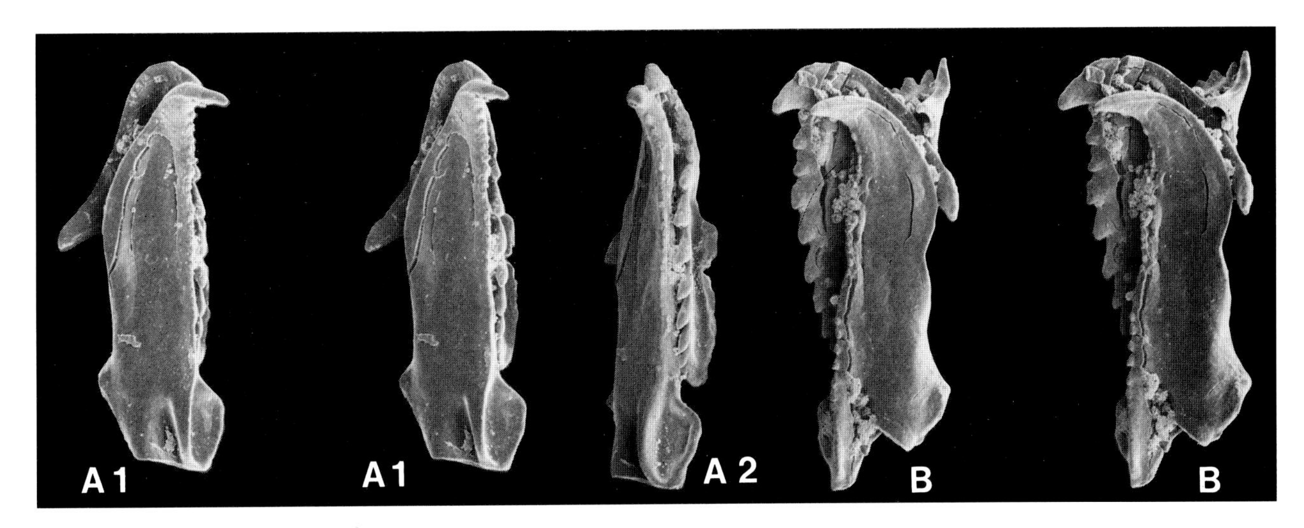

Fig. 49. Kettnerites (A.) sisyphi sisyphi. A and B form main part of a jaw apparatus of fused elements from Amlings 1, Hemse Beds, Hemse Marl, NW part, 75-73CB, ×120. □A1. Left MI and MII, stereopair, dorsal view, LO 5828:3; □A2. Lateral view of the same specimens as A1. □B. Right MI, MII and MIV, stereopair, dorsal view, LO 5826:1.

Left MI, dorsal side: Length 0.36–0.83 mm, width about 1/4 of length. The slender jaw, with a sickle-shaped falx, ends in a fairly small fang with a prominent cutting edge. The largest denticles are at the anteriormost part of the denticulated ridge. They are perpendicular to the extension of the jaw on the almost straight inner margin. The denticles decrease in size to the posterior, ending as small knobs or crenulations in the middle of the denticulated ridge. The posterior denticles are perpendicular to the extension of the jaw or slant slightly towards the posterior. The transition to the low, undenticulated ridge is smooth and gradual. The basal portion is trapezoidal and fairly wide. The inner wing is almost rectangular and rounded, the basal furrow of normal size and wide. The posterior outer margin of the basal portion is almost straight; the basal angle about 30°. The outer margin between the basal portion and falx is slightly concave.

Ventral side: The strongly enclosed myocoele opening is 1/4–1/3 of the jaw length, in other respects it is very similar to the one of the right MI.

Right MII, dorsal side: Length 0.26–0.52 mm, width about half the length. Two small, pre-cuspidal denticles of unequal size project from the anteriormost part of the large cusp. The cusp is oriented almost perpendicular to the length axis of the almost straight denticulated ridge, which is composed of 2–3 minor denticles followed by 6–8 larger ones, slanting towards the posterior, with the largest denticles at the middle. The shank occupies slightly more than half the jaw length, the almost straight inner and outer margins taper to the posterior. The bight is fairly deep, bight angle 60° to almost 90°. The ramus is fairly narrow, the anterior outer margin almost straight. The jaw has a triangular appearance.

Ventral side: The slightly enclosed myocoele opening, 2/3 of the jaw length or somewhat less, is bounded by a fairly wide and somewhat elevated ligament rim to the anterior and along the inner side. The almost straight anterior and inner margins form a corner which protrudes slightly anteriorly.

Left MII, dorsal side: Length 0.22–0.50 mm, width approximately half the length. The large, sickle-shaped, cusp with a distinct cutting edge, is oriented perpendicular to the denticulated ridge and is followed by 4–6 fairly large intermediate denticles. The denticulated ridge is slightly convex with usually 8–9 posteriorly slanting denticles, the largest in the posterior half of the denticulated ridge. The shank tapers to the posterior and occupies about half the jaw length, ending fairly pointed and somewhat bent outwards. The inner wing is of normal size. The bight is deep, the bight angle 60° to almost 90°. The ramus is long and narrow, posteriorly somewhat pointed. The anterior outer margin is sigmoidal because of a pronounced concavity at the base of the ramus.

Ventral side: The slightly enclosed myocoele opening is 1/2–2/3 of the jaw length. The almost straight anterior and inner margins form an almost right angle. The rim is normal and becomes wider at the outer part of the ramus.

Fig. 48. Kettnerites (A.) sisyphi sisyphi. All specimens in dorsal view, ×120. □A. Snoder 2, Hemse Beds, Hemse Marl, NW part, 82-14CB; A1 left MI, LO 5809:1; A2 right MI, LO 5809:2. □B. Värsände 1, Mulde Beds, lowermost part, 75-22CB; B1 left MI, LO 5793:5; B2 right MI, LO 5793:6; B3 left MII, LO 5793:7; B4 right MII, LO 5793:8. □C. Däpps 2, Mulde Beds, upper part, 82-37CB; C1 left MI, LO 5786:1; C2 right MI, LO 5786:2; C3 left MII, LO 5786:3; C4 right MII, LO 5786:4. □D. Gandarve 1, Halla Beds, 71-81LJ; D1 left MI, LO 5836:1; D2, holotype, right MI, LO 5836:2; D3 right MII, LO 5839:1. □E. Robbjäns Kvarn 3, Mulde Beds, lower part, 71-138LJ; E1 left MI, LO 5775:1; E2 right MI, LO 5775:2. □F. Vallstena 2, Slite Beds, *Pentamerus gothlandicus* Beds or slightly older, 77-2CB; F1 left MI, LO 5777:1; F2 right MI, LO 5777:2. □G. Värsände 1, Mulde Beds, lower part, 75-23CB; G1 left MI, LO 5790:1; G2 right MI, LO 5790:2. □H. Slitebrottet 2, Slite Beds, Slite Marl, 83-31LJ; H1 left MI, LO 5773:3; H2 right MI, LO 5773:4. □I. Right MI, LO 5362:3, Buske 1, Lower Visby Beds, unit e, 79-40LJ. □J. Right MI, LO 5363:1, Lickershamn 2, Lower Visby Beds, unit f, 73-53LJ.

Kettnerites (*Aeolus*) *sisyphi* var. valle

Figs. 12K, 18H:3, 50

Material. – Figs. 2 and 3; more than 59 right MI, 57 left MI, 55 right MII, 35 left MII.

Occurrence. – Figs. 4 and 7; Middle Wenlock. Slite Beds *Pentamerus gothlandicus* Beds, Slite Siltstone, Halla Beds, unit b. Gandarve 1, Gerumskanalen 1, Gothemshammar 8, Haganäs 1, Mulde Tegelbruk 1, Klinteenklaven 1, Möllbos 1, Svarvare 1 and 3, Valbybodar 1, Valle 1 and 2.

Diagnosis. – Right MI: Inner and outer margin almost straight and parallel, denticles fairly large and of almost equal size, shank ending bluntly.

Left MI: Inner margin almost straight, outer margin convex. Denticles decrease in size posteriorly. Basal portion angular, large inner wing almost rectangular, parallel with undenticulated ridge.

Right MII: Two pre-cuspidal denticles of normal size on a sickle-shaped anterior margin. Anterior margin slightly angular in the smaller specimens. Ramus fairly small and narrow.

Left MII: One distinct pre-cuspidal denticle; ramus fairly small and narrow.

Description. – Right MI, dorsal side: Length 0.35 to about 2.1 mm, width about ¼ of length. The anterior part of the dorsal surface is almost flat, the posterior part convex. The inner and outer margins are almost straight and almost parallel. The fang is narrow and bent upwards in relation to the surface. The denticulation of the inner margin covers about 0.6–0.7 of the jaw length. The approximately 30 denticles are of almost equal size, though the posterior central and posteriormost ones are somewhat smaller. The anterior 5 or so denticles are slightly spaced and directed forward or obliquely with regard to the direction of the inner margin. The following denticles are closely spaced and slightly slanting. The denticulation ends posterior to the anterior end of the inner wing with some very small denticles. The high, smooth and narrow undenticulated ridge is bent outward in the centre, without reaching the posterior margin. The shank is wide with a blunt end. The bight angle is acute, close to 70°. The inner wing is of normal size, almost rectangular, deeply downfolded; the inner margin is parallel to the undenticulated ridge. The basal furrow is deep and long. The flange is thin-walled and angular, its outer part deeply downfolded and extending anteriorly. The anterior part of the basal portion is folded horizontally, forming a rounded outer wing. The basal angle is 25–30°, with the inner and outer margins of the basal portion more or less parallel.

The basal plate is almost square.

The outer margin of the falx is somewhat flattened on its outer anterior part. Along the outer margin a cutting edge is very distinct for more than half of the jaw length.

Ventral side: The crescent-shaped, strongly enclosed myocoele opening, about ⅓ of the jaw length, is bounded in the anterior and along the inner margin by a fairly high ligament rim which forms a sharp ridge along the inner wing. Along the outer side, the margin, including the rim, is bent over and visible from the dorsal side.

Left MI, dorsal side: Length 0.38–3.0 mm, width about ¼ of length. The dorsal surface is flat to concave along the outer margin. The anterior part of the fang is pointed and narrow. The inner margin is almost straight, denticulated from the anteriormost part of the falcal arch, for about ⅔ of the length of the inner margin. The denticles are of medium size, decreasing toward the posterior end. There is a smooth transition to the long, smooth and narrow undenticulated ridge. The inner wing is large, rectangular, its posterior outer corner prominent, and its outer margin parallel to the undenticulated ridge. The basal furrow is narrow, deep and parallel to the outer margin of the basal portion. The posterior margin is almost straight, sometimes directed obliquely towards the inner margin. The outer margin of the basal portion is slightly downfolded, wide in the anterior and forming an outer wing bordered by a straight ligament rim. The basal portion is almost pentagonal and angular. The basal angle is 25–30°. The outer margin is strongly convex, somewhat flattened in the anterior outer portion and dominated by the cutting edge which runs along the outer margin from the anteriormost part of the falx almost to the basal portion.

Ventral side: The crescent-shaped, strongly enclosed myocoele opening is about ⅓ of the jaw length or slightly less. The ligament rim is of normal width and slightly lower than the jaw surface.

Right MII, dorsal side: Length 0.25–1.33 mm, width slightly more than half the length. There are two pre-cuspidal denticles of normal size on the sickle-shaped anterior margin. Smaller specimens may have a somewhat angular anterior margin. The cusp is of normal size and slants slightly posteriorly, with a prominent cutting edge on its anterior basal part. The denticles slant posteriorly and increases slightly in size from the anterior towards the centre of the dentary. There are no distinct intermediate denticles. The number of denticles is 10–16, more or less correlated with jaw size. The almost straight shank tapers slightly to the posterior and represents slightly more than half the jaw length. The inner wing is very minute, following the inner margin, widest at the posterior. The bight angle varies from acute to almost straight. The ramus is relatively small, particularly in the larger specimens. It is fairly narrow and pointed with a slight basal sinus on the outer, anterior margin.

Ventral side: The slightly enclosed myocoele opening is ⅔–¾ of the jaw length. The ligament rim is fairly narrow,

Fig. 50. Kettnerites (*A.*) *sisyphi* var. valle. I and J are referred to as *K.* (*A.*) *sisyphi* var. cf. valle. Valle 2, Slite Beds, *Pentamerus gothlandicus* Beds, 66-145SL. G, H, L, M, N are ×80, all other specimens ×120. All in dorsal view. □A. Left MII, LO 5770:1. □B. Left MII, LO 5770:2. □C. Left MII, LO 5771:2. □D. Right MII, LO 5771:3. □E. Right MII, LO 5771:4. □F. Right MII, LO 5770:3. □G. Left MII, LO 5770:4. □H. Right MII, LO 5770:5. □I. Left MI, LO 5771:5. □J. Right MI, LO 5771:6. □K. Left MI, LO 5771:7. □L. Right MI, LO 5771:8. □M. Left MI, LO 5771:9. □N. Right MI, LO 5771:10.

A
B
C
D
E
F
G
H
I
J
K
L
M
N

the opening extends in the anterior inner part to form a protruding corner. The margins of the opening are almost straight.

Left MII, dorsal side: Length 0.37–0.83 mm, width about half the length or slightly less. There is one distinct pre-cuspidal denticle with a cutting-edge on the anterior part of the smoothly rounded anterior margin. The cusp is fairly thin and almost perpendicular to the extension of the denticulated ridge. The intermediate dentary is not particularly distinct because of the large size of the denticles. Their number varies from 4 to 6 and they are of equal size. The following denticles are almost of equal size with the exception of the posteriormost ones, and they slant in posterior direction. The shank tapers posteriorly and represents less than half the jaw length. The inner wing is small, with a rounded inner margin. The bight angle is almost 90°. The ramus is pointed and relatively small, with a more or less pronounced sinus at the outer margin of the base.

Ventral side: The crescent-shaped, slightly enclosed myocoele opening extends anteriorly along the inner side for more than half the jaw length. The ligament rim is of normal width, widening in the outer part of the ramus.

Discussion. – This taxon, except the left MII, is very close to *K. (A.) sisyphi sisyphi.* It is only recorded from a few stratigraphically restricted localities. *K. (A.) sisyphi* var. valle occurs together with the nominal subspecies. Further material may indicate whether this taxon should be treated as a species.

Remarks. – A typical element of *K. (A.) sisyphi* var. valle is the right MI, LO 5771:10, Fig. 13N from Valle 2, Slite Beds, Slite Marl, *Pentamerus gothlandicus* Beds.

Kettnerites (Aeolus) sisyphi klasaardensis Bergman 1987

Figs. 12N, 18H:1, 20, 51–53

Synonymy. – □1987 *Kettnerites (Aeolus) sisyphi klasaardensis* n. ssp. – Bergman, p. 83–87, Figs. 12N, 18, 20, 51–53.

Derivation of name. – After the type locality, Klasård 1, where this species is dominant.

Holotype. – LO 5823:13, right MI, Fig. 51P.

Type locality. – Klasård 1, Gotland.

Type stratum. – Hemse Beds, Hemse Marl SE.

Material. – Fig. 3; more than 157 right MI, 131 left MI, 133 right, 139 left MII, 23 MIII, 17 left and right carriers.

Occurrence. – Figs. 4 and 7; Ludlow, Leintwardinian. Hemse Beds, Hemse Marl SE part, Eke Beds basal part. Bodudd 1 and 3, Hallsarve 1, Herrvik 2, Kärne 3, Klasård 1, Stora Kruse 1, Vaktård 2, 4, and 5, Västlaus 1.

Diagnosis. – Right MI: Almost parallel inner and outer margins; inner margin straight, smallest denticles in middle of denticulated ridge. Large angular, almost pentagonal basal portion with fused, square, basal plate.

Left MI: Straight inner margin; denticles decrease in size posteriorly, ending as crenulations without reaching the nearly rectangular inner wing. Basal portion almost pentagonal, angular.

Right MII: One minor, pre-cuspidal denticle, the continuation of which forms, in relation to the large swollen cusp, a characteristic keel along the falx.

Left MII: Prominent cusp with almost straight anterior margin.

MIII: Concavo-convex element with pronounced cusp, preceded anteriorly by three denticles.

Right MIV: Concavo-convex element, in lateral view crescent-shaped, without cusp, second and third denticles from the anterior forming the highest part of the jaw.

Left MIV: Mirror image of the right MIV.

Carriers: A pair of left and right elements, chisel-like, elongated, thin, with pronounced head and shaft with a longitudinal furrow.

Description. – Right MI, dorsal side: Length 0.23–0.96 mm, width approximately 0.3 of length. The anterior part of the fang is slender, bent upward in relation to the almost flat jaw surface. The inner margin is straight, denticulated from the anteriormost part of the falcal arch to the undenticulated ridge. The number of denticles is 18–30, the number increasing more or less regularly with increasing jaw length. The first 2–5 anterior denticles are often large, slightly spaced, some directed forward, or more commonly, perpendicularly to the length axis of the jaw. The middle denticles are smaller and densely spaced. The posterior third of the denticles, again larger, often slanting, end abruptly at the undenticulated ridge with a few small distinct knobs on the margin, which often makes a curve here. The ridge is high, narrow and rounded, slightly bent outwards at the middle, reaching the posterior end of the shank and forming the posteriormost tip of the jaw. The inner wing is strongly downfolded, almost triangular with a rounded anterior inner 'corner'. Almost parallel to and immediately to the right of the undenticulated ridge is the deep and narrow longitudinal basal furrow, partly covered by the basal plate.

The basal plate is square, undenticulated, and confluent with MI on its anterior and inner margins. A low ridge runs along its inner margin, with a small, short furrow immedi-

Fig. 51. Kettnerites (A.) sisyphi klasaardensis. Klasård 1, Hemse Beds, Hemse Marl, SE part, 71-150LJ. A, B, C, D ventral view; J, K, L lateral view; the remaining in dorsal view. □A. Right MI, basal portion, LO 5823:1, ×120. □B. Left MI, basal portion, LO 5823:2, ×120. □C. Right MII, LO 5823:3, ×120. □D. Left MII, LO 5823:4, ×120. □E. Pair of carriers, stereopair, LO 5824:1, ×230. □F. Left MI, LO 5823:4, ×120. □G. Right MI, LO 5823:5, ×120. □H. Left MI, LO 5823:6, ×210. □I. Right MI, LO 5823:7, ×210. □J. Carrier, LO 5824:2, ×210. □K. Pair of carriers, LO 5824:3, ×250. □L. Carrier, LO 5824:4, ×170. □M. Left MII, stereopair, LO 5823:8, ×210. □N. Right MII, stereopair, LO 5823:9, ×210. □O. Left MI, stereopair, LO 5823:10, ×120. □P. Holotype, right MI, stereopair, LO 5823:13, ×120. □Q. Left MII, LO 5823:12, ×120. □R. Right MII, LO 5823:11, ×120.

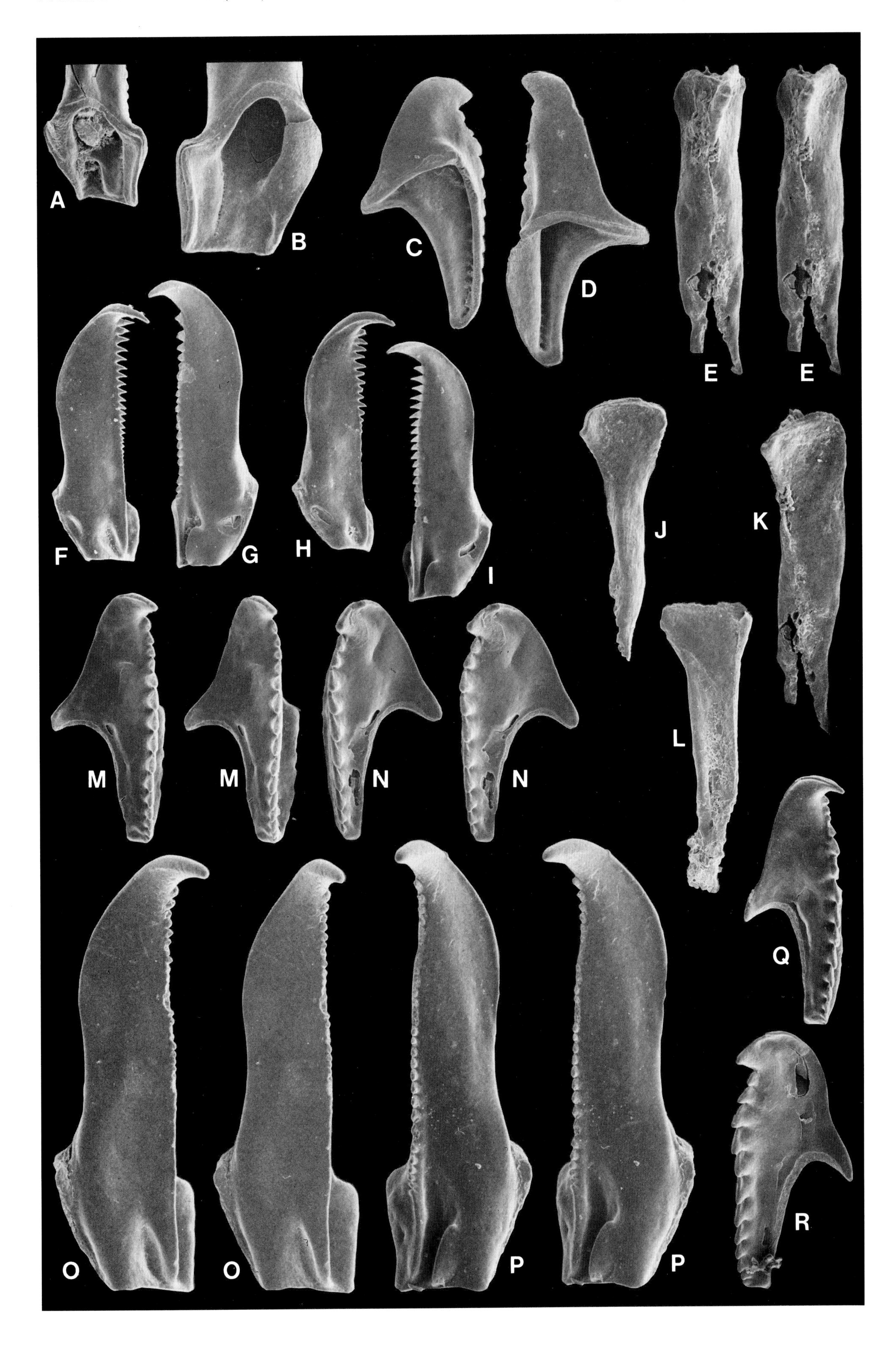
A
B
C
D
E
E
F
G
H
I
J
K
L
M
M
N
N
O
O
P
P
Q
R

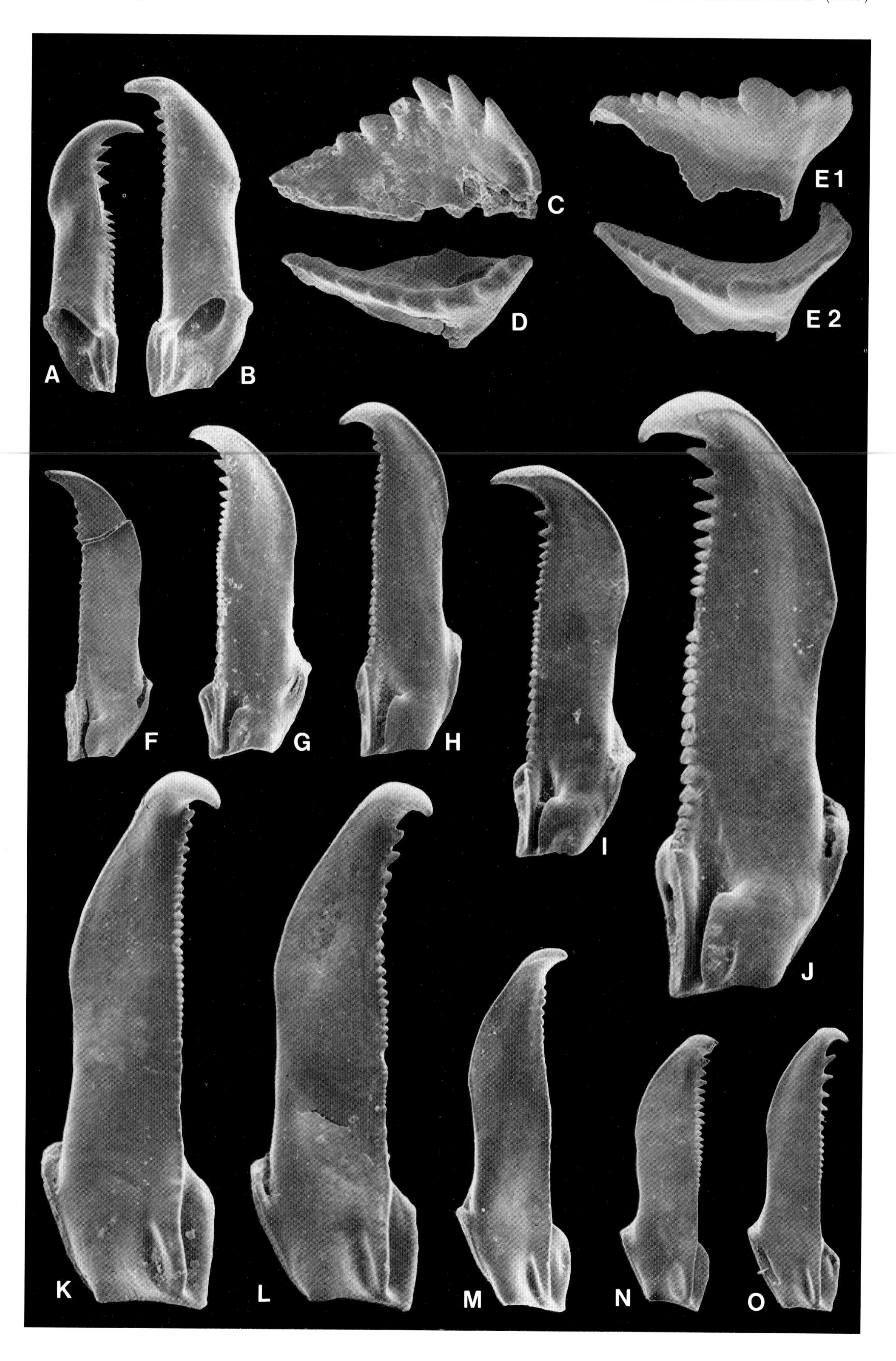
A
B
C
D
E1
E2
F
G
H
I
J
K
L
M
N
O

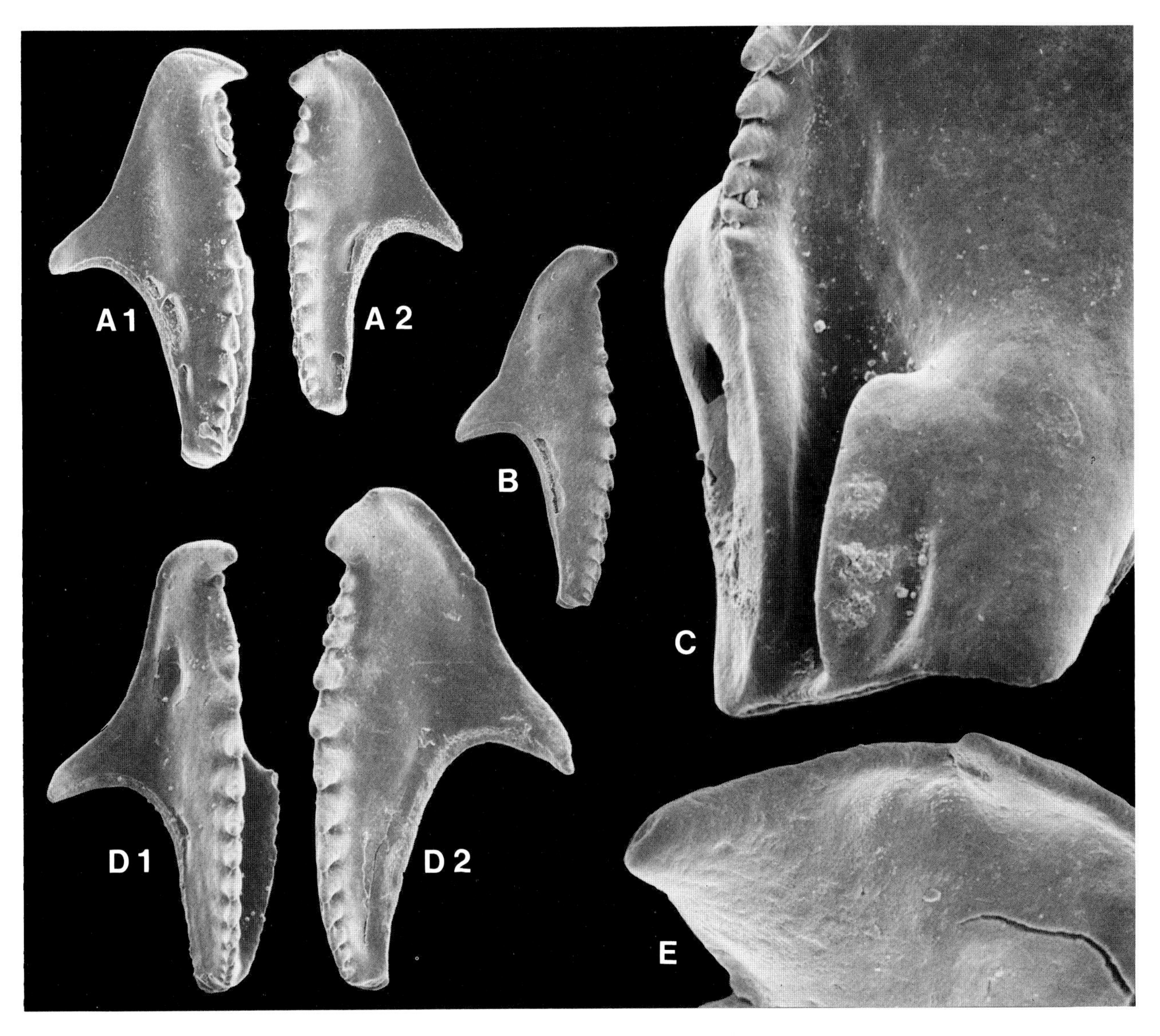

Fig. 53. Kettnerites (A.) sisyphi klasaardensis. All specimens from the Hemse Beds, Hemse Marl, SE part, except B; all in dorsal view. □A. Vaktård 4, 81-35LJ; A1 left MII, LO 5822:3, ×170; A2 right MII, LO 5822:4, ×120. □B. Left MII, LO 5843:3, Kärne 3, Eke Beds, lowermost part, 71-198LJ, ×120. □C. Basal portion of right MI including the basal plate in place, LO 5815:10, Bodudd 3, 71-151LJ, ×280. □D. Bodudd 3, 71-151LJ; D1 left MII, LO 5815:9, ×120; D2 right MII, LO 5815:8, ×120. □E. Precusp and cusp of right MII, LO 5823:17, Klasård 1, 71-150LJ, ×830.

Fig. 52. Kettnerites (A.) sisyphi klasaardensis. F–O are in dorsal position. All ×120 except C, D and E. □A. Right MI, ventral view, LO 5815:7, Bodudd 3, Hemse Beds, Hemse Marl, SE part, 71-151LJ. □B. Left MI, ventral view, LO 5815:6, same sample as A. □C. Left MIV, lateral view, LO 5823:16, Klasård 1, Hemse Beds, Hemse Marl SE part, 71-150LJ, ×250. □D. Right MIV, dorsal view, LO 5823:15, same sample as C, ×250. □E. Same sample as C; E1 MIII, lateral view, LO 5823:14, ×250; E2 same specimen as E1, dorsal view, ×250. □F. Right MI, LO 5843:2, Kärne 3, Eke Beds, lowermost part, 71-198LJ. □G. Right MI, LO 5822:1, Vaktård 4, Hemse Beds, Hemse Marl, SE part, 81-35LJ. □H. Right MI, LO 5804:1, Hallsarve 1, Eke Beds, lowermost part, 69-28LJ. □I. Right MI, LO 5815:5, Bodudd 3, Hemse Beds, Hemse Marl, SE part, 71-151LJ. □J. Right MI, LO 5815:4, same sample as I. □K. Left MI, LO 5815:1, same sample as I. □L. Left MI, LO 5815:2, same sample as I. □ M. Left MI, LO 5822:2, same sample as G. □N. Left MI, LO 5815:3, Bodudd 3, same sample as I. □O. Left MI, LO 5843:1, same sample as F.

ately to the right of its posterior half. The posterior half is slightly concave; the outer margin strongly downfolded.

Anterior to the basal plate, on the basal portion, the ligament rim forms a pronounced outer wing. At a distance about twice the length of the basal plate from the posterior end, the rim drops steeply, forming a characteristic tip at the junction. The boundary between the rim and the rest of the jaw is smooth. The basal angle is about 20–25°. In front of the basal portion the outer margin is concave, swings out and ends in the sickle-shaped falx. The distinct outer cutting-edge of the fang continues along the outer margin for about 1/3 of the jaw length.

Ventral side: The strongly enclosed myocoele opening is approximately 1/3 of the length of the jaw. The basal plate is separated from the myocoele opening by a costa representing the anterior and inner margin of the basal plate.

Left MI, dorsal side: Length 0.15–0.82 mm, width about 1/4–1/5 of length. The inner margin is straight and the outer margin parallel to the inner margin between the basal portion and the falx. The fang is narrow and pointed, fairly strongly bent upwards in relation to surface of the jaw. There are two distinct cutting edges on the fang, one on the inner side continuing onto the denticulated ridge, the other continuing along the outer margin of the falx. The inner margin, denticulated from the falcal arch, extends half the length of the jaw, ending 1½–2 times the length of the inner wing from the posterior margin. The number of denticles increases more or less with jaw length, from about 12 to 32. The anterior 2–4 denticles are often slightly larger, directed forward, more or less widely spaced. The denticles gradually decrease in size posteriorly, ending as crenulations on the inner margin. Thus, the number of denticles is difficult to discern. There is a smooth transition to the undenticulated ridge, which is low, narrow, slightly bent outwards medially, tapering posteriorly, almost reaching the posterior margin as a narrow downbent ridge. The almost rectangular inner wing of normal size, bent down and parallel to the inner margin. The basal portion is angled, irregularly pentagonal, with the slightly concave posterior margin perpendicular to the long axis of the jaw. The basal furrow is short and almost open towards the posterior outer side. The basal angle is about 20–35°. The outer face of the basal portion is large and smoothly convex, downfolded along the posterior outer margin. A more or less pronounced outer wing runs along the outer margin of the basal portion. It is separated from the outer face by a distinct, narrow groove. The wing rises anteriorly, forming a peak about ⅕ of the jaw length from the anteriormost part of the wing, whence it dips down again.

Ventral side: The strongly enclosed, crescent-shaped myocoele opening is about ⅓ of the jaw length. An almost flat, fairly broad ligament rim forms a smaller ridge along the inner margin of the inner wing. On the outer side, the ligament rim continues on the dorsal side forming the outer wing. On the outer ventral side a short posterior ridge tapers and ends at the posterior margin. The undenticulated ridge is marked by a narrow groove.

On the dorsal side, anterior to the basal portion, the outer margin is concave, ending in the large, almost straight falx.

A transverse fold to the left of the basal furrow is found in approximately half of the studied jaws. The fold is often combined with a low elevated flange.

Right MII, dorsal side: Length 0.26–0.68 mm, width about half the length. A fairly small pre-cuspidal denticle forms the anteriormost part of the jaw and continues as a small keel onto the cusp. The cusp is large, in cross-section elongated, with a swollen basal part characteristically continuing into the jaw, and directed almost perpendicularly to the direction of the denticulated inner margin. Along the slightly convex inner margin there are no distinct intermediate denticles; all denticles slanting posteriorly. The number of denticles varies from 9 on a 0.26 mm long jaw to 13 on one that is 0.68 mm long. These denticles are largest in the centre of the dentition. A very narrow inner wing is present on the posterior ¾ of the jaw. The shank occupies somewhat more than half the jaw length, tapering posteriorly, and its posterior extremity is directed somewhat laterally towards the outer side, making the outer margin slightly concave. The angle of the ramal arch is acute to almost straight. The ramus is short, slightly bent upwards and tapers to the fairly sharp tip. The anterior outer margin forms a smooth, characteristic arch which posteriorly continues into a weakly developed sinus at the base of the ramus.

Ventral side: The slightly enclosed myocoele opening occupies the posterior 0.6–0.7 of the jaw length, its anterior part surrounded by a rim which is not visible in ventral view but follows the edge. The opening is enclosed by almost straight borders on two sides, forming an angle of 70–80°. A high, short and narrow ridge is present at the posterior extremity of the shank.

Left MII, dorsal side: Length 0.16–0.64 mm, width somewhat less than half the length. The cusp, with an almost straight anterior margin, and lacking pre-cuspidal denticles, is strong and directed perpendicularly to the length axis of the slightly convex, denticulated ridge. The intermediate dentary, posterior to the cusp, is represented by 3–7 denticles of equal size. These are followed by 9–11 significantly larger denticles (except for the posteriormost ones); the number of denticles varies and does not seem to follow jaw size. The posterior denticles slant; the central ones are the largest. The inner wing is laterally extended, convex and with its widest part at the anterior. It is more than half the total length of the jaw in smaller specimens and less in larger ones. The shank, occupying about half the jaw length, tapers posteriorly, with its posteriormost tip slightly directed towards the outer side. The outer margin, posterior to the ramus, is straight to slightly concave. The ramus is fairly short, somewhat less than half of the width of the jaw, projecting laterally, at nearly right angles and slightly bent upwards. The anterior outer margin is smoothly sigmoidal with an almost rounded anterior part.

Ventral side: The enclosed myocoele opening, surrounded by a rim, is very similar to that of the right MII, except for the large inner wing and the fairly low ridge in the posterior extremity.

MIII, unpaired, dorsal side: Length 0.19–0.31 mm, width about half the length. The narrow, arcuate jaw has a convex inner margin and a concave outer one. The transverse cross-section is high and fairly narrow anterior to the cusp, and low and narrow posterior to it. The ramal angle is about 70°. The ramal extremity is pointed and bent inwards posteriorly. The shank is slender, tapering posteriorly in both lateral and vertical view. The shank extremity is pointed, often slightly bent towards the inner side. The anterior of the cusp, forming a border between the ramus and the shank, is a large node on the inner margin. The denticulation on the ridge forms the upper part of the jaw, a cusp forms the uppermost part. The denticles slant, decreasing in size outwards from the cusp. The cusp is pronounced, slanting slightly towards the posterior. Normally there are three (2–4) pre-cuspidal denticles on the ramus. The number of post-cuspidal denticles on the shank is normally 5 (4–6).

Ventral side: The gaping myocoele on the ventral side is surrounded by a thickened rim.

Right MIV, dorsal side: Length 0.15–0.28 mm, width, including the most basal part, about 0.4 of the length. The narrow, arcuate jaw has a convex inner margin and a concave outer one. The cross-sections along the jaw are almost triangular, high and fairly narrow. No pronounced cusp. Denticles, slanting slightly towards the posterior, usually vary in number between 6 and 8. The second and third anteriormost denticles are usually the largest, and the size of the following denticles decreases to the posterior end.

Left MIV, dorsal side: A mirror image of the right MIV.

Carriers: Chisel-like, flattened, almost bisymmetrical, 3–4 times as long as the maximum width at the pronounced head. The anterior margin of the head is straight and abruptly truncate. The head tends to grow more asymmetrical in later stages of ontogeny, and a shallow furrow on one side of the head, perpendicular to the length of the jaw, becomes more pronounced on some specimens. The head tapers to the posterior and extends into a shaft which in turn tapers posteriorly, ending in a pointed extremity. The inner margin is straight. A longitudinal, shallow furrow is present on the shaft on the opposite side of the furrow in the head.

Remarks. – The transverse fold found on about half of the left MI is often combined with a low elevated flange (Fig. 51O). This indicates that the fold may be caused by compaction, though the overall preservation of the jaws does not entirely support such a conclusion. A less likely explanation is that the fold is an expression of dimorphism.

Comparison. – The MI's show a slight resemblance to the corresponding element of the morph with the straight inner margin of *Kettnerites* (*K.*) *huberti* (see 'Comparison' under *K.* (*K.*) *huberti*).

Kettnerites (*Aeolus*) *microdentatus* Bergman 1987

Figs. 12I, J, Q, R, 18I, 54, 55A, D–F

Synonymy. – □1987 *Kettnerites* (*Aeolus*) *microdentatus* n. sp. – Bergman, p. 89–91, Figs. 12I, J, Q, R, 18, 54, 55A, D–F.

Derivation of name. – Greek *mikros* and Latin *dentatus*, referring to the minute denticles.

Holotype. – LO 5835:2, right MI, Fig 55E:2.

Type locality. – Kärne 3.

Type stratum. – Eke Beds, basal part.

Material. – Figs. 2 and 3; more than 250 right MI, 250 left MI, 250 right MII, 250 left MII.

Occurrence. – Figs. 4 and 8; Late Llandovery to Late Ludlow, Lower Visby Beds to Hamra Beds, unit b. Amlings 1, Aursviken 1, Bjärges 1, Bodudd 1, Bottarve 2, Buske 1, Follingbo 2, Gannor 1 and 3, Gerete 1, Glasskär 1 and 3, Gothemshammar 2, Häftingsklint 4, Hide 1, Källdar 1, Kärne 3, Kroken 2, Kullands 1 and 2, Likmide 2, Lilla Hallvards 4, Närshamn 2 and 3, Nyhamn 5, Slitebrottet 2, Smiss 1, Snäck 1, Snoder 2, Sproge 4, Stora Mafrids 2, Stave 1, Valle 2, Valleviken 1, Valve 3, Vike 2.

Diagnosis. – Right MI: Long slender jaw with about ¾ of the inner margin covered by fine to very fine, comb-shaped, densely spaced denticles. Basal portion wide. Shank pointed.

Left MI: Long slender jaw, inner margin straight, denticulated along about ⅔ of its length. Inner wing almost rectangular, parallel to the undenticulated ridge.

Right MII: One or two small pre-cuspidal denticles, forming a swelling of the cutting edge. Shank and ramus fairly long and shank slender.

Left MII: Fairly large cusp, long and slender shank and ramus, inner wing narrow.

Description. – Right MI, dorsal side: Length 0.36–1.05 mm, width about ¼–⅓ of length. The basal part is wide anterior to the basal portion, but tapers anteriorly into a fang. The sickle-shaped falx has a distinct cutting edge along its outer margin. The inner margin is almost straight, covered for about ¾ by around 40 thin, densely spaced denticles, from the falcal arch and posteriorly. The number, size and extension of the denticles vary, from those of almost normal size (e.g. Upper Visby Beds, Fig. 54F:2) to needle-shaped denticles (e.g. Burgsvik Beds, Fig. 55A:1) covering the whole of the inner margin . The transition to the high and narrow undenticulated ridge is fairly smooth. The ridge ends at the posteriormost part of the shank. The inner wing is strongly downfolded, nearly triangular, widest at its anterior part, with its inner margin almost straight and its anterior corner rounded. The basal portion is fairly wide.

Basal plate not recorded.

The basal furrow is of normal size but fairly deep. The flange is large, thin, posteriorly highly elevated and downfolded along the posterior, outer margin, with a ligament rim along the outer margin of the basal portion. The basal angle is 20–30°. The shank is almost triangular with a pointed proximity. The bight is deep, with a bight angle of almost 90°. The outer margin, anterior to the basal portion and posterior to the falx, has a more or less pronounced concavity.

Ventral side: The crescent-shaped, strongly enclosed myocoele opening is ¼–⅓ of the jaw length. It is bounded along the anterior and inner margins by a fairly wide and slightly raised ligament rim.

Left MI, dorsal side: Length 0.38–1.18 mm, width almost ⅓ to ⅕ of length. The slender jaw tapers to the sickle-shaped falx. The fang, with a pronounced cutting edge, follows the outer margin along the falx. The inner margin is straight, with denticles of small to almost normal size along the anterior ½–⅔ of the inner margin between the falcal arch and the undenticulated ridge. The anteriormost denticles are oriented to the anterior or obliquely to the length axis of the jaw, the posterior ones slightly to the posterior. The denticulation decreases in size to the posterior, ending as small knobs. The transition onto the undenticulated ridge is smooth. The basal portion is angular, irregularly rectangular, with almost straight sides. The inner wing is almost rectangular and parallel to the undenticulated ridge. The basal furrow is fairly deep and of normal size, the basal angle about 20–30°. The ligament rim is visible along the posterior outer margin of the basal

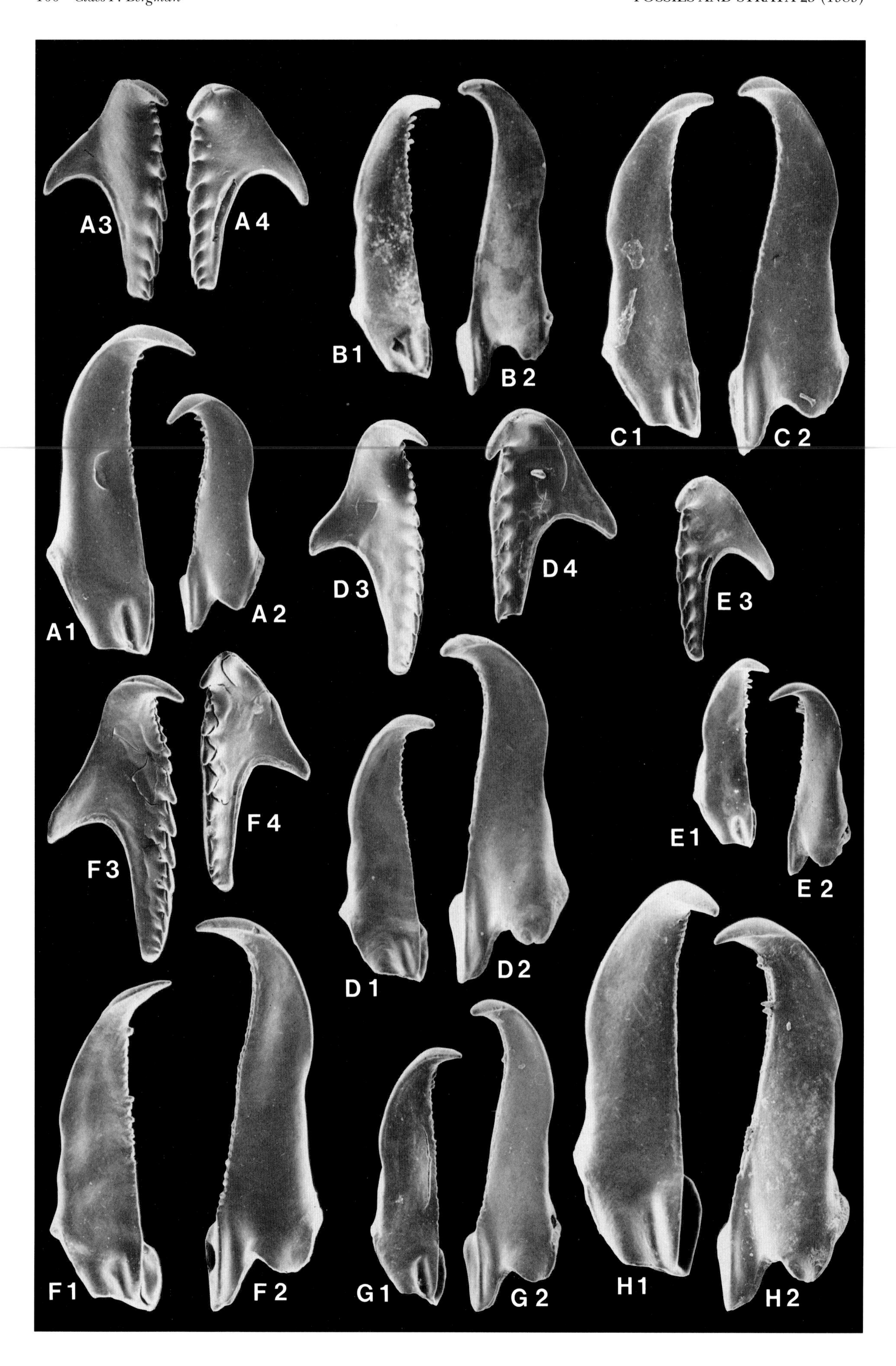
A3
A4
B1
B2
C1
C2
A1
A2
D3
D4
E3
F3
F4
D1
D2
E1
E2
F1
F2
G1
G2
H1
H2

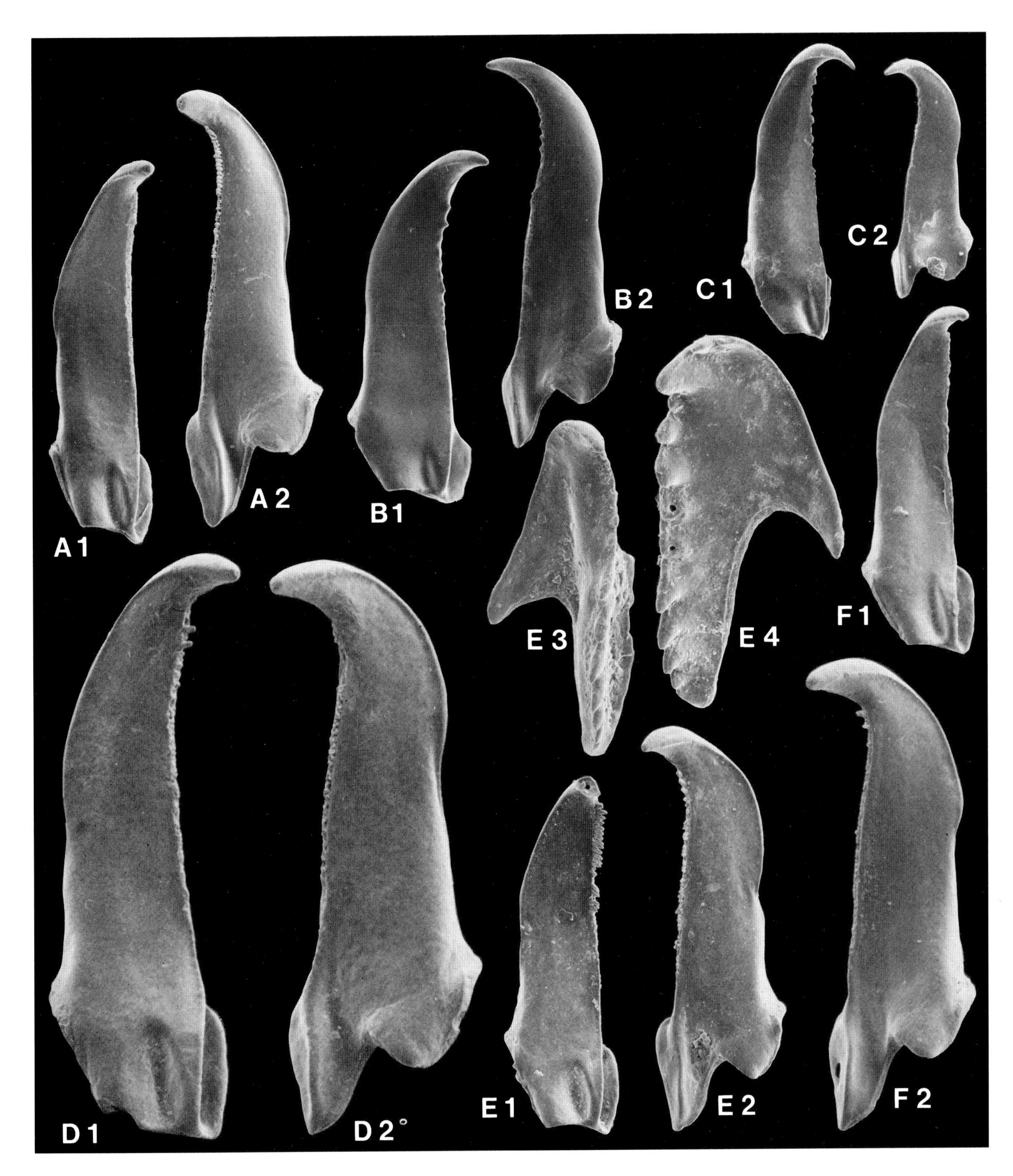

Fig. 55. Kettnerites (A.) microdentatus, except B1–B2, *K. (A.) fjaelensis*, and C1–C2, *K. (A.)* cf. *fjaelensis*. All specimens are in dorsal view, ×120 except A1, A2, E3 and E4. □A. Närshamn 2, Burgsvik Beds, 83-12LJ; A1 left MI, LO 5831:1, ×80; A2 right MI, LO 5831:2, ×80. □B. Kullands 2, Hemse Beds, Hemse Marl, NW part, 84-312DF; B1 left MI, LO 5832:5; B2 right MI, LO 5832:6. □C. Vidfälle 1, Hemse Beds, unit b, 75-41CB; C1 left MI, LO 5825:1; C2 right MI, LO 5825:2. □D. Gannor 1, Eke Beds, basal part, 71-123LJ; D1 left MI, LO 5811:3; D2 right MI, LO 5811:4. □E. Kärne 3, Eke Beds, basal part, 71-199LJ; E1 left MI, LO 5835:1; E2, holotype, right MI, LO 5835:2; E3 left MII, LO 5835:3, ×210; E4 right MII, LO 5835:4, ×210. □F. Gannor 3, Hemse Beds, Hemse Marl, SE part, 71-125LJ; F1 left MI, LO 5813:1; F2 right MI, LO 5813:2.

Fig. 54 (opposite side). *Kettnerites (A.) microdentatus*. All specimens are in dorsal view, ×120. □A. Vattenfallsprofilen 1, Högklint Beds, unit b, 70-6LJ; A1 left MI, LO 5766:1; A2 right MI, LO 5766:2; A3 left MII, LO 5766:3; A4 right MII, LO 5766:4. □B. Lickershamn 2, Lower Visby Beds, unit f, 73-53LJ; B1 left MI, LO 5763:2; B2 right MI, LO 5763:3. □C. Vike 2, Slite Beds, *Pentamerus gothlandicus* Beds or slightly older, 83-4CB; C1 left MI, LO 5781:1; C2 right MI, LO 5781:2. □D. Stave 1, Slite Beds, Slite Marl, central part, 75-11CB; D1 left MI, LO 5772:1; D2 right MI, LO 5772:2; D3 left MII, LO 5772:3; D4 right MII, LO 5772:4. □E. Vattenfallsprofilen 1, Högklint Beds, unit b, 70-6LJ; E1 left MI, LO 5766:5; E2 right MI, LO 5766:6; E3 right MII, LO 5766:7. □F. Häftingsklint 1, Upper Visby Beds, 76-9CB; F1 left MI, LO 5367:1; F2 right MI, LO 5367:3; F3 left MII, LO 5367:2; F4 right MII, LO 5367:4. □G. Vattenfallsprofilen 1, Högklint Beds, unit a, 70-20LJ; G1 left MI, LO 5764:4; G2 right MI, LO 5764:5. □H. Vattenfallsprofilen 1, Högklint Beds, unit d, *Valdaria testudo* level, sample RM; H1 left MI, AN 2717; H2 right MI, AN 2718.

portion. Anterior to the basal portion there is a wide and very shallow concavity on the outer margin.

Ventral side: The crescent-shaped, strongly enclosed myocoele opening is about 1/4–1/3 of the jaw length and resembles the one of the right MI.

Right MII, dorsal side: Length 0.17–0.68 mm, width normally more than half the length. There are two (in some populations probably only one) pre-cuspidal denticles on the dorsal side of the cusp, forming a swelling of the cutting edge, which does not reach the anteriormost part of the jaw. The anterior margin is sickle-shaped, formed by the large cusp which slants towards the posterior. The succeeding denticle is small; the transition to the almost straight denticulated ridge is gradual. Normally there are 6–8 slanting denticles on the ridge; these decrease slightly in size to the posterior. The shank is long and slender, occupying more than half the length of the jaw, and tapering posteriorly. Its extremity is almost pointed and bent slightly ventrally outward. The outer margin of the shank is almost straight. The bight is fairly deep, the bight angle acute to almost straight. The ramus is fairly large and slender, its extremity rounded. The anterior outer margin is almost straight.

Ventral side: The slightly enclosed myocoele opening is 2/3–3/4 of the jaw length. The almost straight anterior and inner margins are surrounded by a ligament rim of normal size. The opening extends anteriorly in the anterior inner part, forming a crescentic shape.

Left MII, dorsal side: Length 0.31–0.53 mm, width approximately half the length. The large cusp forms the sickle-shaped anterior margin. The intermediate dentary is distinct, with 4–7 denticles slightly increasing in size to the posterior. The denticles, normally 6–8, slant to the posterior on the slightly convex, denticulated ridge. The denticles increase in size to a maximum at the middle of the dentary whence they decrease to the posterior, ending fairly abruptly (e.g. in specimens from Snoder 2) or more gradually (e.g. in those from Häftingsklint 1). The shank occupies about half the length of the jaw, tapers posteriorly and ends fairly sharply. The inner wing is very narrow and usually not visible in dorsal view. The outer margin of the the shank is straight to slightly concave. The bight is of moderate size, the bight angle almost straight to slightly acute. The anterior outer margin is smooth and vaguely sigmoidal.

Ventral side: The slightly enclosed myocoele opening is about 1/2–2/3 of the jaw length. The opening is surrounded by a flat ligament rim. The inner and anterior margins are almost obliquely oriented to one another.

Remarks. – The basal portion of the right MI of the population of Närshamn 2 is conspicuously wide with a prominent flange, but whether this is the main trend in the lineage or just an atypical form developed in a population, can at present not be deduced because of the shortage of material.

There are 1–2 pre-cuspidal denticles of the right MII. Normally the dentition formula of the anteriormost denticles of the MII evolves very slowly in a lineage. We may here have an exception, but it is also possible that two different lineages occur in parallel. The difference between them is very modest and may be summarized as follows: In the robust form the left and right MI are somewhat shorter with slightly larger denticles; the right MII has one pre-cuspidal denticle and gives a more short and wide impression than the MII of the slender type.

Comparison. – The left and right MII of *K.* (*A.*) *microdentatus* are very similar to the corresponding element of *K.* (*A.*) *sisyphi sisyphi* and *K.* (*A.*) *fjaelensis.* The MI's however, are diagnostic.

Kettnerites (*Aeolus*) *siaelsoeensis* Bergman 1987

Figs. 12H, 18K, 56

Synonymy. – □1987 *Kettnerites* (*Aeolus*) *siaelsoeensis* n. sp. – Bergman, p. 91–93, Fig. 12H, 18, 56.

Derivation of name. – Latin *Siaelsoeensis,* from Själsö, referring to the type locality.

Holotype. – LO 5360:2, right MI, Fig 56B.

Type locality. – Själsö 1.

Type stratum. – Lower Visby Beds, unit b.

Material. – Fig. 2; more than 80 right MI, 80 left MI, 62 right MII, and 49 left MII.

Occurrence. – Figs. 4 and 7; Late Llandovery. Lower Visby Beds, unit b. Gustavsvik 1, 2, and 3, Ireviken 3, Rönnklint 1, Själsö 1.

Diagnosis. – Right MI: Slender, tapering towards the pointed fang; straight inner margin with thin comb-shaped dentary; shank sharp-ended. Basal plate almost triangular, fused to jaw.

Left MI: Slender, tapering towards anterior end, comb-shaped dentary ending before reaching the low and narrow undenticulated ridge. Basal furrow small and deep, extending somewhat under the undenticulated ridge.

Right MII: Large, sickle-shaped cusp with a fairly small pre-cusp on its anteriormost part. Very slender shank which tapers to the posterior.

Left MII: Large slender cusp, shank slender, tapering to the posterior, diminutive inner wing.

Description. – Right MI, dorsal side: Length 0.34–0.62 mm, width about or slightly less than 1/4 of length. The slender jaw tapers to a thin, sickle-shaped and fairly long fang, slightly bent upwards in relation to the almost flat jaw surface. The inner margin is straight and covered by thin, long, perpendicular denticles (26 on a 0.34 mm long jaw; 42 on a 0.62 mm long one) from the falcal arch to the undenticulated ridge. The latter is straight, high and narrow, almost forming the posteriormost sharp tip of the shank. The inner wing is almost rectangular, strongly downfolded, with an almost straight inner margin. The basal furrow is partly covered by the almost triangular, undenticulated basal plate. The inner margin of the basal plate is smoothly rounded; the posterior, outer part is occupied by a fold. The flange is rounded, the ligament rim exposed as a very indistinct, narrow furrow, to the right of which is a low fold. The basal portion is almost rectangular; the basal angle approximately 20°. Anterior to the basal portion the outer margin is smoothly concave along the posterior half of the jaw, whence it swings out forming the sickle-shaped falx.

Ventral side: The strongly enclosed, rounded myocoele opening is about 1/3 of the jaw length. Along the anterior, outer margin of the basal portion the ligament rim is a smoothly rounded ridge. Along the inner wing the ridge is narrow and surrounded by a groove.

Left MI, dorsal side: Length 0.32–0.61 mm, width 1/4 of length. The slender jaw tapers to a long fang which is slightly bent upwards in relation to dorsal surface. The inner margin is straight, with 20–50 needle-like denticles in the comb-shaped dentary. The denticulation covers most of the inner margin behind the falcal arch, ending as crenulations before reaching the inner wing. There is a smooth transition to the undenticulated ridge which is straight, low and narrow, ending close to the almost straight posterior margin. The inner wing is of normal size, 0.20–0.25 of the jaw length, almost rectangular and moderately downfolded. The basal furrow is short and deep, projecting under the undenticulated ridge. The outer margin of the basal portion is smoothly rounded, slightly downfolded, extending for a quarter of the jaw length, with the ligament rim forming a flat fold posteriorly. The basal portion is almost rectangular and smoothly rounded, the basal angle is about 20°. The outer margin anterior to the basal portion has a fairly deep concavity at the middle of the jaw.

Ventral side: The strongly enclosed myocoele opening is almost a mirror image of that in the right MI.

Right MII, dorsal side: Length 0.27–0.50 mm, width about half the length. A fairly small pre-cuspidal denticle, on the cutting edge of the anterior margin of the sickle-shaped cusp, forms the anteriormost part of the jaw. The cusp points towards the posterior. It is followed by two intermediate denticles, the anteriormost of which is smaller, and about 8–10 denticles of moderate size which decreases towards the posterior. The shank is long, more than half the jaw length, and very slender. The inner wing is diminutive. The inner and outer margins of the shank are convex and concave, respectively, the posteriormost part ends fairly sharply, pointing slightly laterally outwards. The bight is deep, the bight angle acute. The ramus is fairly long and somewhat needle-like, with only a trace of a sinus on the anterior, outer margin.

Ventral side: The slightly enclosed myocoele opening is 2/3–3/4 of the jaw length. The opening extends anteriorly, is delimited to the anterior and inner side by the almost straight margins forming a sharp angle. The ligament rim is narrow and only slightly widened on the outer part of the ramus.

Left MII, dorsal side: Length 0.29–0.49 mm, width approximately half the length. The large, fairly slender, sickle-shaped cusp is followed by 6–7 denticles of different sizes and orientations. The anteriormost ones point anteriorly, the ones behind are posteriorly oriented. The following 7–8 denticles are of moderate size, slanting, and decrease in size towards the posterior. The inner wing is very narrow, about half as long as the jaw. The slender shank, about half the jaw length or slightly less, tapers posteriorly to a fairly sharp end. The bight angle is usually acute, in some cases straight. The ramus is fairly long and slender, with a pronounced sinus at the anterior outer margin.

Ventral side: The crescent-shaped, slightly enclosed myocoele opening is somewhat more than half the jaw length. The opening extends slightly to the anterior inner side of the jaw. The almost straight margins are surrounded by a ligament rim of normal size.

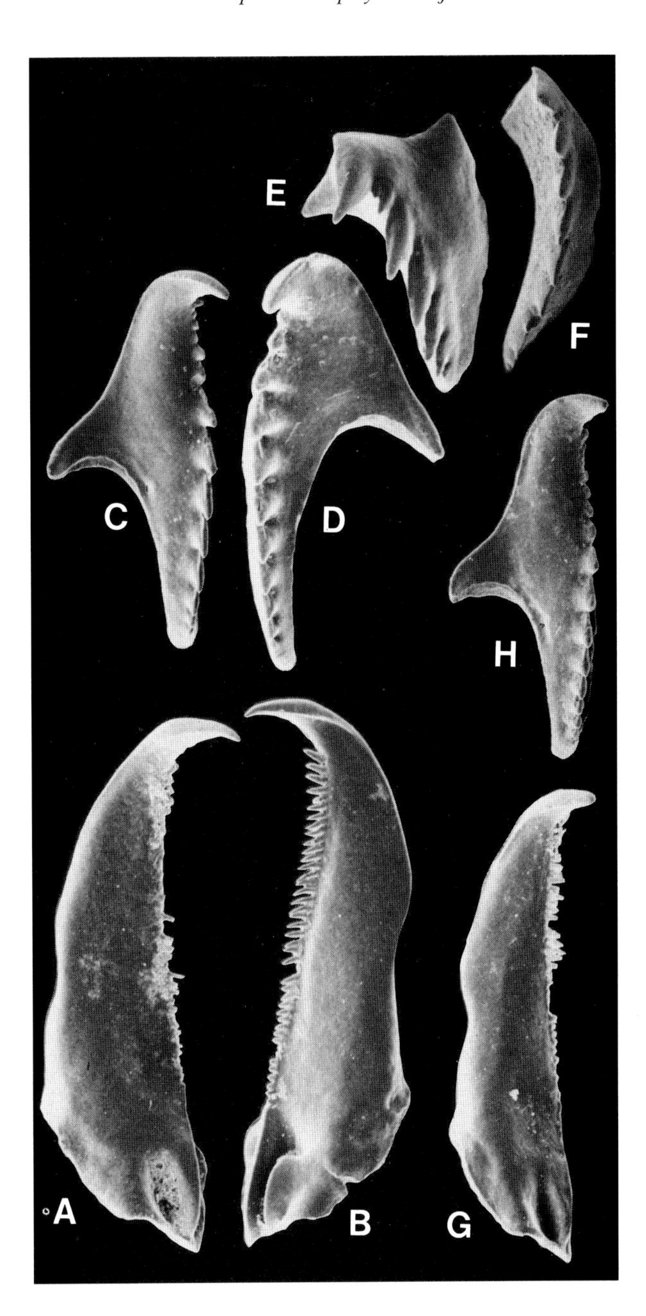

Fig. 56. Kettnerites (A.) siaelsoeensis. All specimens from Själsö 1, Lower Visby Beds, unit b, 79-12LJ. All specimens are in dorsal view, ×120. □A. Left MI, LO 5360:1. □B. Holotype, right MI, LO 5360:2. □C. Left MII, LO 5360:3. □D. Right MII, LO 5360:4. □E. MIII (unpaired), LO 5360:5. □F. Right MIV, LO 5360:6. □G. Left MI, LO 5360:7. □H. Left MII, LO 5360:8.

Remarks. – The good preservation of the type collection, probably due to low water energy and low bioturbation levels, made it possible to make a fairly safe identification of MIII and MIV. These jaws are more or less identical to those of *K. (A.) sisyphus klasaardensis* (see description of this species).

Comparison. – The form and denticulation of the MI are very similar to the corresponding element of *K. (A.) microdentatus*, except for the fused basal plate. The MII of *K. (A.) siaelsoeensis* is usually more slender and has a single pre-cusp on the right MII, while there is usually a double pre-cusp on the right MII element of *K. (A.) microdentatus*.

Kettnerites (Aeolus) fjaelensis Bergman 1987

Figs. 12P, 18J, 55B, C, 57

Synonymy. – □1987 *Kettnerites (Aeolus) fjaelensis* n. sp. – Bergman, p. 93–95, Figs. 12P, 18, 55B, C, 57.

Derivation of name. – Latin *fjaelensis*, referring to the type locality Fjäle 3.

Holotype. – LO 5798:2, right MI, 57D:2.

Type locality. – Fjäle 3.

Type stratum. – Klinteberg Beds, unit e, close to the Wenlock–Ludlow boundary.

Material. – Fig. 3; more than 28 right MI, 30 left MI, 19 right MII, 16 left MII.

Occurrence. – Figs. 4 and 8; Late middle Wenlock to Early middle Ludlow, Halla–Eke Beds. Fjäle 3, Gandarve 1, Herrvik 2, Hide 1, Hörsne 5, Kullands 2, Likmide 2, Lukse 1, Mulde Tegelbruk 1, Snoder 2, Tomtbodarne 1, Vidfälle 1.

Diagnosis. – Right MI: Slender jaw, tapering strongly to the large fang. The anterior two thirds of the inner margin covered by fairly wide-spaced, anteriorly fairly large, almost triangular denticles. Basal portion wide with conspicuously large, rounded flange. Shank pointed.

Left MI: Mirror image of the right MI with the exception of the basal portion (no bight or basal plate). Inner wing almost rectangular.

Right MII: Two very small to fairly small pre-cuspidal denticles. Cusp moderately large, slightly swollen at its base.

Left MII: Single, fairly large cusp. Shank and ramus long and slender.

Description. – Right MI, dorsal side: Length 0.33–0.76 mm, width slightly more than half the length. The jaw is very wide at its base anterior to the basal portion and tapers anteriorly, ending in the fang. The large sickle-shaped falx has a distinct cutting edge along its outer margin. The inner margin is almost straight and covered by about 15–20 slightly spaced, triangular denticles over about ⅔ of the inner margin from the falcal arch. The anteriormost 5–10 denticles are of normal size, but the size decreases to the posterior, where the denticles are ultimately reduced to small knobs. There is a smooth transition to the high and narrow undenticulated ridge which ends at the posteriormost part of the shank. The inner wing is very deeply downfolded, almost triangular, widest at the anterior, its inner margin almost straight and its anterior corner rounded. The basal portion is wide, its outer margin rounded.

Basal plate not recorded.

The basal furrow is of normal size but fairly deep. The flange is very wide, large, thin, posteriorly highly elevated and slightly downfolded along the rounded, posterior, outer margin. The ligament rim along the outer anterior margin of the basal portion is more or less indistinct. The basal angle is about 20–35°. The shank is slender and almost triangular with a pointed extremity, slightly bent to the outside in relation to the denticulated ridge. The bight angle is almost 90°. The transition between the anteriormost part of the basal portion and the outer margin is rounded. A minor concavity is present posterior to the falx.

Ventral side: The crescent-shaped, strongly enclosed myocoele opening is about ⅓ of the jaw length, or slightly less. The ligament rim is fairly distinct and forms a low, sharp ridge on the inner wing.

Left MI, dorsal side: Length 0.42–0.67 mm, width about ¼ of length. The slender jaw tapers to the sickle-shaped falx. The fang is large, with a pronounced cutting edge along the outer margin of the falx. The inner margin is straight, with slightly spaced denticles of normal size on the anterior of the inner margin. The denticles cover slightly more than half of the inner margin between the falcal arch and the undenticulated ridge. They decrease in size posteriorly, ending as small knobs. The anteriormost denticles project anteriorly or perpendicularly to the length axis of the jaw, the posterior ones are more or less perpendicular or point somewhat posteriorly. The transition to the undenticulated ridge is very smooth. The basal portion is wide, irregularly rectangular, with almost straight inner margin, slightly concave posterior side and a slightly convex to rounded outer side. The inner wing is almost rectangular, parallel to the undenticulated ridge. The basal furrow is of normal size, the basal angle about 25–40°. The ligament rim along the posterior, outer margin of the basal portion is indistinct. In some populations an outer wing is developed (Fig. 57D:1, B:1). The outer margin anterior to the basal portion is almost straight, ending in the falx.

Ventral side: The crescent-shaped, strongly enclosed myocoele opening is about ⅓ of the jaw length or slightly less. The opening is similar to that of the right MI.

Fig. 57. Kettnerites (A.) fjaelensis. All specimens in dorsal view, ×120, except B4. □A. Mulde Tegelbruk 1, Mulde Beds, 82-7CB; A1 left MI, LO 5788:2; A2 right MI, LO 5788:1; A3 left MII, LO 5788:3; A4 right MII, LO 5788:4. □B. Snoder 2, Hemse Beds, Hemse Marl NW part, 82-14CB; B1 left MI, LO 5809:3; B2 right MI, LO 5809:4; B3 left MII, LO 5809:5; B4 right MII, LO 5809:6, ×100. □C. Hörsne 5, Halla Beds, unit b, 75-52CB; C1 left MI, LO 5794:1; C2 right MI, LO 5794:2; C3 left MII, LO 5794:3; C4 right MII, LO 5794:4. □D. Fjäle 3, Klinteberg Beds, unit e, 77-7CB; D1 left MI, LO 5798:1; D2, holotype, right MI, LO 5798:2; D3 left MII, LO 5798:3; D4 right MII, LO 5798:4. □E. Gandarve 1, Halla Beds, 71-81CB; E1 left MI, LO 5836:2; E2 right MI, LO 5836:3. □F. Likmide 2, Hemse Beds, Hemse Marl, SE part, 82-28LJ; F1 left MI, LO 5801:1; F2 right MI, LO 5801:2.

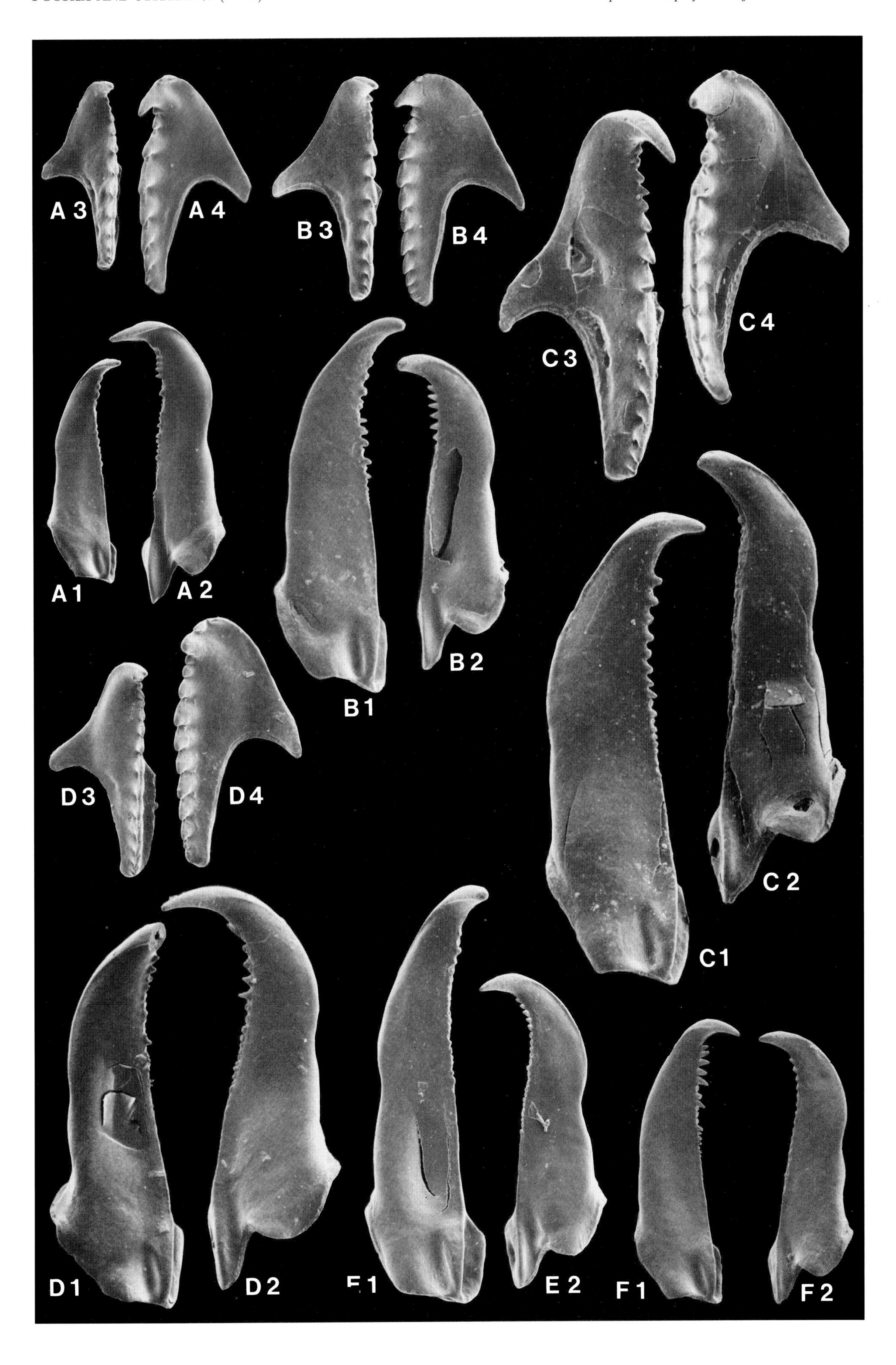
A 3
A 4
B 3
B 4
C 3
C 4
A 1
A 2
B 1
B 2
C 1
C 2
D 3
D 4
D 1
D 2
E 1
E 2
F 1
F 2

Right MII, dorsal side: Length 0.30–0.47 mm, width about half the length or slightly more. Two pre-cuspidal denticles form the anteriormost part of the jaw. Their size varies from very small, indistinct swellings of the cutting edge (e.g. Fig 57C:4) to fairly large, separate pre-cusps (Fig. 57D:4). The anterior margin is more or less rounded, bearing a large cusp with a swollen base. The cusp is directed perpendicular to the length axis of the jaw or slants slightly to the posterior. The following 1–2 denticles are small. There is a gradual transition to the almost straight denticulated ridge, with a post-cuspidal dentary composed of 8–10 slanting denticles, largest at the centre. The shank, about half the length of the jaw, is slender and tapers to the posterior; its extremity is pointed and bent slightly ventrally outward. The bight is of normal size, the bight angle almost 90°. The ramus is large, wide, almost triangular and somewhat pointed. The anterior outer margin is almost straight.

Ventral side: The slightly enclosed myocoele opening is ½–⅔ of the jaw length. The ligament rim is wide with a low relief.

Left MII, dorsal side: Length 0.32–0.47 mm, width approximately half the length. The fairly large cusp with a distinct cutting edge forms the sickle-shaped anterior margin. The intermediate dentary is distinct with 3–5 fairly small denticles of more or less equal size. The usually 8–10 post-cuspidal denticles slant to the posterior, the largest at the middle. The shank, about half (or slightly less) of the jaw length, tapers to a fairly sharp posterior end. The inner wing is very narrow and usually not visible in dorsal view. The outer margin of the shank is straight to slightly concave. The bight is of moderate size, the bight angle almost straight to slightly acute. The ramus is almost triangular, wide at its base. The anterior, outer margin has a slight concavity at the centre.

Ventral side: The crescent-shaped, slightly enclosed myocoele opening is about ⅔ of the jaw length. The opening extends to the anterior on the anterior, inner side of the jaw.

Comparison. – The MI's are easy to identify on the denticulation. The MII's are very similar to the corresponding element of *K.* (*A.*) *sisyphi sisyphi* and *K.* (*A.*) *microdentatus*.

Genus *Lanceolatites* Bergman 1987

Synonymy. – *Lanceolatites* n. gen. – Bergman, p. 96.

Derivation of name. – Latin *lanceolatites*, from lance, referring to the very elongated and sharp-pointed MI.

Type species. – *Lanceolatites gracilis* Bergman 1987

Other species. – *L.* sp. A.

Diagnosis. – Right MI: Long slender jaw, oval cross-section, comb-shaped dentary, pointed shank.

Left MI: Long slender jaw, oval cross-section, comb-shaped dentary, small basal portion, almost rectangular inner wing.

Right MII: Double cusp with one minor denticle between the cusps; shank and ramus long, slender with pointed extremities. Bight deep.

Left MII: Double cusp, shank and ramus long, slender with pointed extremities.

Remarks. – *Lanceolatites* is characterized by the conspicuously long, slender and pointed extremities of its jaws, the slenderness often being combined with an undulating shape, particularly the late Wenlock to Ludlow *Lanceolatites gracilis* specimens. The type species has a variety which is coarser and often slightly smaller than the type form. The *Lanceolatites* sp. A specimens are less slender and have fewer denticles.

Lanceolatites gracilis Bergman 1987

Figs. 14A, B, 18L, 19C, 58, 59, 60B–D

Synonymy. – □1979 *'Oenonites' aspersus* Hinde – Bergman, p. 99, Fig. 28:4C,D (only). □1987 *Lanceolatites gracilis* n. sp. – Bergman, p. 96–100, Figs. 14A, B, 18, 19C, 58, 59, 60B, C, D.

Derivation of name. – Latin *gracilis*, referring to the long slender form of both MI and MII.

Holotype. – LO 5365:4, right MI, Fig. 58D:2.

Type locality. – Nygårdsbäckprofilen 1.

Type stratum. – Lower Visby Beds, unit e.

Material. – Figs. 2 and 3; more than 171 right MI, 158 left MI, 176 right MII, 121 left MII.

Occurrence. – Figs. 4 and 11; Late Llandovery to latest middle Ludlow, Lower Visby Beds, unit e, to Eke Beds, lower–middle part. Ajmunde 1, Amlings 1, Autsarve 1, Bodudd 1, 2, and 3, Buske 1, Fågelhammar 3, Fjäle 3, Gannor 1 and 3, Gardsby 1, Gerete 1, Gerumskanalen 1, Gnisvärd 1 and 2, Gogs 1, Grogarnshuvud 1, Häftingsklint 1 and 4, Hägvide 1, Hummelbosholm 1, Ireviken 3, Källdar 1, Kättelviken 5, Klasård 1, Korpklint 1, Kullands 2, Lickers 1, Lickershamn 2, Likmide 2, Lilla Hallvards 1 and 4, Lukse 1, När 2, Nygårdsbäckprofilen 1, Petsarve 2, Rönnklint 1, Snäckgärdsbaden 1, Snoder 2, Stora Kruse 1, Vaktård 2, 4, and 5, Västlaus 1, Vattenfallsprofilen 1.

Diagnosis. – Right MI: Long, slender, evenly tapering jaw, almost straight inner margin, covered along its entire length by about 50–60 denticles in a comb-shaped dentary. Pointed shank.

Left MI: Apart from the small, angular basal portion with a nearly rectangular inner wing, it is almost a mirror image of right MI.

Right MII: Double cusp with two parts of almost equal size, separated by a minor denticle. Three to six intermediate denticles of equal size. Shank and ramus long, slender, with pointed extremities.

Left MII: Double cusp, parts of equal size. Intermediate denticles covering up to half the length of the denticulated inner margin. Shank and ramus long and slender with pointed extremities.

Description. – Right MI, dorsal side: Length 0.36–1.18 mm, width ⅕ of length. The slender jaw tapers anteriorly, ending in a narrow, pointed fang, which is bent fairly strongly

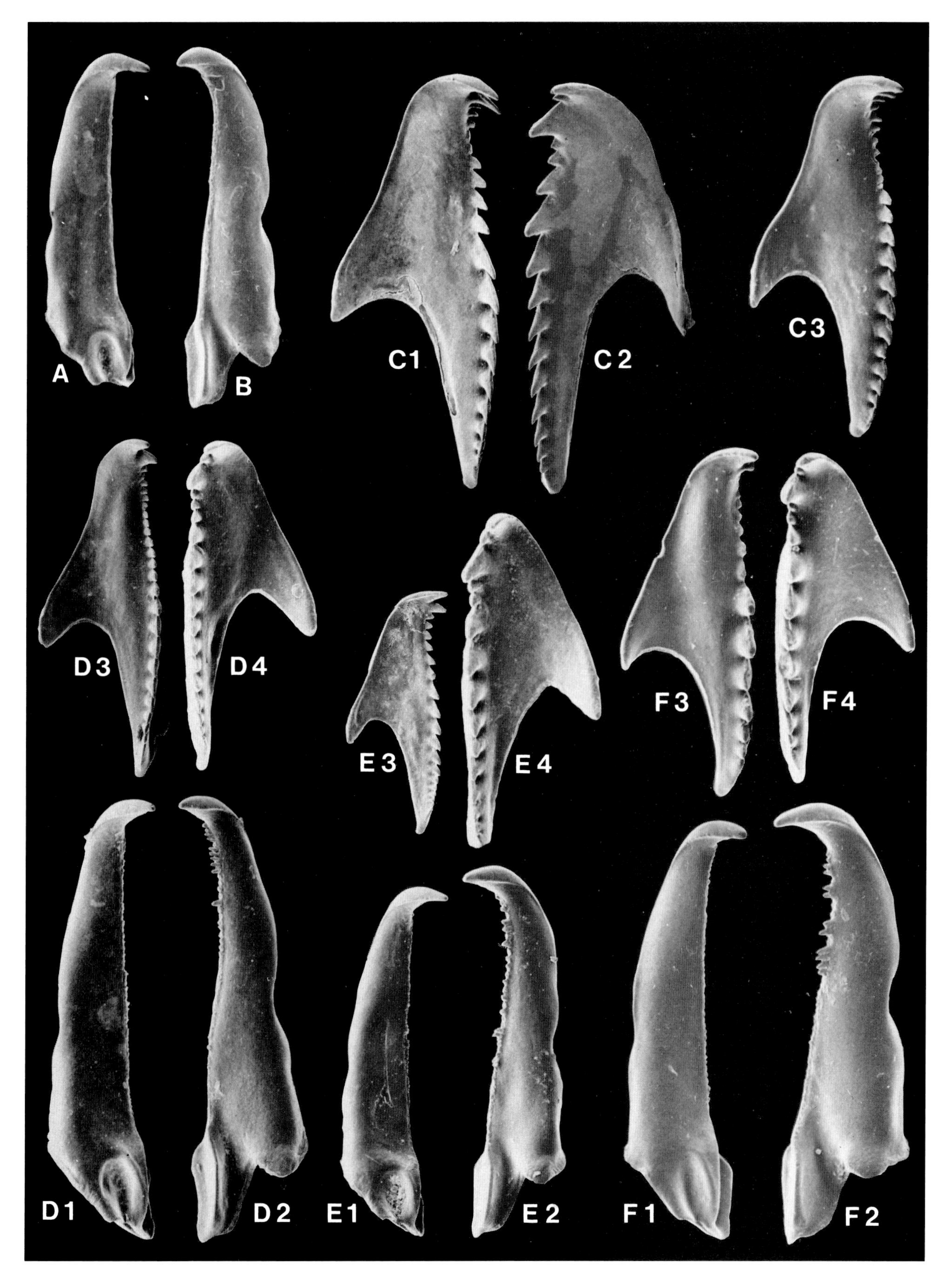

Fig. 58. Lanceolatites gracilis. E and F *L. gracilis* var. visby. All specimens in dorsal view; all ×80 except C. □A. Left MI, LO 5765:2, Lickershamn 2, Högklint Beds, unit a, 73-57LJ. □B. Right MI, LO 5764:3, Vattenfallsprofilen 1, Högklint Beds, unit a, 70-20LJ. □C. Same sample as A; C1 left MII, LO 5765:4, ×120; C2 right MII, LO 5765:5, ×120; C3 left MII, LO 5765:6. □D. Nygårdsbäckprofilen 1, Lower Visby Beds, unit e, 79-42LJ; D1 left MI, LO 5365:3; D2, holotype, right MI, LO 5365:4; D3 left MII, LO 5365:5; D4 right MII, LO 5365:6. □E. Vattenfallsprofilen 1, Lower Visby Beds, unit e(?), 76-6LJ; E1 left MI, LO 5366:1; E2 right MI, LO 5366:2; E3 left MII, LO 5366:3; E4 right MII, LO 5366:4. □F. Vattenfallsprofilen 1, Högklint Beds, unit b, 70-6LJ; F1 left MI, LO 5768:1; F2 right MI, LO 5768:2; F3 left MII, LO 5768:3; F4 right MII, LO 5768:4.

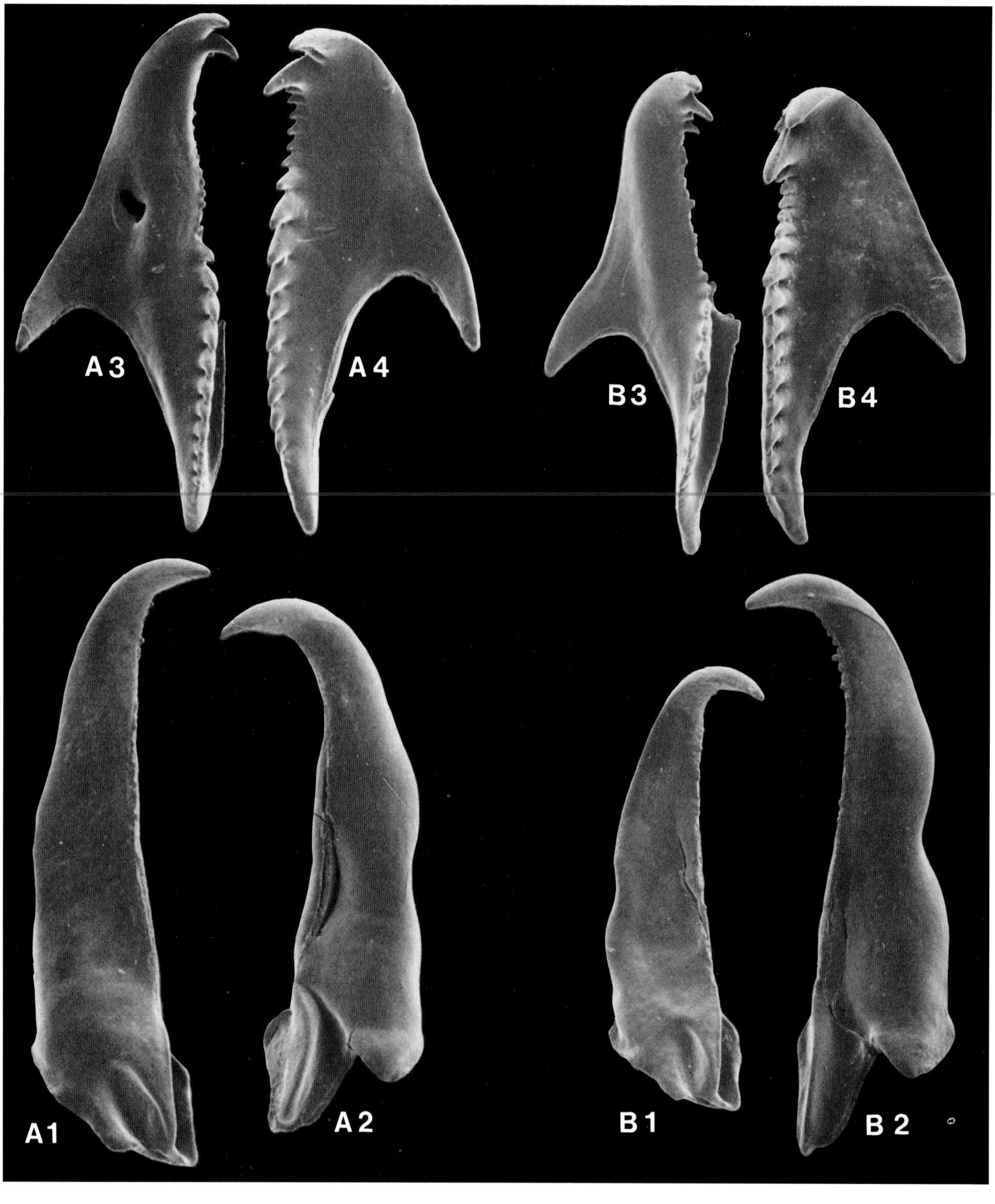

A3 A4 B3 B4 A1 A2 B1 B2

Fig. 59. Lanceolatites gracilis, Hemse Beds, Hemse Marl, NW part, all specimens in dorsal view, ×120. □A. Likmide 2, 82-28LJ; A1 left MI, LO 5802:1; A2 right MI, LO 5802:2; A3 left MII, LO 5802:3; A4 right MII, LO 5802:4. □B. Ajmunde 1, 71-40LJ; B1 left MI, LO 5807:1; B2 right MI, LO 5807:2; B3 left MII, LO 5807:3; B4 right MII, LO 5807:4.

Fig. 60 (opposite page). All specimens in dorsal view, except E1–4, all ×120, except A, D2, D3, E1 and E2. □A. *Lanceolatites* sp. A, Strands 1, Hamra Beds, unit b, 75-14LJ; A1 left MI, LO 5840:5, ×80; A2 right MI, LO 5840:6, ×80. □ B. *L. gracilis* var. visby, Fjäle 3, Klinteberg Beds, unit e, 77-7CB; B1 left MI, LO 5797:1; B2 right MI, LO 5797:2; B3 left MII, LO 5797:3; B4 right MII, LO 5797:4. □C. *L. gracilis* Vaktård 4, Hemse Beds, Hemse Marl, SE part, 81-35LJ; C1 left MI, LO 5821:1; C2 right MI, LO 5821:2; C3 left MII, LO 5821:3; C4 right MII, LO 5821:4. □D. *L. gracilis* Källdar 1, Hemse Beds, Hemse Marl, NW part, 75-75CB; D1 right MI, LO 5810:2; D2 basal portion, with basal plate in place, of right MI, same specimen as D1, ×600; D3 anterior part of left MI, LO 5810:1, ×600. □*L. gracilis*, Västlaus 1, Hemse Beds, Hemse Marl, SE part, ventral view, 82-31LJ; E1, right MI, LO 5806:1, ×210; E2, left MI, LO 5806:2, ×210; E3, right MII, LO 5806:3; E4, left MII, LO 5806:4.

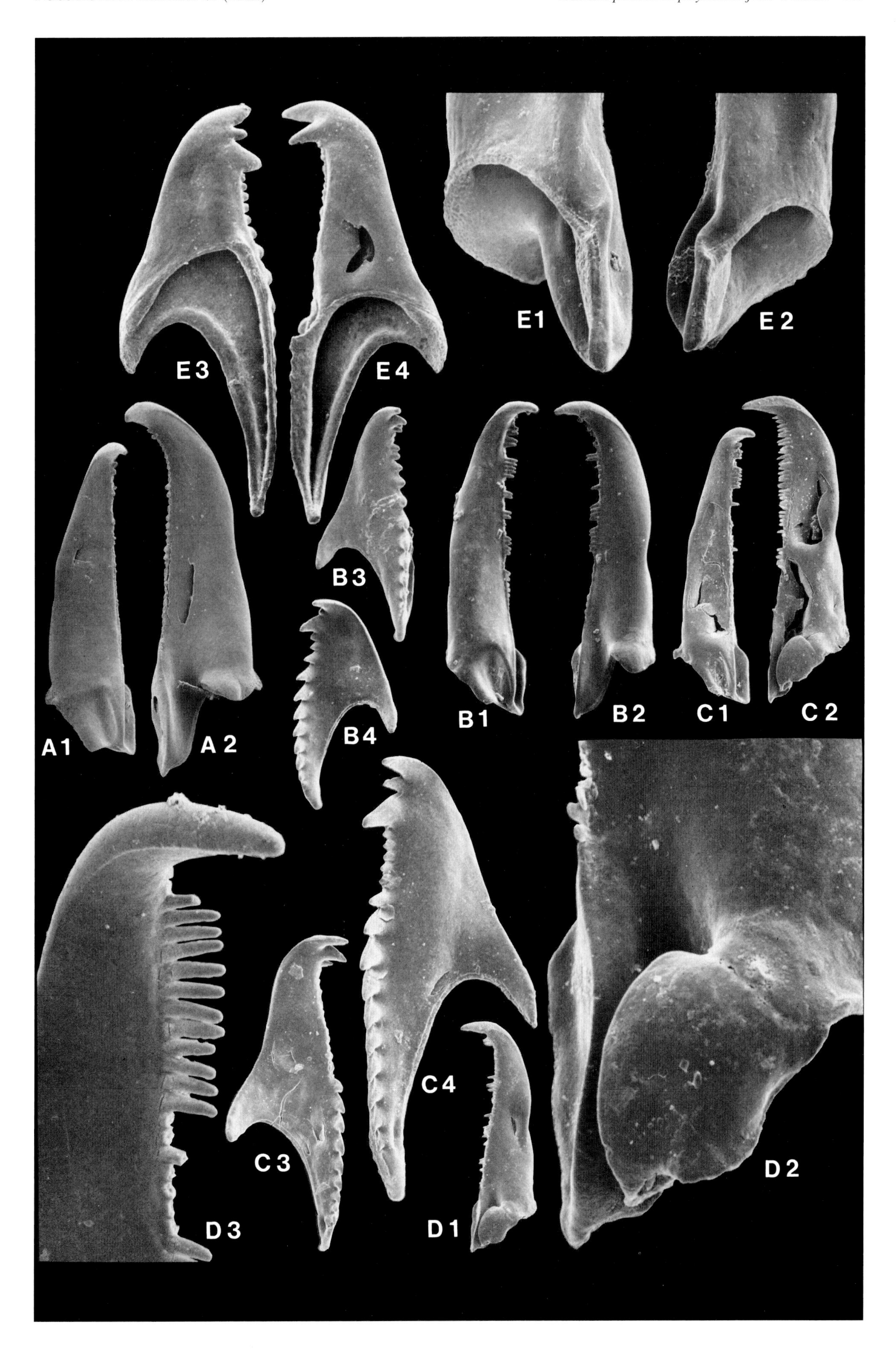
E3
E4
E1
E2
A1
A2
B3
B4
B1
B2
C1
C2
D3
C3
C4
D1
D2

upwards in relation to the convex dorsal surface. The distinct cutting-edge runs transversely to the extension of the jaw along the fang. In the Late Llandovery and Early Wenlock specimens the long (0.7 of the jaw length), almost straight inner margin has about 50–60 thin, needle-shaped denticles of equal size along its entire length (Fig. 58D:2). In Late Wenlock and Early Ludlow specimens the number of denticles is smaller, about 50 or even lower (Fig. 58B:2). The denticulation and the inner margin end before reaching the inner wing. There is an abrupt transition to the high, smoothly convex and very narrow, undenticulated ridge. The posterior part of the ridge is somewhat broadened, slightly bent outward at the middle, extending down and forming the posteriormost part of the shank together with the inner wing. The inner wing is usually downbent, the anterior outer corner is rounded, the inner margin medially concave. The shank is wide posteriorly, fairly blunt-ended, though the undenticulated ridge forms a conspicuously pointed tip. The basal furrow is long, wide and fairly deep. The posterior part of the elevated flange is fairly thin-walled. There is a short and fairly narrow fold in the outer, anterior margin of the basal portion. The basal angle is 30–40°.

The basal plate which is fused to the jaw, is recorded only from specimens of late Wenlock and Ludlow age (Fig. 60D:2, C:2). It is almost triangular, with a semicircular, rounded margin along the inner side.

The outer margin, anterior to the fairly small basal portion is smoothly wavy with a distinct median concavity. The falx is sickle-shaped, very short and indistinct.

Ventral side: The strongly enclosed myocoele opening, about ⅓ of the jaw length, is surrounded by a narrow, high and fairly sharp ridge, the ligament rim. The opening is anteriorly crescent-shaped, laterally following the inner and outer margins (Fig. 60E:1).

Left MI, dorsal side: Length 0.40–1.18 mm, width ⅕ of length. The jaw tapers evenly towards the anterior, ending in a finely pointed fang, only slightly bent upwards in relation to the convex, dorsal surface. The fang has a distinct, almost transverse cutting-edge. In the Late Llandovery and Early Wenlock specimens, the inner margin is almost straight, densely denticulated by around 50–60 needle-shaped denticles of equal size, oriented perpendicular to the jaw axis. In Early Ludlow specimens the number of denticles is about 50 or less. The comb-shaped dentary, about 0.7 of the jaw length, completely covers the inner margin and does not reach the inner wing. The transition to the undenticulated ridge is abrupt. The ridge is slightly elevated and rounded in longitudinal view, fairly narrow, thinning posteriorly. The posteriormost part of the ridge is bent inwards along the posterior margin of the deeply downfolded inner wing. The basal furrow is of normal length, wide and fairly deep. The posterior half of outer margin of the basal portion is slightly downfolded. The anterior half of the ligament rim is visible. The basal portion is almost triangular with a very small outer face. The basal angle is about 40–50°. The outer margin, anterior to the basal portion, is smooth, rounded in the anterior-posterior direction, normally with a concavity at middle. The falx is short and sickle-shaped.

Ventral side: The strongly enclosed myocoele opening is about ¼–⅓ of the jaw length. The shape of the opening is very similar to that of the right MI (Fig. 60E:2).

Right MII, dorsal side: Length 0.31–0.93 mm, width somewhat less than half the length. The almost equal parts of the double cusp are separated by a minor denticle. The posteriormost part slants slightly more to the posterior. In the Late Llandovery and Early Wenlock specimens, the intermediate dentary is well developed with about three equal denticles of normal size; whereas specimens from latest Wenlock to Ludlow Beds have more denticles, about 5–6. The post-cuspidal dentary is represented by 10–12 fairly slender, slightly slanting denticles. The shank tapers to the posterior, the posterior 0.3 being without denticles, the extremity very pointed and often characteristically bent to the right. The shank occupies slightly more than half the jaw length. The bight is deep with an acute bight angle. The ramus is long, often slender and pointed. The anterior outer margin has a wide and shallow sinus.

Ventral side: The parabolic, slightly enclosed myocoele opening extends anteriorly on the inner side for about ⅔ of the jaw length. The ligament rim is narrow (Fig. 60E:3).

Left MII, dorsal side: Length 0.47–1.40 mm, width less than half the length. The double cusp has parts of equal and moderate size. The posterior part slants slightly to the posterior, and the anterior one is perpendicular to the length axis of the jaw. The intermediate dentary is represented by about 12–15 small denticles. There are about ten post-cuspidal, fairly small, slightly slanting denticles decreasing in size to the posterior. The denticulation ends abruptly at the posterior end of the inner wing, leaving the posterior 0.2 of the shank without denticles. The fairly slender shank, slightly less than half the jaw length, tapers distinctly to the posterior, its posterior proximity sharply pointed. The bight is fairly deep, the bight angle acute. The ramus is usually fairly slender and long with a pointed proximity. The anterior, outer margin has a wide, shallow sinus.

Ventral side: The slightly enclosed myocoele opening, somewhat more than half the jaw length, is slightly curved and surrounded by a fairly narrow rim along the anterior and the inner side (Fig. 60E:4).

Remarks. – The transition from the intermediate to the post-cuspidal dentary of the MII is gradual in some populations, making the count of intermediate denticles somewhat uncertain.

Lanceolatites gracilis var. visby

Figs. 14C, D, 18L, 58E, F, 60B, D

Occurrence. – Early Wenlock to Wenlock–Ludlow boundary, Lower Visby Beds, unit e, to Hemse Beds, Hemse Marl NW part. Ajmunde 1, Buske 1, Fjäle 3, Gnisvärd 1, Häftingsklint 4, Ireviken 3, Korpklint 1, Lickers 1, Nygårdsbäckprofilen 1, Nyhamn 4, Rönnklint 1, Snäckgärdsbaden 1, Vattenfallsprofilen 1.

Diagnosis. – Right MI: Fairly long, slender jaw, tapering evenly to the anterior. Densely denticulated, straight inner margin with around 40 denticles.

Left MI: Fairly long, slender jaw, tapering evenly to the anterior. Almost straight inner margin densely denticulated with around 40 needle-shaped denticles.

Right MII: Double cusp, parts of equal size, separated by a denticle. Two, more rarely three, intermediate denticles. Shank fairly slender.

Left MII: Double cusp, parts of almost equal size, followed by 6–8 intermediate denticles. Shank fairly slender.

Description. – Right MI, dorsal side: Length 0.42–0.95 mm, width about 1/5 of length. The inner margin is straight, strongly denticulated with approximately 40 densely spaced, erect, needle-shaped denticles, slightly decreasing in size posteriorly. In elements from the Lower Visby Beds, the denticles cover the complete inner margin, 0.6–0.7 of the jaw length (Fig. 58E:2). In those from the Klinteberg Beds the denticulation covers about 0.6 of the jaw length, ending before reaching the inner wing (Fig. 60B:2).

Ventral side: Similar to the nominal species.

Left MI, dorsal side: Length 0.49–1.01 mm, width about 1/4–1/5 of length. The number of denticles is about 40 (Fig. 60B:1). Except for the basal part, this is almost a mirror image of the right MI.

Ventral side: Similar to the nominal species.

Right MII, dorsal side: Length 0.42–0.70 mm, width less than half the length. The double cusp has parts of almost equal size and separated by a denticle. The anterior part, bent upwards in relation to the dorsal surface, is perpendicular to the length axis of the jaw; the posterior part slants slightly to the posterior. Two denticles of fairly small and equal size form the intermediate dentary, sometimes with a slightly larger denticle forming the transition to the post-cuspidal dentary which has 7–10 fairly large, posteriorly slanting denticles (Fig. 60B:4). The shank is slender and tapers to the posterior, with a pointed extremity, sometimes slightly bent outwards. The shank occupies about half the jaw length. The bight is fairly deep, the bight angle acute; the ramus is almost triangular. The anterior outer margin is almost straight, ending in a pointed anterior end.

Ventral side: Similar to the nominal species.

Left MII, dorsal side: Length 0.40–0.75 mm, width less than half the jaw length. The double cusp has parts of almost equal size and varying orientation: usually the anterior part is oriented slightly anteriorly (Fig. 58E:3) or perpendicularly to the length axis of the jaw, and the posterior one slightly posteriorly or perpendicularly to the length axis. There are 6–9 intermediate denticles and a somewhat gradual transition to the post-cuspidal dentary with 6–11 posteriorly slanting denticles. The shank, slightly less than half the jaw length, tapers to a pointed posterior end. The inner wing is narrow. The bight is fairly deep, the bight angle acute. The ramus is almost triangular, with a somewhat pointed extremity. The anterior outer margin is almost straight or has a very weak, wide sinus. The anterior end is almost sickle-shaped.

Ventral side: Similar to the nominal species.

Discussion. – The MI's are very similar to those of *L. gracilis*, but the latter are slightly more slender and have a larger number of denticles. The MII's differ more, both in general morphology (*L. gracilis* being more slender) and in the number of denticles.

The species evolved slowly; changes of denticle formulae of the MII element are fairly easy to observe. There is no verifiable size difference between them, such as would be expected if they had been ontogenetic variations of the same species. Also, they are only occasionally found together. *L. gracilis* has a longer stratigraphical range than the variety, so they are not likely to be sexual dimorphs. The variety seems always to be less frequent than the nominal species.

Lanceolatites sp. A

Figs. 14E, 60A

Material. – Nine right MI, 12 left MI, 13 right MII, 8 left MII.

Occurrence. – Fig. 4 and 11; Late Ludlow, Hamra Beds, unit b. Bankvät 1, Strands 1.

Diagnosis. – Right MI: Jaw tapers anteriorly; width about 1/4–1/3 of jaw length. Fairly finely denticulated. Shank large and pointed, flange highly elevated.

Left MI: Jaw tapers anteriorly, fairly finely denticulated. Basal portion skew, angular; inner wing almost rectangular.

Right and left MII: Similar to the corresponding element of the *L. gracilis* var. visby.

Description. – Right MI, dorsal view: Length 0.78–1.1 mm, width about 1/4–1/3 of length. The jaw tapers evenly to the anterior (Fig. 60A:2), ending in a fairly slender fang. The inner margin is straight, covered by about 30 fairly small denticles from the anteriormost part of the falcal arch, ending slightly anterior to the deeply downfolded inner wing.

Basal plate is not recorded.

The basal portion is of typical *Lanceolatites* shape, with a large, almost triangular, sharp-ended shank, representing about 0.2 of the jaw length. The flange is highly elevated, smoothly rounded with a spur-like extension in the anterior, outer part of the basal portion. The outer margin anterior to the basal portion is smoothly undulated, with a sickle-shaped falx.

Ventral side: Similar to *L. gracilis*.

Left MI, dorsal view: Length 0.49–0.99 mm, width about 1/4–1/3 of length. The jaw tapers evenly (Fig. 60A:1) to the anterior, ending in a fairly slender fang. The inner margin is straight, covered by about 30 or more fairly small denticles. The dentition ends before reaching the almost rectangular inner wing. The basal portion is irregularly rectangular, with almost straight sides and a spur-like extension of its anterior outer margin. There is a large basal furrow, and the outer margin anterior to the basal portion is almost crescent-shaped.

Ventral side: Similar to *L. gracilis*.

The right and left MII are similar to the corresponding element of *Lanceolatites gracilis* var. visby, with the following addition: The posterior dentary of the right MII (probably also of the left MII) has smaller denticles. The number of denticles increases with increased size of the jaws. Eleven post-cuspidal denticles have been recorded on a broken, at least 0.88 mm long right MII and eight denticles on a 0.40 mm long right MII. The length of the right MII's is 0.40–0.95 mm, and of left MII's 0.61–1.03 mm.

Discussion. – *Lanceolatites* sp. A is referred to under open nomenclature on two grounds. First, the very few specimens encountered do not permit a study of the variation. Second, the stratigraphical gap between *L. gracilis* and *L.* sp. A makes it difficult to determine whether or not they are of the same lineage. On the basis of the MII morphology, a close relation between the two seems probable.

Remarks. – A typical specimen of the taxon is the right MI (Fig. 60A:1, LO 5840:6) from Strands 1, Hamra Beds, unit b, of Late Ludlow age.

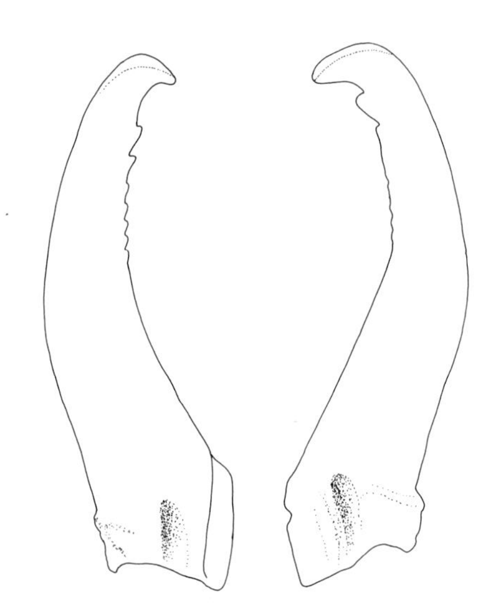

Fig. 61. Langeites glaber Kielan-Jaworowska 1966, camera-lucida drawing of the type specimen, right MI (Z. Pal. No. O.466/13b), ×39.

Genus *Langeites* Kielan-Jaworowska 1966

Type and only species. – *Langeites glaber* Kielan-Jaworowska 1966.

Emended diagnosis. – Right MI: More or less arcuate, convex outwards, almost parallel inner and outer margins, rounded cross section. A few, wide, widely-spaced denticles along the anterior part of the inner margin.

Left MI: Mirror image of right MI except for basal portion.

Right MII: Two almost equal-sized, large, pre-cuspidal denticles followed by a fairly large, slanting cusp. Shank fairly short. Bight angle sharply acute, ramus long, needle-shaped.

Left MII: Double cusp oriented to the anterior, very short shank. Bight angle sharply acute, ramus long and needle-shaped.

Langeites glaber Kielan-Jaworowska 1966

Figs. 14Q–S, 61, 62

Synonymy. – □1980 *Langeites glaber* Kielan-Jaworowska 1966 – Wolf, pp. 86–87, Pl. 10:89–94. □1987 *Langeites glaber* Kielan-Jaworowska 1966 – Bergman, p. 101–103, Figs. 14Q–S, 61, 62.

Holotype. – No. O.466/13b, right MI, Fig. 30:2.

Type locality. – Unknown, erratic boulder.

Type stratum. – Ordovician or Silurian

Material from Gotland. – Four right MI, 6 left MI, 2 right MII, 2 left MII.

Occurrence on Gotland. – Figs. 4 and 8; Middle–Late Ludlow, Hemse Beds, Hemse Marl, SE part, and Sundre Beds. Stora Kruse 1, Vaktård 4, and Holmhällar 1.

Diagnosis. – As for the genus.

Additional description. Right MI, dorsal side: Length 0.62–>3.1 mm (anteriormost part broken off and not included in measurement), width about ⅕ of length. The large, slender jaw has almost parallel to anteriorly slightly tapering sides. The cross-section is almost rounded. The inner margin is nearly straight to pronouncedly concave with about 10–16 large (wide), anteriorly directed, knoblike denticles, largest in the anterior part. A very short spur exists in the middle Ludlow specimens but not in the Late Ludlow ones.

The basal plate is rectangular and recorded in place in the Late Ludlow specimens but not on the late Early Ludlow ones. The latter specimens have a small flange which is almost right-angled, fairly thick, uplifted in the posterior part and slightly downfolded along the outer margin.

Left MI, dorsal side: Length 0.89–>4.1 mm (anteriormost and posteriormost parts broken off), width about ⅕ of length. The large slender jaw has almost parallel to anteriorly slightly tapering sides. The cross-section is almost rounded. The inner margin is almost straight to pronouncedly concave, with about 10 large (worn and wide), anteriorly directed, knoblike denticles, largest on the anterior. The basal portion is almost quadratic, particularly in the Late Ludlow specimens. A very short spur exists in the Hemse specimens but is not seen in dorsal view on the Late Ludlow ones. The basal furrow is pronounced, short and fairly deep.

Right MII, dorsal side: Length about 1.5–2.5 mm (reconstructed from two fragmented jaws), width about half the length. Two very large pre-cuspidal denticles of almost equal size are followed by a large cusp, which is in turn followed by fairly large denticles. The dentary with the pre-cusps forms a weak convex arch. The shank is fairly short with a sharply acute bight angle. The ramus is needle-shaped, long and narrow.

Left MII, dorsal side: Length 1.55–2.75 mm, width about half the length. There is a large double cusp but no pronounced intermediate dentary. The dentary along the inner margin is about half the jaw length. It ends conspicuously early, leaving the posterior 0.15–0.2 of the jaw undenticulated. The bight angle is sharply acute. The shank is wide, short, and bluntly ended; the ramus is narrow.

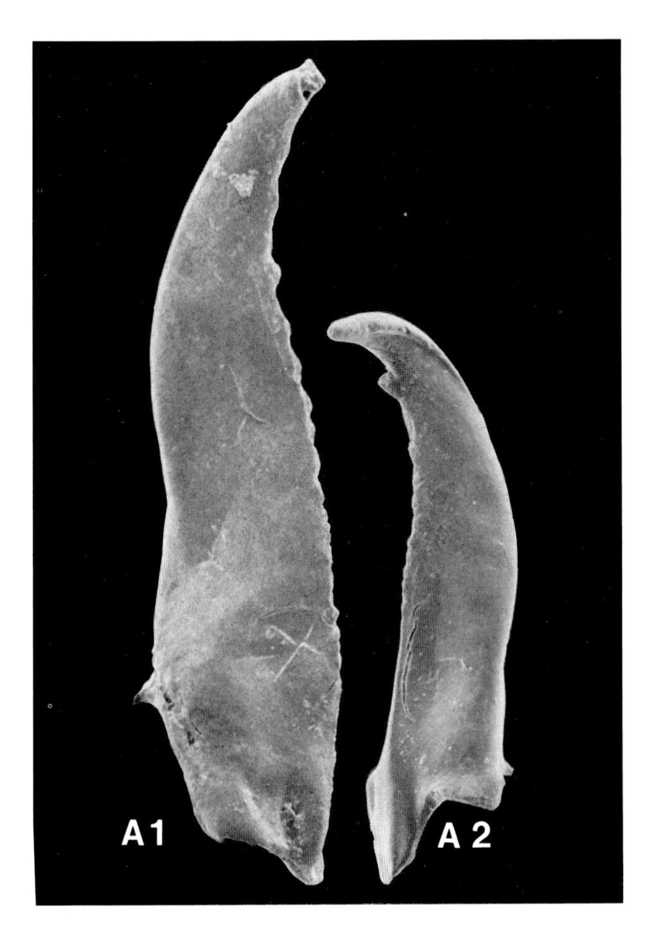

Fig. 62. Langeites glaber. Specimens are in dorsal view, ×120. From Vaktård 4, Hemse Beds, Hemse Marl SE part, 81-35LJ; A1 left MI, LO 5820:1; A2 right MI, LO 5820:2.

Discussion. – Jaws ascribed to *Langeites glaber* have been encountered in the late Early and Late Ludlow strata on Gotland viz. Hemse Beds, Hemse Marl SE part and Sundre Beds, middle upper part.

Due to the very limited material at hand, the relation between the specimens from the Hemse and Sundre Beds is somewhat dubious, though it is probable that they represent the same lineage. The Late Ludlow specimens from Gotland generally agree with Kielan-Jaworowska's description, but the late Early Ludlow jaws differ, particularly in being almost straight to slightly convex outwards.

Remarks. – It is notable that no MII elements have been described from Kielan-Jaworowska's collection. The assignments here of some particular MII elements to the same species are less well-based than in reconstructions of other Gotland apparatuses. The material derives from one locality (Holmhällar 1).

The few jaws and their bad state of preservation also explain the brevity of the description and the absence of variation analysis. The Holmhällar 1 sample was collected by Anders Martinsson as an ostracode marl sample. It has been processed with a coarse sieve, and the residue has been picked over for ostracodes. I have only had access to the picked out scolecodonts, and so do not know if these are representative for the sample.

References

Agterberg, F. P. 1958: An undulation of the rate of sedimentation in southern Gotland. *Geologie en Mijnbouw (NW) 20,* 253–260.

Åkesson, B. 1973: Morphology and life history of *Ophryotrocha maculata* sp. n. (Polychaeta, Dorvilleidae). *Zoologica Scripta 2,* 141–144.

Aldridge, R. J., Dorning, K. J., Hill, P. J., Richardson, J. B. & Siveter, D. J. 1979: Microfossil distribution in the Silurian of Britain and Ireland. *In* Harris, A. L. (ed.): The Caledonides of the British Isles; reviewed. *Geological Society of London, Special Publications 8,* 433–438.

Aldridge, R. J. & Jeppsson, L. 1984: Ecological specialists among Silurian conodonts. *In* Bassett, M. G. & Lawson, J. D. (eds.): Autecology of Silurian organisms. *Special Papers in Palaeontology, The Palaeontological Association 32,* 141–149.

[Angelin, N. P. (year unknown): *Palaeontologica Scandinavica.* Tab. 53:20–26. (Unpublished plates).]

Barnes, C. R., Fredholm, Doris & Jeppsson, L. 1985: Improved techniques for picking of microfossils. *In* Austin, R. L. (ed.): *Conodonts: Investigative Techniques and Applications,* 55, 74–76. Ellis Horwood Ltd., Chichester, England.

Barnes, R. S. K. & Head, S. M. 1977: Variation in paragnath number in some British populations of the estuarine polychaete *Nereis diversicolor. Estuarine and Coastal Marine Science 5,* 771–781.

Bassett, M. G. 1984: Life strategies of Silurian brachiopods. *In* Bassett, M. G. & Lawson, J. D. (eds.): Autecology of Silurian organisms. *Special papers in Palaeontology 32,* 237–263.

Bassett, M. G. & Cocks, L. R. M. 1974: A review of Silurian brachiopods from Gotland. *Fossils and Strata 3,* 1–56.

Bengtson, S. 1981: *Atractosella,* a Silurian alcyonacean octocoral. *Journal of Paleontology 55,* 281–294.

Bengtson, S. 1986: Preparing clean backgrounds in published photographic illustrations. *Lethaia 19,* 193–194.

Bergman, C. F. 1979a: Polychaete jaws. *In* Jaanusson, V., Laufeld, S. & Skoglund, R. (eds.): Lower Wenlock faunal and floral dynamics – Vattenfallet section, Gotland. *Sveriges Geologiska Undersökning C 762,* 92–102.

Bergman, C. F. 1979b: Ripple marks in the Silurian of Gotland, Sweden. *Geologiska Föreningens i Stockholm Förhandlingar 101,* 217–222.

Bergman, C. F. 1980a: Macrofossils of the Wenlockian Slite Siltstone of Gotland. *Geologiska Föreningens i Stockholm Förhandlingar 102,* 13–25.

Bergman, C. F. 1980b: Lower Wenlock polychaete fauna from Gotland, Sweden. *Fifth International Palynological Conference, Abstracts,* 38. Cambridge 1980.

Bergman, C. F. 1981a: Palaeocurrents, wave marks and reefs; a palaeogeographical instrument applied to the Silurian of Gotland. – The whole environment – Slite Beds, Gotland. *In* Laufeld, S. (ed.): Proceedings of Project Ecostratigraphy Plenary Meeting, Gotland, 1981. *Sveriges Geologiska Undersökning, Rapporter och Meddelanden 25,* 6.

Bergman, C. F. 1981b: Dispersed polychaete jaws and clusters in the Högklint Beds, Gotland. *In* Laufeld, S. (ed.): Proceedings of Project Ecostratigraphy Plenary Meeting, Gotland, 1981. *Sveriges Geologiska Undersökning, Rapporter och Meddelanden 25,* 5–6.

Bergman, C. F. 1984: Palaeocurrents in the Silurian Slite Marl of Fårö, Sweden. *Geologiska Föreningens i Stockholm Förhandlingar 105,* 169–179.

Bergström, S. M. & Sweet, W. C. 1966: Conodonts from the Lexington Limestone (Middle Ordovician) of Kentucky and its lateral equivalents in Ohio and Indiana. *Bulletins of American Paleontology 50-229,* 269–441.

von Bitter, P. 1972: Environmental control of conodont distribution in the Shawnee Group (Upper Pennsylvanian) of Eastern Kansas. *The University of Kansas Paleontological Contributions 59,* 1–105.

Bonnier, J. 1893: Notes sur les Annelides du Boulonnais. I. *Ophryotrocha puerilis* Clap. et Metschn. et son appareil maxillaire. *Bulletin Scientifique France et Belgique 25,* 198–226.

Boucot, A. J. 1975: *Evolution and extinction rate controls.* 427 pp. Elsevier Publishing Co.

Boyer, P. S. 1975: Polychaete jaw apparatus from the Devonian of central Ohio. *Acta Palaeontologica Polonica 20:3,* 425–435.

Boyer, P. S. 1981: Calcite in the mandibles of a marine polychaete. *Review of Palaeobotany and Palynology 34,* 247–250.

Brenchley, G. A. 1979: Post-mortem transport and population longevity recorded in scolecodont death assemblages. *Palaeogeography, Palaeoclimatology, Palaeoecology 28,* 297–314.

Brood, K. 1982: *Gotländska fossil.* 95 pp. P. A. Nordstedt & Söners Förlag. Stockholm.

Brood, K. 1985: Bryozoans from the Rönnklint section, Gotland. *Geologiska Föreningens i Stockholm Förhandlingar 107,* 71–75.

Cain, A. J. 1963: *Animal Species and their Evolution.* 190 pp. Hutchinson and Co. London.

Charletta, A. C. & Boyer, P. S. 1974: Scolecodonts from Cretaceous greensand of the New Jersey coastal plain. *Micropaleontology 20,* 354–366.

Cherns, L. 1981: An enigmatic Silurian metazoan from Gotland. *Palaeontology 24,* 195–202.

Cherns, L. 1982: Palaeokarst, tidal erosion surfaces and stromatolites in the Silurian Eke Formation of Gotland, Sweden. *Sedimentology 29,* 819–833.

Cherns, L. 1983: The Hemse–Eke boundary: facies relationships in the Ludlow Series of Gotland, Sweden. *Sveriges Geologiska Undersökning C 800,* 1–45.

Claesson, C. 1979: Early Palaeozoic geomagnetism of Gotland. *Geologiska Föreningens i Stockholm Förhandlingar 101,* 149–155.

Claparède, E. 1870: Annélides chétopodes du golfe de Naples. *Mémoires de la Société de physique et d'historie natureile de Geneve 19-20.*

Clarke, J. M. 1887: Annelid teeth from the lower portion of the Hamilton group and from the Naples shales of Ontario Country, New York. *Report of the State Geologist. New York State, Annual Report 6,* 30–33.

Colbath, G. K. 1986a: Jaw mineralogy in eunicean polychaetes. *Micropaleontology 32,* 186–189.

Colbath, G. K. 1986b: Taphonomic implications of polychaete jaws recovered from gut contents of the queen triggerfish. *Lethaia 19,* 339–342.

Colbath, G. K. 1987a: Evidence for shedding of maxillary jaws in eunicoid polychaetes. *Journal of Natural History 21,* 443–447.

Colbath, G. K. 1987b: An articulated polychaete jaw apparatus from the Carboniferous Kittanning Formation, western Pennsylvania, U.S.A. *Paläontologische Zeitschrift 61,* 81–86.

Colbath, G. K. & Larson, S. K. 1980: On the chemical composition of fossil polychaete jaws. *Journal of Paleontology 54,* 485–488.

Corradini, D. & Serpagli, E. 1968: Preliminary report on the discovery and initial study of large amounts of 'scolecodonts' and polychaete jaw apparatuses from Mesozoic formations. *Bollettino della Societa Paleontologica Italiana 7,* 3–5.

Corradini, D. & Olivieri, R. 1974: *Langeites siciliensis* n. sp., a polychaete jaw apparatus from the Permo-Carboniferous of northwestern Sicily. *Bollettino della Societa Paleontologica Italiana 13,* 156–163.

Corradini, D., Russo, F. & Serpagli, E. 1974: Ultrastructure of some fossil and recent polychaete jaws (scolecodonts). *Bolletino della Societa Paleontologica Italiana 13:1–2,* 122–134.

Croneis, C. 1941: Micropaleontology – past and future. *Bulletin of the American Association of Petroleum Geologists 25,* 1208–1255.

Croneis, C. & Scott, H. W. 1933: Scolecodonts. *Bulletin of the Geological Society of America 44,* 207.

Crossland, C. 1924: Polychaeta of Tropical East Africa, the Red Sea and Cape Verde Islands, collected by Cyril Crossland, and of the Maladive Archipelago collected by Professor Stanley Gardiner, M.A., F.R.S. *Proceedings of the Zoological Society (London) 1,* 1–106.

Dorning, K. J. 1984: Acritarch, Chitinozoa, conodont, dinocyst, graptolite, miospore and scolecodont coloration: multiple indicators of the thermal alteration index. *Abstracts of the 6th IPC, Calgary 1984.*

Edgar, D. R. 1984: Polychaetes of the Lower and Middle Paleozoic: A multi-element analysis and a phylogenetic outline. *Review of Palaeobotany and Palynology 43,* 255–285.

Ehlers, E. 1864–1868: *Die Borstenwürmer, nach systematischen und anatomischen Untersuchungen dargestellt (Annelida, Chaetopoda) 1/2.* 748 pp. Leipzig.

Ehlers, E. 1868a: Ueber eine fossile Eunicee aus Solenhofen (*Eunicites avitus*), nebst Bemerkungen über fossile Würmer überhaubt. *Zeitschrift für wissenschaftliche Zoologie 18,* 421–443.

Ehlers, E. 1868b: Ueber fossile Würmer aus dem lithographischen Schiefer in Bayern. *Palaeontographica 17,* 145–175.

Eichwald, E. 1854: Die Grauwackenschichten von Liev- und Esthland. *Bulletin de la Société Impériale des Naturalistes de Moscou 27,* 1–111.

Eisenack, A. 1939: Einige neue Annelidenreste aus dem Silur und dem Jura des Baltikums. *Zeitschrift für Geschiebeforschung und Flachlandsgeologie 15,* 153–176.

Eisenack, A. 1970: Mikrofossilien aus dem Silur Estlands und der Insel Ösel. *Geologiska Föreningens i Stockholm Förhandlingar 92,* 302–322.

Eisenack, A. 1975: Beiträge zur Anneliden-Forschung, I. *Neues Jahrbuch für Geologie und Paläontologie Abhandlungen 150,* 227–252.

Eller, E. R. 1933: An articulated annelid jaw from the Devonian of New York. *American Midland Naturalist 14,* 186.

Eller, E. R. 1934a: Annelid jaws from the Upper Devonian of New York. *Annals of the Carnegie Museum 22,* 303–316.

Eller, E. R. 1934b: Annelid jaws from the Hamilton Group of Ontario County, New York. *Annals of the Carnegie Museum 24,* 51–56.

Eller, E. R. 1936: A new scolecodont genus, *Idraites,* from the Upper Devonian of New York. *Annals of the Carnegie Museum 25,* 73–76.

Eller, E. R. 1938: Scolecodonts from the Potter Farm Formation of the Devonian of Michigan. *Annals of the Carnegie Museum 27,* 275–286.

Eller, E. R. 1940: New Silurian scolecodonts from the Albion Beds of the Niagara Gorge, New York. *Annals of the Carnegie Museum 28,* 9–46.

Eller, E. R. 1941: Scolecodonts from the Windom, Middle Devonian, of Western New York. *Annals of the Carnegie Museum 28,* 323–340.

Eller, E. R. 1942: Scolecodonts from the Erindale, Upper Ordovician, at Streetsville, Ontario. *Annals of the Carnegie Museum 29,* 241–270.

Eller, E. R. 1945: Scolecodonts from the Trenton Series (Ordovician) of Ontario, Quebec, and New York. *Annals of the Carnegie Museum 30,* 119–212.

Eller, E. R. 1955: Additional scolecodonts from the Potter Farm Formation of the Devonian of Michigan. *Annals of the Carnegie Museum 33,* 347–386.

Eller, E. R. 1963a: Scolecodonts from the Dundee, Devonian of Michigan. *Annals of the Carnegie Museum 36,* 173–180.

Eller, E. R. 1963b: Scolecodonts from the Sheffield Shale, Upper Devonian of Iowa. *Annals of the Carnegie Museum 36,* 159–172.

Eller, E. R. 1964: Scolecodonts of the Delaware Limestone, Devonian of Ohio and Ontario. *Annals of the Carnegie Museum 36,* 229–275.

Eller, E. R. 1969: Scolecodonts from well cores of the Maquoketa Shale Upper Ordovician, Ellsworth County, Kansas. *Annals of the Carnegie Museum 41,* 1–17.

Epstein, A., G., Epstein J. B. & Harris, L. D. 1977: Conodont color alteration – An index to organic metamorphism. *U.S. Geological Survey Professional Paper 995,* 1–27.

Fauchald, K. 1970: Polychaetous annelids of the families Eunicidae, Lumbrineridae, Iphitimidae, Arabellidae, Lysaretidae and Dorvilleidae from western Mexico. *Allan Hancock Monographs in Marine Biology 5,* 1–335.

Fauvel, P. 1919: Annélides Polychetes de la Guyane Française. *Bulletin du Museum national d'histoire natureile, Paris, 25,* 472–479.

Fauvel, P. 1953: Annelida Polychaeta. *In* Sewell, S.: *The Fauna of India, including Pakistan, Ceylon, Burma and Malaya. I–XII,* 1–507, Allahabad.

Flodén, T. 1980: Seismic stratigraphy and bedrock geology of the central Baltic. *Stockholm Contributions in Geology 35*, 1–240.

Ford, E. B. 1965: *Genetic Polymorphism.* 110 pp. Massachusetts Institute of Technology Press. Cambridge, Massachusetts.

Franzén, C.1977: Crinoid holdfasts from the Silurian of Gotland. *Lethaia 10*, 219–234.

Franzén, C. 1983: Ecology and taxonomy of Silurian crinoids from Gotland. *Acta Universitatis Upsaliensis 665*, 1–31.

Fredholm, D. 1988: Vertebrate biostratigraphy of the Ludlovian Hemse Beds of Gotland, Sweden. *Geologiska Föreningens i Stockholm Förhandlingar 110*, 237–253.

Frykman, P. 1985: Subaerial exposure and cement stratigraphy of a Silurian bioherm in the Klinteberg Beds, Gotland, Sweden. *Geologiska Föreningens i Stockholm Förhandlingar 107*, 77–88.

Frykman, P. 1986: Diagenesis of Silurian Bioherms in the Klinteberg Formation, Gotland, Sweden. *In* Schroeder, J. H. & Purser, B. H. (eds.): *Reef Diagenesis*, 399–423. Springer-Verlag.

Frykman, P. 1989: Carbonate ramp facies of the Klinteberg Formation, Wenlock–Ludlow transition on Gotland, Sweden. *Sveriges Geologiska Undersökning C 820*, 1–79.

Germeraad, J. H. 1980: Dispersed scolecodonts from the Cainozoic strata of Jamaica. *Scripta Geologica 54*, 1–24.

Gray, J., Laufeld, S. & Boucot, A, J. 1974: Silurian trilete spores and spore tetrads from Gotland: their implications for land plant evolution. *Science 185*, 260–263.

Grinnell, G. B. 1877: Notice of a new genus of annelids from the Lower Silurian. *American Journal of Science 3:14*, 229–230.

Hadding, A. 1913: Undre Dicellograptusskiffern i Skåne jämte några därmed ekvivalenta bildningar. *Lunds Universitets Årsskrift, N.F. 2,9:15*, 1–90. [Also in *Meddelande från Lunds Geologiska Fältklubb (B) 6*, 90 pp.]

Hadding, A. 1915: Undre och mellersta dicellograptusskiffern i Skåne och å Bornholm. *Meddelelser fra Dansk geologisk Forening 4*, 361–382.

Hadding, A. 1941: The Pre-Quaternary sedimentary rocks of Sweden. VI. Reef Limestones. *Lunds Universitets Årsskrift N.F. 2, 37:10*, 1–137.

Hadding, A. 1959: Silurian algal limestones of Gotland. Indicators of shallow waters and elevation of land. Some reflections on their lithological character and origin. *Lunds Universitets Årsskrift N.F. 2, 56:7*, 1–26.

Hartman, Olga: 1944: Polychaetous Annelids. Part V: Eunicea. *Allan Hancock Pacific Expeditions. 10*, 1–200. Los Angeles.

Hede, J. E. 1917: Faunan i kalksandstenens märgliga bottenlager söder om Klintehamn på Gottland. *Sveriges Geologiska Undersökning C 281*. 1–32.

Hede, J. E. 1919: Djupborrningen vid Burgsvik på Gottland 1915. Paleontologisk-stratigrafiska resultat. *Sveriges Geologiska Undersökning C 298*, 1–59.

Hede, J. E. 1921: Gottlands silurstratigrafi. *Sveriges Geologiska Undersökning C 305*. 1–100.

Hede, J. E. 1925a: Berggrunden (Silursystemet). *In* Munthe, H., Hede, J. E. & von Post, L.: Beskrivning till kartbladet Ronehamn. *Sveriges Geologiska Undersökning Aa 156*, 14–51.

Hede, J. E. 1925b: Beskrivning av Gotlands silurlager. *In* Munthe, H., Hede, J. E. & von Post, L.: Gotlands geologi, en översikt. *Sveriges Geologiska Undersökning C 331*, 13–30.

Hede, J. E. 1927a: Berggrunden (Silursystemet). *In* Munthe, H., Hede, J. E. & Lundqvist, G.: Beskrivning till kartbladet Klintehamn. *Sveriges Geologiska Undersökning Aa 160*, 12–54.

Hede, J. E. 1927b: Berggrunden (Silursystemet). *In* Munthe, H., Hede, J. E. & von Post, L.: Beskrivning till kartbladet Hemse. *Sveriges Geologiska Undersökning Aa 164*, 15–56.

Hede, J. E. 1928: Berggrunden (Silursystemet). *In* Munthe, H., Hede, J. E. & Lundqvist, G. 1928: Beskrivning till kartbladet Slite. *Sveriges Geologiska Undersökning Aa 169*, 13–65.

Hede, J. E. 1929: Berggrunden (Silursystemet). *In* Munthe, H., Hede, J. E. & Lundqvist, G.: Beskrivning till kartbladet Katthammarsvik. *Sveriges Geologiska Undersökning Aa 170*, 14–57.

Hede, J. E. 1933: Berggrunden (Silursystemet). *In* Munthe, H., Hede, J. E. & Lundqvist, G.: Beskrivning till kartbladet Kappelshamn. *Sveriges Geologiska Undersökning Aa 171*, 10–59.

Hede, J. E. 1936: Berggrunden (Silursystemet). *In* Munthe, H., Hede, J. E. & Lundqvist, G.: Beskrivning till kartbladet Fårö. *Sveriges Geologiska Undersökning Aa 180*, 11–42.

Hede, J. E. 1940: Berggrunden (Silursystemet). *In* Munthe, H., Hede, J. E. & Sundius, N.: Beskrivning till kartbladet Visby och Lummelunda. *Sveriges Geologiska Undersökning Aa 183*, 9–68.

Hede, J. E. 1942: On the correlation of the Silurian of Gotland. *Lunds Geologiska Fältklubb 1892–1942.* [Also in *Meddelanden från Lunds Geologisk-Mineralogiska Institution 101*, 1–25.]

Hede, J. E. 1960: The Silurian of Gotland. *In* Regnéll, G. & Hede, J.E.: The Lower Palaeozoic of Scania. The Silurian of Gotland. *International Geological Congress, XXI Session, Norden, Guidebook d, Sweden*, 44–87.

Hedström, H. 1904: Detaljprofil från skorpionfyndorten – *Pterygotus*- lagret – i siluren strax S om Visby. *Geologiska Föreningens i Stockholm Förhandlingar 26*, 93–96.

Hedström, H. 1910: The Stratigraphy of the Silurian strata of the Visby district. *Geologiska Föreningens i Stockholm Förhandlingar 32*, 1455–1484.

Hedström, H. 1923: Remarks on some fossils from the diamond boring of the Visby cement factory. *Sveriges Geologiska Undersökning C 314*, 1–26.

Heider, K. 1922: Uber Zahnwechsel bei polychaeten Anneliden. *Sitzungsberichte der Deutschen Akademie der Wissenschaften zu Berlin (1922)*, 488–491.

Herpin, R. 1926: Recherches biologiques sur la reproduction et le développement de quelques Annélides Polychetes. *Bulletin de la Société des sciences naturelles de l'ouest de la France (4) 5*, 1–250.

Hinde, G. J. 1879: On annelid jaws from the Cambro-Silurian, Silurian and Devonian Formations in Canada and from the Lower Carboniferous in Scotland. *The Quarterly Journal of the Geological Society of London 35*, 370–389.

Hinde, G. J. 1880: On annelid jaws from the Wenlock and Ludlow Formations of the West of England. *The Quarterly Journal of the Geological Society of London 36*, 368–378.

Hinde, G. J. 1882: On annelid remains from the Silurian strata of the Isle of Gotland. *Bihang till Kungliga Vetenskapsakademiens Handlingar (7) 5*, 3–28..

Hinde, G. J. 1896: On the jaw-apparatus of an annelid (*Eunicites reidiae* sp. nov.) from the Lower Carboniferous of Halkin Mountain, Flintshire. *The Quarterly Journal of the Geological Society of London 52*, 438–450.

Howell, B. F. 1962: Worms. *In* Moore, R. C. (ed.): *Treatise on Invertebrate Paleontology, Part W, Miscellanea*, 3–65. The University of Kansas Press.

International Code of Zoological Nomenclature, third edition, 1985. 338 pp. International Trust for Zoological Nomenclature, London.

Jaanusson, V., Laufeld, S. & Skoglund, R. 1979 (eds.): Lower Wenlock faunal and floral dynamics – Vattenfallet section, Gotland. *Sveriges Geologiska Undersökning C 762*. 1–294.

Jaeger, H. 1981: Comments on the graptolite chronology of Gotland. *In* Laufeld, S. (ed.): Proceedings of Project Ecostratigraphy Plenary Meeting, Gotland 1981. *Sveriges Geologiska Undersökning, Rapporter och meddelanden 25*, 12.

Jansonius, J. & Craig, J. H. 1971: Scolecodonts: I. Descriptive terminology and revision of systematic nomenclature; II. Lectotypes, new names for homonyms, index of species. *Bulletin of Canadian Petroleum Geology 19*, 251–302.

Jansonius, J. & Craig, J. H. 1974: Some scolecodonts in organic association from the Devonian strata of western Canada. *Geoscience and Man 9*, 15–26.

Janvier, Ph. 1978: On the oldest known teleostome fish *Andreolepis hedei* Gross (Ludlow of Gotland), and the systematic position of the lophosteids. *Eesti NSV Teaduste Akadeemia Toimetised. Keemia Geoloogia 27:3*, 88–95.

Jeppsson, L. 1969: Notes of some Upper Silurian multielement conodonts. *Geologiska Föreningens i Stockholm Förhandlingar 91*, 12–24.

Jeppsson, L. 1972: Some Silurian conodont apparatuses and possible conodont dimorphism. *Geologica et Palaeontologica 6,* 51–69.

Jeppsson, L. 1974: Aspects of late Silurian conodonts. *Fossils and Strata 6.* 54 pp.

Jeppsson, L. 1976: Autecology of Late Silurian conodonts. *In* Barnes, C. R. (ed.): Conodont Palaeoecology. *The Geological Association of Canada Special Paper 15,* 105–118.

Jeppsson, L. 1979: Conodont element function. *Lethaia 12,* 153–171.

Jeppsson, L. 1982: Third European Conodont Symposium (ECOS III). Guide to excursion. *Publications from the Institutes of Mineralogy, Palaeontology and Quaternary Geology, University of Lund, Sweden, 239.* 1–32.

Jeppsson, L. 1983: Silurian conodont faunas from Gotland. *Fossils and Strata 15,* 121–144.

Jeppsson, L. 1987: Some thought about future improvments in conodont extraction. *In:* Austin, R. L. (ed.): *Conodonts: Investigative Techniques and Applications,* 45–51. Ellis Horwood Ltd., Chichester, England.

Jeppsson, L., Fredholm, D. & Mattiasson, B. 1985: Acetic acid and phosphatic fossils – a warning. *Journal of Paleontology 59,* 952–956.

Jeppsson, L. & Fredholm, D. 1987: Temperature dependence of limestone dissolution in conodont extraction. *In:* Austin, R. L. (ed.): *Conodonts: Investigative Techniques and Applications,* 39–42. Ellis Horwood Ltd., Chichester, England.

Jumars, P. A. 1974: A generic revision of the Dorvilleidae (Polychaeta), with six new species from the deep North Pacific. *Zoological Journal of the Linnean Society 54,* 101–135. London.

Kershaw, S. 1981: Stromatoporoid growth form and taxonomy in a Silurian biostrome, Gotland. *Journal of Paleontology 55,* 1284–1295.

Kershaw, S. 1987: Stromatoporoid-coral intergrowths in a Silurian biostrome. *Lethaia 20,* 371–380.

Kershaw, S. & Riding, R. 1978: Parameterization of stromatoporoid shape. *Lethaia 11,* 233–242.

Kielan-Jaworowska, Z. 1961: On two Ordovician polychaete jaw apparatuses. *Acta Palaeontologica Polonica 6,* 237–254.

Kielan-Jaworowska, Z. 1962: New Ordovician genera of polychaete jaw apparatuses. *Acta Palaeontologica Polonica 7,* 291–325.

Kielan-Jaworowska, Z. 1966: Polychaete jaw apparatuses from the Ordovician and Silurian of Poland and comparison with modern forms. *Palaeontologia Polonica 16,* 1–152 pp.

Kielan-Jaworowska, Z. 1968: Scolecodonts versus jaw apparatuses. *Lethaia 1,* 39–49.

Korschelet, E. 1893: Uber *Ophryotrocha puerilis* Clap.-Metschn. und die polytrochen Larven eines anderen Anneliden (*Harpochaeta cingulata* nov. gen. nov. spec.). *Zeitschrift für wissenschaftliche Zoologie 57,* 224–289.

Kozlowski, R. 1956: Sur quelques appareils masticateurs des Annélides polychetes ordoviciens. *Acta Palaeontologica Polonica 1,* 165–205.

Kozur, H. 1970: Zur Klassifikation und phylogenetischen Entwicklung der fossilen Phyllodocida und Eunicida (Polychaeta). *Freiberger Forschungshefte C 260,* 35–81.

Kozur, H. 1971: Die Eunicida und Phyllodocida des Mesosoikums. *Freiberger Forschungshefte C 267,* 73–111.

Kozur, H. 1972: Die Bedeutung der triassichen Scolecodonten insbesondere für die Taxonomie und Phylogenie der fossilen Eunidica. Hat sich die Synthese vom 'orthotaxonomischen' und 'parataxonomischen' System in der Praxis bewährt?. *Mitteilungen der Gesellschaft der Geologie- und Bergbaustudenten 21,* 745–776.

Lange, F. W. 1947: Annelidos poliquetos dos folhelhos devonianos do Parana. *Arquivos do Museu Paranaense 6,* 161–230. [English translation 1949: Polychaete annelids from the Devonian of Parana, Brazil. *Bulletins of American Paleontology 33-134,* 5–104.]

Lange, F. W. 1950: Um novo escolecodonte dos folhelhos Ponta Grossa. *Arquivos do Museu Paranaese 8,* 189–213.

Larsson, K. 1979: Silurian tentaculitids from Gotland and Scania. *Fossils and Strata 11.* 180 pp.

Laufeld, S. 1974a: Silurian Chitinozoa from Gotland. *Fossils and Strata 5.* 130 pp.

Laufeld, S. 1974b: Reference localities for palaeontology and geology in the Silurian of Gotland. *Sveriges Geologiska Undersökning C 705,* 1–172.

Laufeld, S. 1974c: Preferred orientation of orthoconic nautiloids in the Ludlovian Hemse Beds of Gotland. *Geologiska Föreningens i Stockholm Förhandlingar 96,* 157–162.

Laufeld, S. 1975: Paleoecology of Silurian polychaetes and chitinozoans in a reef-controlled sedimentary regime. *The Geological Society of America, Abstracts with Programs,* 804–805.

Laufeld, S. & Bassett, M. 1981: Gotland: The anatomy of a Silurian carbonate platform. *Episodes 1981:2,* 23–27.

Laufeld, S. & Jeppsson, L. 1976: Silicification and bentonites in the Silurian of Gotland. *Geologiska Föreningens i Stockholm Förhandlingar 98,* 31–44.

Laufeld, S. & Martinsson, A. 1981: *Reefs and Ultrashallow Environments. Guidebook to the Field Excursions in the Silurian of Gotland, Project Ecostratigraphy Plenary Meeting 22nd–28th August, 1981.* 24pp. Museum Department, Geological Survey of Sweden.

Laufeld, S., Sundquist, B. & Sjöström, H. 1978: Megapolygonal bedrock structures in the Silurian of Gotland. *Sveriges Geologiska Undersökning C 759,* 1–26.

Liljedahl, L. 1981: Silicified bivalves from the Silurian of Gotland. *In* Laufeld, S. (ed.): Proceedings of Project Ecostratigraphy Plenary Meeting, Gotland, 1981. *Sveriges Geologiska Undersökning, Rapporter och meddelanden 25,* 22.

Liljedahl, L. 1983: Two silicified Silurian bivalves from Gotland. *Sveriges Geologiska Undersökning C 799.* 1–51.

Liljedahl, L. 1984: Silurian silicified bivalves from Gotland. *Sveriges Geologiska Undersökning C 804,* 1–82.

Liljedahl, L. 1985: Ecological aspects of a silicified bivalve fauna from the Silurian of Gotland. *Lethaia 18,* 53–66.

Liljedahl, L. 1986: Endolithic micro-organisms and silicification of a bivalve fauna from the Silurian of Gotland. *Lethaia 19,* 267–278.

Lindström, G. 1885: *List of the fossils of the Upper Silurian Formation of Gotland.* 20 pp. Stockholm.

Lindström, G. 1888: *List of the fossil faunas of Sweden.* 29 pp. Stockholm.

Lindström, M. 1955: The conodonts described by A. R. Hadding, 1913. *Journal of Paleontology 29,* 105–111.

Männil, R. M. & Zaslavskaya, N. M. 1985a: Silurian polychaetes from the northern Sibiria. *In:* Gudina V. I. & Kanigin A. V. (eds): Phanerozoic microfauna from Sibiria with surroundings. *Trudy Instituta Geologii i Geofiziki 615,* 98–119, 127–130. Nauka, Novosibirsk.

Männil, R. M. & Zaslavskaya, N. M. 1985b: Finds of Middle Palaeozoic polychaetes in the southeastern part of the West Sibirian Platform. *In:* Dubatolov, V. N. & Kanigin, A. V. (eds): Palaeozoic biostratigraphy of western Sibiria. *Trudy Instituta Geologii i Geofiziki SO AN SSSR 619,* 69–72, 199–200 and 213. Nauka, Novosibirsk.

Manten, A. A. 1971: *Silurian reefs of Gotland.* 539 pp. Elsevier Publishing Company.

Martinsson, A. 1958: Deep boring on Gotska Sandön. I: The Submarine Morphology of the Baltic Cambro-Silurian Area. *Bulletin of the Geological Institutions of the University of Uppsala 38,* 11–35.

Martinsson, A. 1960: Two assemblages of polychaete jaws from the Silurian of Gotland. *Bulletin of the Geological Institutions of the University of Uppsala 39:2,* 1–8.

Martinsson, A. 1962: Ostracodes of the family Beyrichiidae from the Silurian of Gotland. *Bulletin of the Geological Institutions of the University of Uppsala 41,* 1–369.

Martinsson, A. 1967: The succession and correlation of ostracode faunas in the Silurian of Gotland. *Geologiska Föreningens i Stockholm Förhandlingar 89,* 350–386.

Martinsson, A. 1972: Review of Manten, A. A.: Silurian reefs of Gotland. *Geologiska Föreningens i Stockholm Förhandlingar 94,* 128–129.

Massalongo, 1855: *Monographia delle nereidi fossili del Monte Bolca.* Verona.

Mierzejewska, G. & Mierzejewski, P. 1974: The ultrastructure of some fossil invertebrate skeletons. *Annals of the Medical Sections of the Polish Academy of Sciences 19 (2)*, 133–135.

Mierzejewski, P. 1978a: Molting of the jaws of the Early Paleozoic Eunicida (annelida, polychaeta). *Acta Palaeontologica Polonica 23*, 73–88.

Mierzejewski, P. 1978b: New placognath Eunicida (Polychaeta) from the Ordovician and Silurian of Poland. *Acta Palaeontologica Polonica 28*, 273–281.

Mierzejewski, P. 1984: *Synclinophora synclinalis* Eisenack – the oldest arabellid polychaete. *Review of Palaeobotany and Palynology 43*, 285–292.

Mierzejewski, P. & Mierzejewska, G. 1975: Xenognath type of polychaete jaw apparatuses. *Acta Palaeontologica Polonica 20*, 437–444.

Mierzejewski, P. & Mierzejewska, G. 1977: Preliminary transmission electron microscopy studies on pharate jaws of Palaeozoic Eunicida. *Acta Med. Pol. 18:4*, 347–348.

Mori, K. 1970: Stromatoporoids from the Silurian of Gotland. Part 2. *Stockholm Contributions in Geology 22*, 1–152.

Munthe, H. 1921: Beskrivning till kartbladet Burgsvik jämte Hoburgen och Ytterholmen. *Sveriges Geologiska Undersökning Aa 152*, 1–172.

Murphy, M. A., Matti, J. C. & Walliser O. H. 1981: Biostratigraphy and evolution of the *Ozarcodina remscheidensis – Eognathodus sulcatus* Lineage (Lower Devonian) in Germany and central Nevada. *Journal of Paleontology 55*, 747–772.

Muus, B. J. 1967: The fauna of Danish estuaries and lagoons. *Meddelelser fra Danmarks Fiskeri- og Havundersøgelser (Ny Serie) 5*. 316 pp.

Nathorst, A. G. 1888–1894: *Jordens historia*, part II 1892, 610–611. F. & G. Beijer, Stockholm.

Nathorst, A. G. 1894: *Sveriges geologi*, 90–91. F. & G. Beijer, Stockholm.

Odin, G. S., Spjeldnaes N., Jeppsson, L. & Nielsen Thorshoj, A. 1984: Field meeting in Scandinavia. *Bulletin de Liaison et Informations 3*, 6–23.

Odin, G. S., Hunziker, J. C., Jeppsson, L. & Spjeldnaes, N. 1986: Ages radiometriques K–Ar de biotites pyroclastiques sedimentaires dans le Wenlock de Gotland (Suede). *In*: Odin, G. S. (guest ed.): Calibration of the Phanerozoic Time Scale. *Chemical Geology (Isotope Geoscience Section) 59*, 117–125.

Olive, P. J. W. 1977: The life history and population structure of the polychaetes *Nephtys caeca* and *Nephtys hombergii* with special reference to the growth rings in the teeth. *Journal of Marine Biological Association of the United Kingdom 57*, 133–150.

Olive, P. J. W. 1980: Growth lines in polychaete jaws (teeth). *In* Rhoads, D. C. & Lutz, R. A. (eds.): *Skeletal Growth of Aquatic Organisms, Biological Records of Environmental Change*, 561–626. Plenum Press, New York.

Olive, P. J. W. & Clark, R. B. 1978: Physiology of reproduction. *In* Mill, P. J. (ed.): *Physiology of Annelids*, 271–368. Academic Press, New York.

Pander, C. H. 1856: *Monographie der fossilen Fische des Silurischen Systems der Russisch-Baltischen Gouvernements.* Kaiserliche Akademie Wissenschaften St. Petersburg. 91 pp.

Paxton, H. 1980: Jaw growth and replacement in Polychaeta. *Journal of Natural History 14*, 543–546.

Pfannenstiel, H. D. 1977: Experimental analysis of the 'paarkultureffekt' in the protandric polychaete *Ophryotrocha puerilis* Clap. Mecz. *Journal of Experimental Marine Biology and Ecology 28*, 31–40.

Poulsen, K. D., Saxov, S., Balling, N. & Kristiansen, J. I. 1982: Thermal conductivity measurements of Silurian limestones from the Island of Gotland, Sweden. *Geologiska Föreningens i Stockholm Förhandlingar 103*, 349–356.

Ramsköld, L. 1983: Silurian cheirurid trilobites from Gotland. *Palaeontology 26*, 175–210.

Ramsköld, L. 1984: Silurian odontopleurid trilobites from Gotland. *Palaeontology 27*, 239–264.

Ramsköld, L. 1985a: Silurian phacopid and dalmanitid trilobites from Gotland. *Stockholm Contribution in Geology 40*, 1–62.

Ramsköld, L. 1985b: *Studies on Silurian trilobites from Gotland, Sweden.* 24 pp. Department of Geology, University of Stockholm, and Department of Palaeozoology, Swedish Museum of Natural History.

Ramsköld, L. 1986: Silurian encrinurid trilobites from Gotland and Dalarna, Sweden. *Palaeontology 29*, 527–575.

Regnéll, G. 1951: Centenary of 'Palaeontologia Svecica' – with a sketch of the work and life of N. P. Angelin. *Geologiska Föreningens i Stockholm Förhandlingar 73*, 619–629.

Riding, R. 1979: Calcareous algae. *In* Jaanusson, V., Laufeld, S. & Skoglund, R. (eds.): Lower Wenlock faunal and floral dynamics – Vattenfallet section, Gotland. *Sveriges Geologiska Undersökning C 762*, 54–60.

Riding, R. 1981: Composition, structure and environmental setting of Silurian bioherms and biostroms in Northern Europe. *In* Toomey, D. F. (ed.): European fossil reef models. *The Society of Economic Paleontologists and Mineralogists, Special Publication 30*, 41–83.

Sandford, J. T. & Mosher, R. E. 1985: Insoluble residues and geochemistry of some Llandoverian and Wenlockian rocks from Gotland. *Sveriges Geologiska Undersökning C 811*, 1–31.

Schäfer, W. 1972: *Ecology and Palaeoecology of Marine Environments.* 578 pp. Oliver & Boyd, Edinburgh.

Schallreuter, R. 1982: Mikrofossilien aus Geschieben II. Scolecodonten. *Der Geschiebe-Sammler. Mitteilungsheft der Sammlergruppe für Geschiebekunde 16 (1)*, 1–23. Hamburg.

Schwab, K. W. 1966: Microstructure of some fossil and recent scolecodonts. *Journal of Paleontology 40*, 416–423.

Serpagli, E. 1967: I conodonti dell'Ordoviciano superiore (Ashgilliano) delle Alpi Carniche. *Bollettino della Societa Paleontologica Italiana 6*, 30–111.

Simpson, G. G. 1961: *Principles of Animal Taxonomy.* 247 pp. Columbia University Press, New York.

Sivhed, U. 1976: Sedimentological studies of the Wenlockian Slite Siltstone of Gotland. *Geologiska Föreningens i Stockholm Förhandlingar 98*, 59–64.

Snajdr, M. 1951: On Errant Polychaeta from the Lower Paleozoic of Bohemia. *Sbornik of the Geological Survey of Czechoslovakia, Paleontology 18*, 241–296.

Stauffer, C. R. 1933: Middle Ordovician Polychaeta from Minnesota. *Bulletin of the Geological Society of America 44*, 1173–1218.

Stauffer, C. R. 1939: Middle Devonian Polychaeta from the lake Eire district. *Journal of Paleontology 13*, 500–511.

Strauch, F. 1973: Die Feinstruktur einiger Scolecodonten. *Senckenbergiana Lethaea 54*, 1–19.

Stridsberg, S. 1981a: Apertural constrictions in some oncocerid cephalopods. *Lethaia 14*, 269–276.

Stridsberg, S. 1981b: Silurian oncocerid nautiloids from Gotland. *In* Laufeld, S. (ed.): Proceedings of Project Ecostratigraphy Plenary Meeting, Gotland, 1981. *Sveriges Geologiska Undersökning, Rapporter och Meddelanden 25*, 32.

Stridsberg, S. 1985: Silurian oncocerid cephalopods from Gotland. *Fossils and Strata 18.* 65 pp.

Sundquist, B. 1981: The whole environment – Silurian Slite Beds, Gotland. – Petrography. *In* Laufeld, S. (ed.): Proceedings of Project Ecostratigraphy Plenary Meeting, Gotland, 1981. *Sveriges Geologiska Undersökning, Rapporter och Meddelanden 25*, 33.

Sundquist, B. 1982a: Carbonate petrography of the Wenlockian Slite Beds at Haganäs, Gotland. *Sveriges Geologiska Undersökning C 796*, 1–79.

Sundquist, B. 1982b: Wackestone petrography and bipolar orientation of cephalopods as indicators of littoral sedimentation in the Ludlovian of Gotland. *Geologiska Föreningens i Stockholm Förhandlingar 104*, 81–90.

Sundquist, B. 1982c: Palaeobathymetric interpretation of wave ripple-marks in a Ludlovian grainstone of Gotland. *Geologiska Föreningens i Stockholm Förhandlingar 104*, 157–166.

Sylvester, R. K. 1959: Scolecodonts from central Missouri. *Journal of Paleontology 33*, 33–49.

Szaniawski, H. 1968: Three new polychaete apparatuses from the Upper Permian of Poland. *Acta Palaeontologica Polonica 13*, 255–281.

Szaniawski, H. 1970: Jaw apparatuses of the Ordovician and Silurian polychaetes from the Mielnik borehole. *Acta Palaeontologica Polonica 15*, 445–472.

Szaniawski, H. 1974: Some Mesozoic scolecodonts congeneric with recent forms. *Acta Palaeontologica Polonica 19*, 179–199.

Szaniawski, H. & Wrona, R. 1973: Polychaete jaw apparatuses and scolecodonts from the Upper Devonian of Poland. *Acta Palaeontologica Polonica 18*, 223–262.

Szaniawski, H. & Gazdzicki, A. 1978: A reconstruction of three Jurassic polychaete jaw apparatuses. *Acta Palaeontologica Polonica 23*, 3–29.

Tasch, P. & Shaffer, B. L. 1961: Study of scolecodonts by transmitted light. *Micropaleontology 7*, 369–371.

Tasch, P. & Stude, J. R. 1965: A scolecodont natural assemblage from the Kansas Permian. *Transactions of the Kansas Academy of Science 67*, 646–658.

Taugourdeau, Ph. 1968: Propositions concernant l'établissement de formules dentaires pour l'étude des Scolécodontes. *Proceedings IPU, 23 International geological congress*, 437–442.

Taugourdeau, Ph. 1972: Débris cuticulaires d'annelides associés aux scolécodontes. *Review of Palaeobotany and Palynology 13*, 233–252.

Taugourdeau, Ph. 1976: Les schistes et calcaires Eodevoniens de Saint-Céneré (Massif Armoricain, France). *Mémoires de la Société géologique et minéralogique de Bretagne 19*, 135–141.

Taugourdeau, Ph. 1978: Les scolécodontes dispersés. *Cahiers de Micropaléontologie 2(1)*. 1–104.

Thorell, T. & Lindström, G. 1885: On a Silurian scorpion from Gotland. *Kungliga Svenska Vetenskaps-Akademiens Handlingar 21:9*. 33pp. Stockholm.

Voss-Foucart, M. F., Fonze-Vignaux, M. T. & Jeuniaux, C. 1973: Systematic characters of some annelid Polychaetes at the level of the chemical composition of the jaws. *Biochemical Systematics and Ecology 1*, 119–122.

Walmsley, V. G. & Boucot, A. J. 1975: The phylogeny, taxonomy and biogeography of Silurian and Early to Mid Devonian Isorthinae (Brachiopoda). *Palaeontographica A 148*, 34–108.

Watkins, R. 1975: Silurian brachiopods in a stromatoporoid bioherm. *Lethaia 8*, 53–61.

[Watts, N. R. 1981: Sedimentology and diagenesis of the Högklint reefs and their associated sediments, lower Silurian, Gotland, Sweden. 406 pp. University of Wales, Cardiff. Unpublished Ph. D. thesis.]

Watts, N. R. 1982: Early carbonate facies in Silurian reefs on Gotland, Sweden. *Bulletin of the American Association of Petroleum Geologists 66*, 640.

Webers, G. F. 1966: The Middle and Upper Ordovician conodont faunas of Minnesota. *Minnesota Geological Survey, Special Publication Series, SP-4*, 1–133.

Westergård, A. H. 1909: Studier öfver dictyograptusskiffern och dess gränslager med särskild hänsyn till i Skåne förekommande bildningar. *Meddelande från Lunds Geologiska Fältklubb (B) 4*. 71 pp. [Also as *Kongliga Fysiografiska Sällskapets i Lund Handlingar (N.F.) 20:3*. 71 pp. Also as *Lunds Universitets Årsskrift (N.F.) 2:5:3*. 71 pp.]

Wiman, C. 1893: Ueber das Silurgebiet des Bottnischen Meeres. *Bulletin of the Geological Institution of the University of Upsala 1*, 65–75.

Wolf, G. 1976: *Bau und Funktion der Kieferorgane von Polychaeten.* [*On the morphology and function of polychaete jaws.*] 55 pp. Zoologisches Institut und Museum, Universität Hamburg.

Wolf, G. 1980: Morphologische Untersuchungen an den Kieferapparaten einiger rezenter und fossiler Eunicoidea (Polychaeta). *Senckenbergiana maritima 12:1/4*, 1–182.

Zebera, K. 1935: Les Conodontes et les Scolécodontes du Barrandien. *Bulletin international de l'Academie des Sciences de Boheme 36*, 88–96.

Appendix: Localities

In the list of references to each locality, an asterisk marks the publication in which a major description of the locality is found. Other references are given only if not included in 'Reference localities for palaeontology and geology in the Silurian of Gotland' (Laufeld 1974b). The grid references refer to the Swedish National Grid system measured from the 2nd topographical map edition and (within parentheses) the UTM system from the 1st edition.

The sample codes consist of the two last figures of the sampling year, a sample number for that year, and the initials of the collector. Apart from my own samples I have received a large number from Lennart Jeppsson (including a few samples from 'Project Silurian Silicified Fossils from Gotland' marked PSSFG) and some further samples from the Swedish Museum of Natural History, Stockholm, Kent Larsson, Lund, Anders Martinsson, Uppsala, Doris Fredholm, Lund, Sven Laufeld, Uppsala, and Sven Stridsberg, Lund.

The neutral term 'Beds' has been used for designating Hede's major stratigraphical units.

Only part of the jawed annelid fauna (the paulinitids) is described in this paper. Paulinitids are reported by specific name, all other species of jawed annelid as 'annelid jaws'. If only fragments and no identifiable annelid remains are found, 'fragments' is noted.

AJMUNDE 1. Hemse beds, Hemse Marl NW part. *Age:* Ludlow, probably Bringewoodian.
References: Laufeld 1974a, b*; Laufeld & Jeppsson 1976; Larsson 1979; Jeppsson 1983; Ramsköld 1985; Frykman 1989.
Samples/jawed annelids: 71-40LJ, bottom of the ditch, *Lanceolatites gracilis, Kettnerites (K.)* cf. *martinssonii, K. (A.) sisyphi, K. (K.)* cf. *burgensis*, annelid jaws; 75-77CB, bottom of the ditch, fragments; 77-36LJ, bottom of the ditch, few annelid jaws.

AJSTUDDEN 1. 639599 167863 (CJ 6866 9246), ca 4250 m SSE of Boge church. Topographical map sheet 6 J Roma NV & NO. Geological map sheet Aa 169 Slite.
Shore exposure comprising about 100–150 m north of the small point along the shore and some tens of meters out to sea.
Slite Beds, Slite Marl, probably slightly older than *Pentamerus gothlandicus. Age:* Wenlock.
Samples/jawed annelids: 83-6CB, fossiliferous grey calcarenite immediately below the mean water level, *Kettnerites (A.) sisyphi, Gotlandites slitensis*, annelid jaws; 83-7CB, grey dense calcarenite about 10 m beyond the shoreline and 0.2 m below the mean water level covered by a few mm of quaternary sediment, *Gotlandites slitensis*, other paulinitid jaws, annelid jaws.

ALBY 1. Slite Beds, Slite Marl. *Age:* Wenlock.
References: Laufeld 1974b*; Larsson 1979.
Samples/jawed annelids: 67-20LJ, *Kettnerites (A.) sisyphi*, annelid jaws.

ALBY 2. Slite Beds, Slite Marl. *Age:* Wenlock.
References: Laufeld 1974a, b*; Ramsköld 1983.
Samples/jawed annelids: 75-17CB, excavated material, annelid jaws.

ALBY 4. Slite Beds, Slite Marl. *Age:* Wenlock.
References: Laufeld 1974a, b*.
Samples/jawed annelids: 77-16CB, uppermost bed about 0.75 m below ground level, annelid jaws.

AMLINGS 1. Hemse Beds, Hemse Marl NW part. *Age:* Ludlow; probably Late Bringewoodian, possibly earliest Leintwardinian.
References: Laufeld 1974a, b*; Laufeld & Jeppsson 1976; Larsson 1979; Ramsköld 1983; Jeppsson 1983*; Fredholm 1988.
Samples/jawed annelids: 71-111LJ, top of the section, *Kettnerites (K.) huberti, K. (A.) microdentatus*, annelid jaws; 75-73CB, 0.10 m above the bottom of the ditch, same sample level as 66-170SL (Laufeld 1974a), *Kettnerites (K.) polonensis, K. (A.) sisyphi* aff. *klasaardensis, K. (A.) microdentatus*, annelid jaws; 75-74CB, 0.15 m above the bottom of the ditch, same level as sample G66-170SL (Laufeld 1974a), *Lanceolatites gracilis, Kettnerites (K.) huberti, K. (A.) sisyphi*, annelid jaws; 77-39PSSFG, *Kettnerites (K.) martinssonii, K. (K.)* cf. *burgensis*, annelid jaws.

ÄNGMANS 2. Hemse Beds, unit b. *Age:* Ludlow, Eltonian.
Reference: Fredholm 1988*.
Samples/jawed annelids: 83-50LJ, excavated material, *Kettnerites (K.) bankvaetensis, K. (K.) huberti*, annelid jaws.

ANSARVE 1. Högklint Beds, southwestern facies, upper part. *Age:* Early Wenlock.
References: Laufeld 1974b*, Larsson 1979.
Samples/jawed annelids: 79-46LJ, bottom of the ditch, *Kettnerites (K.) martinssonii, K. (A.) sisyphi, K. (K.) abraham isaac, K. (A.) sisyphi* var. valle, *Hindenites angustus*, annelid jaws.

AR 1. Högklint Beds, upper part. *Age:* Early Wenlock.
Reference: Larsson 1979*.
Samples/jawed annelids: 84-13LJ 0.25 to 0.5 m above base of the section, *Kettnerites (K.) martinssonii, K. (K.) abraham isaac, Hindenites* cf. *gladiatus*, annelid jaws.

ASKRYGGEN 2. Slite Beds, Slite Siltstone, top. *Age:* Wenlock.
References: Laufeld 1974a, b*; Bergman 1980a.
Samples/jawed annelids: 71-52LJ, no annelid jaws.

AURSVIKEN 1. Högklint Beds, lower–middle part. *Age:* Early Wenlock.
References: Laufeld 1974a, b*.
Samples/jawed annelids: 77-25CB, surface exposure, *Kettnerites (K.) martinssonii, K. (A.) microdentatus, Hindenites* cf. *gladiatus*, annelid jaws.

AUTSARVE 1. Hemse Marl NW part. *Age:* Early Middle Ludlow.
References: Laufeld 1974a, b*, Jeppsson 1983*; Fredholm 1988; Frykman 1989.
Sample/jawed annelids: 71-43LJ, surface exposure, *Lanceolatites gracilis, Kettnerites (K.) martinssonii, K. (K.) huberti, K. (A.) sisyphi*, annelid jaws; 75-60CB, loose slab, no annelid jaws.

BAJU 1. 637823 167773 (CJ 6639 7480), ca 4100 m ENE of Anga church. Topographical map sheet 6 J Roma NV & NO. Geological map sheet Aa 170 Katthammarsvik.
Low cliff at the shore at the southernmost tip of the point about 50 m south of the pier.
Klinteberg Beds(?). *Age:* Late Wenlock – Early Ludlow.
Samples/jawed annelids: 78-12CB, 0.15 m below top of the exposure, *Kettnerites (K.)* cf. *bankvaetensis, K. (K.) martinssonii, K. (K.) polonensis*, annelid jaws.

BANDLUNDE 1. Eke Beds, lower part. *Age:* Ludlow, Late Leintwardinian.
References: Laufeld 1974a, b*.
Samples/jawed annelids: 76-12CB, loose slab from a temporary ditch excavation, *Kettnerites (K.) martinssonii, K. (K.)* cf. *huberti*, annelid jaws.

BANKVÄT 1. Hamra Beds, unit b. *Age:* Latest Ludlow.
References: Laufeld 1974a, b*; Larsson 1979; Jeppsson 1982, 1983.
Samples/jawed annelids: 81-39LJ, some 10 m N of the lake (existing only during wet periods) and slightly south of sample location of 82-33CB, *Kettnerites (K.) bankvaetensis, K. (K.) huberti*, annelid jaws; Samples 82-27CB to 32CB: About 155 m NE of the stone fence and 10–15 m N of the beach, in a shallow shore exposure, 82-27CB, bed 0.12–0.20 m below the uppermost Silurian rock surface, undulating, competent, brownish-grey calcarenite between soft calcilutite, *Kettnerites (K.) bankvaetensis*, annelid jaws; 82-28CB, 20–50 mm thick bed and about 0.10 m below surface, *Kettnerites (K.) bankvaetensis, K. (K.) huberti*, annelid jaws; 82-29CB, 10–30 mm thick bed and about 0.08 m below surface, *Kettnerites (K.) bankvaetensis*, annelid jaws; 82-30CB, 30–40 mm thick bed about 0.05 m below surface, *Kettnerites (K.) bankvaetensis, K. (K.) huberti*, annelid jaws; 82-31CB, 10–30 mm thick bed and about 0.03 m below surface, *Kettnerites (K.) bankvaetensis*, annelid jaws; 82-32CB, about 30 mm thick, comprising the uppermost bed, *Kettnerites (K.) bankvaetensis, K. (K.) huberti*, annelid jaws; 82-33CB, about 150 m NW of samples 82-27CB to 82-32CB and 100 m SE of the 82-34CB, about 0.30–0.40 m below the soil surface, *Kettnerites (K.) bankvaetensis, K. (K.) huberti*, annelid jaws; 82-34CB, about 250 m NW of the beach parallel to and 100 m NE of the stone fence, 0.1 m below ground surface, *Kettnerites (K.) bankvaetensis, K. (K.) huberti*, annelid jaws; 83-15CB, ca 50 m E of the small ruin adjacent to the stone fence, the second erratic boulder from the ruin, about 1 m in diameter, sample from the stromatoporoid-rich bed, the topmost bed, *Kettnerites (K.) bankvaetensis*, annelid jaws; 83-16CB, calcarenite with calcareous algae, 0.15–0.20 m below surface, *Kettnerites (K.) bankvaetensis*, annelid jaws; 83-17CB, ca 300 m N of the ruin and 75 m E of the stone fence, 0.25 m below soil surface, calcarenitic nodules in soft calcilutite, *Lanceolatites* sp. A, *Kettnerites (K.) bankvaetensis, K. (K.) polonensis*, annelid jaws.

BARA 1. Slite Beds, Halla Beds, oolite. *Age:* Wenlock.
References: Laufeld 1974a, b*; Larsson 1979; Jeppsson 1982, 1983; Ramsköld 1983; Frykman 1989.
Samples/jawed annelids: Slite Beds: 70-12LJ, on the slope above the quarry, no annelid jaws; Halla Beds: 67-24LJ, in the quarry, no annelid jaws; 78-4CB, 0.75 m above the bottom in the section, no annelid jaws; 70-13LJ, on the slope near the telephone pole above the quarry, no annelid jaws.

BARKARVEÅRD 1. Hamra Beds, unit c. *Age:* Late Ludlow.
References: Laufeld 1974a, b*; Larsson 1979.
Samples/jawed annelids: 76-14CB, shore exposure, *Kettnerites (K.) huberti, K. (K.) polonensis*, annelid jaws.

BARSHAGEUDD 2. Hamra Beds, unit c. *Age:* Late Ludlow.
References: Laufeld 1974a, b*.
Samples/jawed annelids: 76-26CB, shore exposure, paulinitid fragment, annelid jaws.

BÅTA 1. Slite Beds, unit f (*Rhipidium tenuistriatum* Beds). *Age:* Early Wenlock.
References: Laufeld 1974a, b*; Larsson 1979; Jeppsson 1983.
Samples/jawed annelids: 77-27CB, 0.15 m below the uppermost bed in the ditch, no annelid jaws.

BÅTELS 1. Klinteberg Beds, unit c. *Age:* Late Wenlock.
References: Laufeld 1974a, b*; Jeppsson 1983; Frykman 1989.
Samples/jawed annelids: 71-89LJ, the lowest exposed bed, no annelid jaws; 71-90LJ, 0.90 m above the lowest exposed bed, no annelid jaws; 71-91LJ, 2.10 m above the base of the section, top of the section, no annelid jaws.

BÅTELS 2. Klinteberg Beds, unit b. *Age:* Late Wenlock.
References: Laufeld 1974a, b*; Frykman 1989.
Samples/jawed annelids: 71-92LJ, top of the section, no annelid jaws.

BINGERS KVARN 2. Tofta Beds. *Age:* Early Wenlock.
References: Laufeld 1974a, b*.
Samples/jawed annelids: 75-27CB, uppermost bed of the road exposure, no annelid jaws.

BJÄRGES 1. Mulde Beds, upper part. *Age:* Late Wenlock.
References: Laufeld 1974a, b*; Larsson 1979; Frykman 1989.
Samples/jawed annelids: 71-9LJ, excavated material, *Kettnerites (A.) microdentatus*, paulinitid sp., annelid jaws; 75-68CB, about 0.5 m below ground level, *Kettnerites (A.) sisyphi*, annelid jaws.

BJÄRGES 2. 635275 164223 (CJ 2911 5214), ca 2725 m WSW of Eksta church. Topographical map sheet 6 I Visby SO. Geological map sheet Aa 164 Hemse.
Ditch exposure north of and along the private road, where a pipe underlies the road, about 350 m SW of Ajvide and about 175 m W of the small houses to the left of the H in Hägur.
Mulde Beds, upper part. *Age:* Late Wenlock.
Samples/jawed annelids: 79-11LJ, bottom of the ditch, *Kettnerites (A.) sisyphi*, annelid jaws.

BLÅHÄLL 1. Mulde Beds, lower part. *Age:* Late Wenlock.
References: Laufeld 1974a, b*; Claesson 1979; Larsson 1979; Poulsen *et al.* 1982; Jeppsson 1983; Ramsköld 1984, 1985; Frykman 1989.
Reference level: About 2 m above water level, there is a protruding bed that can be followed for at least 50–100 m. The lower surface of this bed is designated as reference level.
Samples/jawed annelids: 71-140LJ, 1.50 m below reference level, *Kettnerites (K.) martinssonii* var. mulde, *K. (A.) sisyphi*, annelid jaws; 75-65CB, 0.50 m below the top of the section, *Kettnerites (K.) martinssonii* var. mulde, *K. (A.) sisyphi*, annelid jaws; 75-66CB, 1.2 m below the top of the section, *Kettnerites (K.) martinssonii* var. mulde, *K. (A.) sisyphi*, annelid jaws.

BODUDD 1. Hemse Beds, Hemse Marl uppermost part and Eke Beds, basal part. *Age:* Ludlow, Late Leintwardinian.
References: Laufeld 1974a, b*, c; Larsson 1979; Jeppsson 1982, 1983*; Cherns 1983*; Fredholm 1988.
Samples/jawed annelids: Hemse Beds, uppermost part: 75-100CB, just above mean water level, *Kettnerites (A.) sisyphi klasaardensis*, annelid jaws; 81-30LJ, in the southeasternmost part of the exposure, at mean water level, in the water, the lowest bed about 5 m NW of the sea-wall, annelid jaws, 81-31LJ, about 0.3 m higher than 81-30, 2 m NW of the sea-wall, fragments; 81-32LJ, topmost Hemse Beds, *Kettnerites (A.) sisyphi klasaardensis*, annelid jaws; lowest Eke Beds: 81-33LJ, about 100 m NW of the tail of erratic boulders in the sea and 0.3 m below the uppermost exposed bed, *Kettnerites (K.) martinssonii, K. (A.) sisyphi klasaardensis, K. (K.) huberti, K. (A.)* cf. *microdentatus*, annelid jaws; 81-34LJ, the uppermost exposed bed, *Lanceolatites gracilis* var. visby, *Kettnerites (A.) sisyphi klasaardensis*, annelid jaws.

BODUDD 2. Eke Beds, lowest part. *Age:* Ludlow, Late Leintwardinian.
References: Laufeld 1974a, b*; Larsson 1979.
Samples/jawed annelids: 75-101CB, from the shoreline, *Lanceolatites gracilis, Kettnerites (K.) huberti*, annelid jaws.

BODUDD 3. Hemse Beds, Hemse Marl uppermost part. *Age:* Ludlow, Late Leintwardinian.
References: Hede 1919:21, Loc. 9; Hede 1942:20, Loc 1C; Jeppsson 1983*.
Samples/jawed annelids: 71-151LJ, shallow shore exposure available at low water, *Lanceolatites gracilis, Kettnerites (K.) martinssonii, K. (A.) sisyphi klasaardensis, K. (K.) huberti, K. (K.)* cf. *burgensis*, annelid jaws.

BOFRIDE 1. Klinteberg Beds, upper part. *Age:* Late Wenlock to Early Ludlow.
References: Laufeld 1974a, b*; Larsson 1979; Frykman 1989.
Samples/jawed annelids: 70-32LJ, 0.0–0.1 m above the lower marly bed, *Kettnerites (K.)* cf. *bankvaetensis*, annelid jaws; 75-70CB, 1.25 m below reference level, *Kettnerites (K.)* cf. *martins-*

sonii, K. (A.) sisyphi cf. var. valle, annelid jaws; 75-71CB, 2.0 m below reference level, *Kettnerites (K.)* cf. *bankvaetensis*, annelid jaws; 75-72CB, 1.0 m above reference level, no annelid jaws.

BOFRIDE 2. Klinteberg Beds, upper part. *Age:* Early Ludlow.
References: Laufeld 1974a, b*; Frykman 1989.
Samples/jawed annelids: 75-69CB, 1.5 m from top of the section, no annelid jaws.

BOTTARVE 1. Hamra Beds, unit b. *Age:* Ludlow, Whitcliffian.
References: Laufeld 1974a, b*; Laufeld & Jeppsson 1976.
Samples/jawed annelids: 77-32LJ, shallow ditch, *Kettnerites (K.) bankvaetensis*, annelid jaws; 83-21CB, shallow ditch, paulinitid fragments, annelid jaws.

BOTTARVE 2. Hamra Beds, unit b. *Age:* Ludlow, Whitcliffian.
References: Laufeld 1974a, b*; Laufeld & Jeppsson 1976.
Samples/jawed annelids: 76-21CB, excavated material, *Kettnerites (K.)* cf. *bankvaetensis*, paulinitid fragments, annelid jaws; 76-22CB, excavated material, *Kettnerites (K.)* cf. *bankvaetensis*, annelid jaws; 83-19CB, ca. 20 m W of the sharp bend of the ditch which continues in an E–W direction, bottom bed, *Kettnerites (K.) martinssonii, K. (A.) microdentatus*, annelid jaws; 83-20CB, immediately above 83-19CB, paulinitid fragments.

BOTVALDE 1. Klinteberg Beds, unit e. *Age:* Wenlock, Late Homerian.
Reference: Jeppsson 1983*; Frykman 1989.
Samples/jawed annelids: 75-37LJ, small quarry, annelid jaws.

BOTVIDE 1. Hemse Marl uppermost part and Eke Beds basal part. *Age:* Ludlow, Late Leintwardinian.
References: Laufeld 1974a, b*; Jeppsson 1974, p. 10, 1982, 1983; Larsson 1979; Laufeld & Martinsson 1981; Cherns 1982*, 1983*.
Samples/jawed annelids: Hemse Beds: 69-18LJ, 0.60–0.70 m below the conglomerate level, *Kettnerites (K.)* cf. *polonensis, K. (K.)* cf. *burgensis*; 69-19LJ, 0.05–0.10 m below the conglomerate level, paulinitid sp. IMI, annelid jaws; 69-20LJ, 0.0–0.05 below the conglomerate level, *Kettnerites (K.) huberti*, annelid jaws; Eke Beds: 69-22LJ, 0.0–0.05 above the conglomerate level, *Kettnerites (K.) huberti, K. (K.) burgensis*; 69-23LJ, 0.10–0.15 m above the conglomerate level, no annelid jaws; 69-24LJ, 0.25 above the conglomerate level, no annelid jaws; 69-25LJ, on the northern slope 0.5 m below top of the isolated sea stack, no annelid jaws.

BRINGES 1. Sundre Beds, lower part. *Age:* Latest Ludlow.
References: Laufeld 1974a, b*.
Samples/jawed annelids: 71-186LJ, no annelid jaws.

BRINGES 3. 632195 165562 (CJ 4006 2043), ca 4020 m NE of Hamra church. Topographical map sheet 5 I Hoburgen SO & 5 J Hemse SV. Geological map sheet Aa 152 Burgsvik.
Abandoned, shallow quarry, 400 m SSE of the manor house at Bringes.
Sundre Beds, lower part. *Age:* Latest Ludlow.
Samples/jawed annelids: 71-187a LJ, *Kettnerites (K.)* cf. *bankvaetensis*, annelid jaws.

BROA 2. Slite Beds, Ryssnäs Limestone. *Age:* Wenlock, middle part.
References: Laufeld 1974a, b*.
Samples/jawed annelids: 78-16CB, sample from the reference level, *Kettnerites (K.) jacobi, K. (A.) sisyphi*, annelid jaws.

BROTRÄSKKRÖKEN 1. Hemse Beds, middle upper part. *Age:* Ludlow.
References: Laufeld 1974a, b*; Larsson 1979.
Samples/jawed annelids: 71-44LJ, lowest exposed bed, no annelid jaws.

BURGEN 1. Burgsvik Beds, upper part. *Age:* Late Ludlow, possibly Whitcliffian.
References: Laufeld 1974a, b*; Larsson 1979.
Samples/jawed annelid jaws: 82-21CB, about 0.85 m below the boundary to the crinoid-rich beds, conglomeratic balls, no annelid jaws; 82-22CB, from the same conglomeratic level as the former sample but also including the oolitic limestone which embeds the balls, ca 0.85 m below the bed rich in crinoids, paulinitid fragments; 82-23CB, immediately below the crinoid-rich bed about 2.35 above the bottom of the quarry, no annelid jaws; 82-24CB, probably Hamra Beds, the lowest crinoid-rich layer about 2.40 m above the quarry floor, no annelid jaws.

BURGEN 5. Eke Beds, lower part. *Age:* Ludlow, possibly Late Leintwardinian.
References: Laufeld 1974a, b*; Larsson 1979.
Samples/jawed annelids: 77-32CB, ditch section, brownish blue-grey calcibiorudite, no annelid jaws.

BURGEN 9. Burgsvik Beds, Burgsvik Oolite. *Age:* Late Ludlow.
References: Larsson 1979*.
Samples/jawed annelids: 71-132LJ, northern wall, low in the section, no annelid jaws; 71-133LJ, 3–4 m above 71-132, upper part of the section, no annelid jaws.

BUSKE 1. Lower Visby Beds, Upper Visby Beds. *Age:* Early Wenlock.
References: Laufeld 1974a, b*; Larsson 1979; Odin *et al.* 1984*.
Auxiliary reference level: The boundary between the Lower and Upper Visby Beds has been reported at 0.75 m above the base of the section (Hede 1940:13, Martinsson 1962:47, Laufeld 1974b), but according to Lennart Jeppsson (personal communication, 1986) and Odin *et al.* (1984) it is about 2.8 m above the base. Martinsson's level cannot be identified with the necessary accuracy, thus it has not been used as reference level by Lennart Jeppsson. Instead, the sampled levels have been measured from the best visible bentonite, which is about 1.6 m above the base of the section and about 1.2 m below the level abundant in large solitary rugose corals.
Samples/jawed annelids: Lower Visby Beds, unit e: 79-40LJ, 1.3–1.1 m below the auxiliary reference level, *Kettnerites (K.) versabilis* form A, B, and C, *Lanceolatites gracilis, Kettnerites (K.) martinssonii, K. (A.)* cf. *microdentatus, K. (K.) abraham abraham*, annelid jaws; 79-41LJ, 0.2–0.25 m above the auxiliary reference level, *Kettnerites (K.) versabilis* form A and B, *K. (K.) martinssonii, K. (K.) abraham abraham*, annelid jaws.

DACKER 1. Slite Beds, unit e. *Age:* Wenlock.
References: Laufeld 1974b*; Claesson 1979.
Samples/jawed annelids: 67-14LJ, top of the section, no annelid jaws.

DÄPPS 1. Mulde Beds, upper part. *Age:* Late Wenlock.
References: Laufeld 1974a, b*; Larsson 1979; Frykman 1989.
Samples/jawed annelids: 81-56LJ, about 15 m SE of the road, 1 m section exposed, sample from the 0.15 m thick bed, *Kettnerites (K.) bankvaetensis*, other paulinitid taxa, annelid jaws; 82-45CB, 1.48–1.55 m below the surface of the SW corner of a concrete ramp, built for the former railway, *Kettnerites (K.) martinssonii* var. mulde *K. (A.) sisyphi sisyphi*, annelid jaws.

DÄPPS 2. Mulde Beds, upper part. *Age:* Late Wenlock.
References: Laufeld 1974a, b*; Larsson 1979.
Samples/jawed annelids: 75-25CB, 0.5 m below the top of the section, *Kettnerites (K.) martinssonii, K. (A.) sisyphi sisyphi*, annelid jaws; 82-37CB, top of the section, 20–30 mm thick bed, *Hindenites gladiatus, Kettnerites (K.) martinssonii, K. (A.) sisyphi sisyphi*, annelid jaws; 82-38CB, 0.60 m below the top of the section and 1.0 m above railway embankment, about 50 mm thick bed composed of bluish-grey, dense limestone, *Kettnerites (K.) martinssonii, K. (A.) sisyphi sisyphi*, annelid jaws; 82-39CB, 30 m north and about 1.70 above the railway embankment, *Kettnerites (K.) martinssonii, K. (A.) sisyphi sisyphi*, annelid jaws, 82-40CB, 0.98 m below the lower surface of 82-39CB, *Kettnerites (K.) martinssonii, K. (A.) sisyphi sisyphi*, annelid jaws.

DIGRANS 1. Hamra Beds, lower part. *Age:* Ludlow, Whitcliffian.
References: Laufeld 1974a, b*.
Samples/jawed annelids: 75-15LJ, ditch section, paulinitid jaw, annelid jaws.

DJAUPVIKSUDDEN 4. 637500 167953 (CJ 6796 7149), ca 6040 m NW of Östergarn church. Topographical map sheet 6 J Roma SO. Geological map sheet Aa 170 Katthammarsvik.
About 100 m NE of Djaupviksudden 1 (described in the geological map description).
Hemse Beds, unit b and c. *Age:* Ludlow.
Samples/jawed annelids: 71-155LJ, 0–10 mm, immediately above the corroded surface, annelid jaws; 71-157LJ, from the lower part of the three levels in the section, described in Hede's 1960 guide, paulinitid fragments, annelid jaws; 71-159LJ, from the uppermost level, *Kettnerites (K.) huberti, K. (K.) polonensis*, annelid jaws.

DJUPVIK 1. Mulde Beds, lower part. *Age:* Late Wenlock.
References: Laufeld 1974a, b*; Larsson 1979, Jeppsson 1982, 1983; Odin et al. 1984, 1986.
Samples/jawed annelids: 75-67CB, middle part of the section, *Kettnerites (K.) martinssonii* var. mulde, *K. (A.) sisyphi sisyphi*, annelid jaws; 84-21LJ, the calcarenitic bed between the bentonite bed selected as reference level by Jeppsson (1983:140) and the bentonite 0.2 m above it, *Kettnerites (K.) martinssonii, K. (A.) sisyphi sisyphi*, annelid jaws.

DJUPVIK 2. Mulde Beds, lower part. *Age:* Late Wenlock.
Reference: Jeppsson 1983*; Odin et al. 1984*.
Reference level: 1.05 m above the distinct marl horizon (erroneously given as 1.50 m in Jeppsson 1983; L. Jeppsson, personal communication, 1986).
Samples/jawed annelids: 70-25LJ, 0.0–0.2 m above the reference level, *Kettnerites (A.) sisyphi sisyphi*, annelid jaws; 70-26LJ, 1.05–0.95 m below the reference level, *Kettnerites (K.) martinssonii, K. (A.) sisyphi sisyphi*, annelid jaws; 70-27LJ, 1.07–1.20 m above the reference level, *Kettnerites* cf. *(K.) martinssonii, K. (A.) sisyphi sisyphi*, annelid jaws; 70-28LJ, 2.24 m above the reference level, top of the section, *Kettnerites (K.)* cf. *martinssonii, K. (A.) sisyphi sisyphi*, annelid jaws; 70-29LJ, 1.95–2.05 m below the reference level, *Kettnerites (K.) martinssonii, K. (A.)* cf. *sisyphi*, annelid jaws.

DJUPVIK 3. 635577 164096 (CJ 2804 5525), ca 4275 m SW of Fröjel church. Topographical map sheet 6 I Visby SO. Geological map sheet Aa 164 Hemse.
From the solitary house west of the road to the fault, about 50 m SW of the house. Djupvik 1 starts beyond the fault. A section of about 7.2 m tilted strata is exposed.
Mulde Beds. Age: Late Wenlock.
Samples/jawed annelids: 84-18LJ, the lowest exposed bed at low water, *Kettnerites (K.) martinssonii* mulde type, *K. (A.) sisyphi sisyphi*, annelid jaws; 84-19LJ, the uppermost exposed bed, *Kettnerites (K.) martinssonii* mulde type, annelid jaws.

DJUPVIK 4. 635576 164106 (CJ 2813 5523), ca 4425 WNW of Eksta church. Topographical map sheet 6 I Visby SO. Geological map sheet Aa 164 Hemse.
Rivulet section in the gully just NE of the solitary house west of the road and up stream some ten metres beyond the road.
Mulde Beds. Age: Late Wenlock.
Samples/jawed annelids: 84-20LJ, uppermost bed in the section, *Kettnerites (K.) martinssonii* mulde type, annelid jaws.

DRAKARVE 1. Hemse Beds, Hemse Marl, uppermost part. *Age:* Ludlow, probably Leintwardinian.
References: Laufeld 1974a, b*; Larsson 1979.
Samples/jawed annelids: 75-99CB, about 0.5 m below the soil surface and 0.7 m above the base of the normally water filled hollow, no annelid jaws.

ENHOLMEN 2. 640083 168001 (CJ 7037 9722), ca 2100 m SE of Slite church. Topographical map sheet 7 J Fårösund SO & NO. Geological map sheet Aa 169 Slite.
Shore exposure on the NE side of the Enholmen island.
Slite Beds, unit g. *Age:* Wenlock.
References: Hede 1928, p. 41, lines 24–26; Martinsson 1962, p. 52 (references to Enholmen island in general).
Samples/jawed annelids: 75-9CB, shore exposure, no annelid jaws.

FÅGELHAMMAR 3. 636173 167652 (CJ 6392 5847), ca 7550 m SW of Gammelgarn church. Topographical map sheet 6 J Roma SO. Geological map sheet Aa 170 Katthammarsvik.
Area with loose slabs, SW of the sea stack field, SSW of Fågelhammar 1, about 50 m SW of the triangulation point at Folhammar.
Hemse Beds, upper part. *Age:* Ludlow.
References: Hede 1929, p. 53, line 17, to p. 54, line 4.
Samples/jawed annelids: 71-218LJ, loose conglomeratic flat slabs, some 10 m SW of the southernmost sea stack on the beach, *Lanceolatites gracilis, Kettnerites (K.) polonensis*, annelid jaws.

FAKLE 1. Hemse Beds, unit c. *Age:* Ludlow, probably Bringewoodian.
References: Laufeld 1974a, b*; Larsson 1979; Fredholm 1988.
Samples/jawed annelids: 71-162LJ, the lowest brownish bed, *Kettnerites (K.) polonensis*, paulinitid fragments, annelid jaws.

FALUDDEN 1. Sundre Beds, lower part. *Age:* Latest Ludlow.
References: Laufeld 1974a, b*; Larsson 1979.
Samples/jawed annelids: 76-15CB, uppermost bed between stromatoporoids, no annelid jaws.

FALUDDEN 2. Hamra Beds and Sundre Beds. *Age:* Latest Ludlow.
References: Laufeld 1974a, b*.
Samples/jawed annelids: Hamra Beds, unit c: 76-16CB, uppermost bed, *Kettnerites (K.) bankvaetensis, K. (K.) huberti, K. (K.) polonensis*, annelid jaws.

FARDUME 1. 641275 168555 (CK 7681 0861), ca 3650 m S of Rute church. Topographical map sheet 7 J Fårösund SO & NO. Geological map sheet Aa 171 Kappelshamn.
Quarry along the road about 100–150 m NNW of Fardume ruin.
Slite Beds, unit g. *Age:* Wenlock.
Reference: Hede 1933:44.
Samples/jawed annelids: 84-17LJ, 1.5 m above quarry floor, thin limestone slabs in a deeply eroded lutitic bed, *Gotlandites slitensis, Kettnerites (K.) martinssonii, K. (K.) polonensis* var. gandarve, *K. (A.)* cf. *sisyphi, Hindenites gladiatus*, annelid jaws.

FÅRÖ SKOLA 1. Slite Beds, Slite Marl. *Age:* Wenlock.
References: Laufeld 1974b*; Larsson 1979.
Samples/jawed annelids: 78-15CB, uppermost bed, *Kettnerites (A.) sisyphi sisyphi*, annelid jaws.

FIE 3. Hemse Beds, Hemse Marl SE part. *Age:* Ludlow, Leintwardinian(?).
Reference: Jeppsson 1983; Fredholm 1988.
Samples/jawed annelids: 71-128LJ, bottom of the ditch, *Kettnerites (K.) bankvaetensis, K. (K.) martinssonii, K. (K.) huberti, K. (K.)* cf. *burgensis*, annelid jaws.

FJÄLE 2. Klinteberg Beds, unit d. *Age:* Latest Wenlock – earliest Ludlow.
References: Laufeld 1974a, b*; Frykman 1989.
Samples/jawed annelids: 77-6CB, shallow ditch, 0.25 m below ground level, *Kettnerites (K.) bankvaetensis*, annelid jaws.

FJÄLE 3. Klinteberg Beds, unit e. *Age:* Latest Wenlock – earliest Ludlow.
References: Laufeld 1974a, b*; Frykman 1989.
Samples/jawed annelids: 73-31LJ, from the 0.3 m exposed, fragments; 77-7CB, 0.50 m below ground level, *Lanceolatites gracilis, L. gracilis* var. visby, *Kettnerites (K.) bankvaetensis, K. (K.) martinssonii, K. (A.) fjaelensis, Hindenites* sp., annelid jaws.

FJÄRDINGE 1. Klinteberg Beds, unit b. *Age:* Latest Wenlock – earliest Ludlow.
References: Laufeld 1974a, b*; Larsson 1979; Frykman 1989.
Samples/jawed annelids: 75-33LJ, top of accessible section, no annelid jaws; 75-34LJ, lowest accessible bed about 1 m below 75-33, *Kettnerites (K.)* cf. *bankvaetensis, K. (K.)* cf. *polonensis*, annelid jaws; 77-4CB, uppermost bed of the section, no annelid jaws; 77-5CB, loose slabs probably excavated from the ditch, *Kettnerites (K.) bankvaetensis, K. (K.) martinssonii*, annelid jaws.

FOLLINGBO 2. Slite Beds, Slite Marl, NW part. *Age:* Wenlock.
References: Laufeld 1974a, b*; Larsson 1979.
Samples/jawed annelids: 75-10CB, sample from the bottom of the ditch, *Kettnerites (K.) martinssonii, K. (A.) microdentatus*, annelid jaws.

FOLLINGBO 3. Slite Beds, Slite Marl, NW part. *Age:* Wenlock.
References: Laufeld 1974a, b*; Larsson 1979.
Samples/jawed annelids: 71-68LJ, lowest accessible bed, *Kettnerites (A.) sisyphi*, fragments.

FOLLINGBO 4. Slite Beds, unit g. *Age:* Wenlock.
References: Laufeld 1974a, b*; Claesson 1979; Larsson 1979.
Samples/jawed annelids: 71-71LJ, 0.5 m below the road surface, fragments.

FOLLINGBO 12. 638848 165401 (CJ 4353 8683), ca 1430 N of Follingbo church. Topographical map sheet 6 J Visby NV & NO. Geological Map sheet Aa 183 Visby & Lummelunda.
Temporary excavation about 50 m west of the almost N–S directed road (the road between the main roads Visby–Endre and Visby–Roma) and some 10 m left of the small road to the sanatorium, at the first right turn on the small road.
Slite Beds, Slite Marl, northwesternmost part. *Age:* Wenlock.
Samples/jawed annelids: 80-1CB, loose slab from excavation, *Kettnerites (K.) martinssonii, Hindenites angustus*, annelid jaws.

FRÖJEL 2. Mulde Beds, uppermost part. *Age:* Late Wenlock.
References: Laufeld 1974a, b*; Frykman 1989.
Samples/jawed annelids: 70-30LJ, shallow ditch, paulinitid fragments, annelid jaws.

GALGBERGET 1. Högklint, unit c, and Tofta Beds. *Age:* Early Wenlock.
References: Laufeld l974a, b*.
Samples/jawed annelids: 67-4LJ, lowermost Tofta bed, fragments; 67-5LJ, uppermost Högklint, unit c, no annelid jaws.

GAMLA HAMN 1. Högklint Beds, unit b. *Age:* Early Wenlock.
Reference: Laufeld l974b*.
Samples/jawed annelids: 77-23CB 0.5 m below the uppermost part of the stratified bed between the sea stacks, *Kettnerites (K.) martinssonii*, annelid jaws.

GAMLA HAMN 2. Högklint Beds, unit c. *Age:* Early Wenlock.
Reference: Laufeld 1974b*.
Samples/jawed annelids: Högklint Beds, unit c: 67-18LJ, annelid jaws; Slite Beds, unit c: 67-19LJ, about 2 m above 67-18, no annelid jaws.

GANDARVE 1. Halla Beds. *Age:* Late Wenlock.
References: Laufeld 1974a, b*; Larsson 1979.
Samples/jawed annelids: 71-81LJ, *Hindenites gladiatus, Kettnerites (K.) martinssonii, K. (K.) polonensis* var. gandarve, *K. (A.) sisyphi sisyphi, K. (A.) sisyphi* var. valle, annelid jaws; 75-13CB, base of the section, no annelid jaws.

GANDARVE 2. Halla Beds. *Age:* Late Wenlock.
References: Laufeld 1974a, b*.
Samples/jawed annelids: 75-12CB, uppermost prominent calciruditic bed, fragments; 84-113CB, third bed from the top in the section, *Kettnerites (K.) martinssonii, K. (A.) sisyphi sisyphi*, annelid jaws.

GANE 2. Slite Beds, Slite Marl. *Age:* Wenlock.
References: Hede 1928*, 1942 locality 21; Larsson 1979*; Ramsköld 1983.
Samples/jawed annelids: 83-29LJ, 0.5 m exposed, sample from the middle, eroded part, 200 m SW of the private road along the former railway line, *Kettnerites (K.) bankvaetensis*, annelid jaws.

GANNES 2. Hemse Beds, Hemse Marl, unit c or d(?). *Age:* Ludlow, Bringewoodian or Early Leintwardinian.
References: Laufeld 1974a, b*.
Samples/jawed annelids: 71-96LJ, about 3 m below the floor of the quarry, *Kettnerites (K.)* cf. *polonensis*, annelid jaws.

GANNES 3. Hemse Beds, unit d. *Age:* Ludlow.
Reference: Laufeld 1974b*; Sundquist 1982c*; Fredholm 1988.
Sample/jawed annelid: 82-13CB, northernmost part of the bed with large wave marks, no annelid jaws.

GANNOR 1. Hemse Beds, Hemse Marl uppermost part and Eke Beds, basal part. *Age:* Ludlow, Late Leintwardinian.
References: Laufeld 1974a, b*; Eisenack 1975, Fig, 30; Laufeld & Jeppsson 1976; Larsson 1979; Claesson 1979; Cherns 1983*; Jeppsson 1983; Ramsköld 1984, 1985b; Fredholm 1988.
Samples/jawed annelids: Hemse Beds: 71-26a LJ, loose slab rich in *Dayia navicula* (*Dayia* flags), representing the uppermost bed in the Hemse Beds, no annelid jaws; 71-26b LJ, loose slab, (*Dayia* flags), annelid jaws; 71-121LJ, about 25 m NW of the reference point, 0.0–0.02 m below the Hemse–Eke boundary, *Kettnerites (K.) martinssonii, K. (K.) huberti*, annelid jaws; 75-87CB, 0.30 m below the boundary, no annelid jaws; Eke Beds: 71-122LJ, lowest bed of Eke Beds, no annelid jaws; 75-88CB, the bed immediately above the boundary, no annelid jaws; 71-123LJ, calcarenite poor in fossils, from the uppermost part of layer D^3 in Munthe (1924, Fig 2) and Hede (1925, p. 36 and 38–39. lines 33–38 and 1–18, respectively), *Lanceolatites gracilis, Kettnerites (K.) bankvaetensis, K. (K.) huberti, K. (K.) polonensis, K. (A.) microdentatus*, annelid jaws; 71-124LJ, calcarenitic bed in the calcilutitic layer E in Munthe and Hede op. cit., *Kettnerites (K.) martinssonii, K. (K.) huberti, K. (K.) polonensis, K. (A.) microdentatus*, annelid jaws.

GANNOR 3. Hemse Beds, Hemse Marl SE part. *Age:* Ludlow, Leintwardinian.
References: Laufeld 1974a, b*; Jeppsson 1983; Fredholm 1988.
Samples/jawed annelids: 71-125LJ, 90 to 100 m SE of the reference point, 0.2 to 0.3 m above the bottom of the ditch, *Lanceolatites gracilis, Kettnerites (K.) martinssonii, K. (K.) polonensis, K. (K.) huberti, K. (A.)* cf. *sisyphi*, annelid jaws; 71-126LJ, uppermost part of the lutitic calcarenite, 0.3 m above the bottom of the small ditch which empties into the main drainage ditch, *Kettnerites (K.) huberti, K. (K.) polonensis, K. (A.) sisyphi sisyphi, K. (A.) microdentatus*; annelid jaws.

GARDE 1. Hemse Beds, unit e. *Age:* Ludlow, Leintwardinian.
Reference: Jeppsson 1983*; Fredholm 1988.
Samples/jawed annelids: 82-34LJ, 1.64–1.70 m below the reference level, no annelid jaws; 82-35LJ, 0.33–0.18 m below the reference level, no annelid jaws.

GARDRUNGS 1. Slite Beds, unit f – *Rhipidium tenuistriatum* Beds.
Age: Early Wenlock.
References: Laufeld 1974a, b*; Larsson 1979.
Samples/jawed annelids: 73-19LJ, the deepest eroded bed in the section, fragments; 73-20LJ, 1.40 m above 73-19, annelid jaws.

GARDSBY 1. Hemse Beds, Hemse Marl NW part. *Age:* Ludlow, Bringewoodian or Leintwardinian.
Reference: Jeppsson 1983*; Fredholm 1988.
Samples/jawed annelids: 82-27LJ, very shallow ditch, *Lanceolatites gracilis, Kettnerites (A.) sisyphi sisyphi*, annelid jaws.

GARNUDDEN 1. Hemse Beds unit a. *Age:* Ludlow, probably Eltonian.
References: Laufeld 1974a, b*; Frykman 1989.
Samples/jawed annelids: 75-47CB, 0.75 m above mean water level and 40 m west of 75-46CB, *Kettnerites (K.)* cf. *polonensis*, paulinitid fragments, annelid jaws.

GARNUDDEN 3. 637575 167865 (CJ 6712 7227) ca 4710 m ESE of Anga church. Topographical map sheet 6 J Roma NV & NO. Geological map sheet Aa 170 Katthammarsvik.
Low cliff in a small bay at the northern sea shore, ca. 440 m SW of the triangulation point at Garnudden.
Hemse Beds, unit a. *Age:* Early Ludlow, possibly Eltonian.
Samples/jawed annelids: 82-25CB, the second to lowest bed accessible at mean water level, *Kettnerites (K.) bankvaetensis, K. (K.) polonensis*, annelid jaws; 82-26aCB, 0.29–0.32 m above 82-25CB, *Kettnerites (K.) huberti, K. (K.) polonensis*, annelid jaws; above former sample, 82-26b CB, ca. 40 mm thick bed with smooth eroded hardground surface containing some corals, *Kettnerites (K.)* cf. *martinssonii, K. (K.) huberti, K. (K.) polonensis*, annelid jaws.

GARNUDDEN 4. 637596 167937 (CJ6788 7246), ca 5400 m E of Anga church. Topographical map sheet 6 J Roma NV & NO. Geological map sheet Aa 170 Katthammarsvik.
Shore exposure at the southernmost tip, 125 m NE of the end of the path marked on the topographical map and ca 150 m SE of the triangulation point at Garnudden.
Hemse Beds, unit a. *Age:* Early Ludlow, possibly Eltonian.
Samples/jawed annelids: 78-13CB, bed rich in *Sphaerorhynchia, Kettnerites (K.) bankvaetensis, K. (K.) martinssonii, K. (K.) huberti, K. (K.) polonensis*, annelid jaws; 78-14CB, bed below 78-13CB, annelid jaws.

GERETE 1. Hemse Beds, Hemse Marl NW part. *Age:* Ludlow, Bringewoodian or Leintwardinian.
References: Laufeld 1974a, b*; Franzén 1977; Larsson 1979; Ramsköld 1983; Jeppsson 1983; Fredholm 1988.
Samples/jawed annelids: 75-80CB, 20 m east of the bridge and about 1 m above the bottom of the drainage ditch from the lowest of the exposed (1975) beds *Lanceolatites gracilis, Kettnerites (A.)* cf. *microdentatus*, annelid jaws. 75-81CB, 0.50 m above 75-80CB, *Lanceolatites gracilis, Kettnerites (K.) martinssonii, K. (A.)* cf. *sisyphi*, annelid jaws.

GERUMSKANALEN 1. Hemse Beds, Hemse Marl NW part. *Age:* Ludlow probably Late Bringewoodian, possibly earliest Early Leintwardinian.
References: Laufeld 1974a, b*; Laufeld & Jeppsson 1976; Jeppsson 1982, 1983*; Fredholm 1988.
Samples/jawed annelids: 71-39LJ, bottom of the ditch, *Kettnerites (K.) martinssonii*. 75-78CB, 2 m W of the bridge, bottom of the ditch, *Lanceolatites gracilis, Kettnerites (K.) martinssonii, K. (K.)* cf. *burgensis, K. (A.) sisyphi* cf. var. valle, annelid jaws.

GISLE 1. Hamra Beds, unit a. *Age:* Ludlow, (Whitcliffian).
References: Laufeld 1974a, b*; Larsson 1979; Ramsköld 1984.
Samples/jawed annelids: 76-18CB, uppermost bed, surface exposure, *Kettnerites (K.) bankvaetensis*, annelid jaws.

GISSLAUSE 1. Slite Beds, unit g. *Age:* Wenlock.
References: Laufeld 1974a, b*; Larsson 1979.
Samples/jawed annelids: 77-13CB, about 1.0 m below road surface of the bridge, fragments.

GLASSKÄR 1. Burgsvik Beds, lower part. *Age:* Ludlow, Whitcliffian.
References: Laufeld 1974a, B*; Larsson 1979; Jeppsson 1983.
Samples/jawed annelids: 72-18LJ, the about 50 mm thick topmost limestone bed, *Kettnerites (K.) polonensis*, annelid jaws; 82-15CB, grey calcarenite, 0.14–0.20 m below ground level, *Kettnerites (K.)* cf. *martinssonii, K. (K.) polonensis, K. (A.) microdentatus, Hindenites naerensis*, annelid jaws; 82-16CB, 0.08–0.14 m below topmost limestone bed, soft calcilutite with calcarenitic nodules, *Kettnerites (K.) polonensis*, annelid jaws, 82-17CB, top bed 20–50 mm thick, *Kettnerites (K.)* cf. *martinssonii, K. (K.) polonensis* including two other varieties, viz. a finely and a coarsely denticulated wide type, *Hindenites naerensis*, annelid jaws; 83-10LJ, topmost bed, *Kettnerites (K.) martinssonii, K. (K.) polonensis* and the variety with fine denticles, *Hindenites naerensis*, annelid jaws.

GLASSKÄR 2. Burgsvik Beds, lower part. *Age:* Ludlow, Whitcliffian.
References: Laufeld 1974a, b*.
Samples/jawed annelids: 72-19LJ, the about 50 mm thick topmost limestone bed, *Kettnerites (K.) polonensis*, annelid jaws.

GLASSKÄR 3. Eke Beds and Burgsviks Beds. *Age:* Ludlow, Whitcliffian.
References: Laufeld 1974a, b*.
Samples/jawed annelids: Eke Beds: 72-20LJ, 0–50 mm above the Eke–Burgsvik boundary, *Kettnerites (K.) polonensis*, annelid jaws; 82-20CB, 0.10–0.07 m below ground level, *Kettnerites (K.) polonensis*, annelid jaws; 82-19CB, the about 30 mm thick soft calcilutite, separating the two competent beds 82-20 and 18, *Kettnerites (K.) polonensis*, annelid jaws; 82-18CB, the about 40 mm thick, uppermost bed, *Kettnerites (K.)* cf. *martinssonii, K. (K.) polonensis, K. (A.) microdentatus*, annelid jaws.

GLASSKÄR 4. 634905 167510 (CJ 6154 4596), ca. 11960 m EESE of Burs church. Topographical map sheet 5 I Hoburgen NO & 5 J Hemse NV. Geological map sheet Aa 156 Ronehamn.
Shore exposure on the easternmost point at Glasskär, ca. 2200 m NNE of the house adjacent to the light house at När. Stratified, crinoid-rich limestone with isolated patches of stromatoporoids.
Burgsvik Beds. Age: Late Ludlow.
Samples/jawed annelids: 83-13CB, the uppermost bed containing stromatoporoids, no annelid jaws; 83-14CB, the uppermost bed 10 m ENE of 83-13CB, coarse calcarenite, rich in crinoidal fragments, no annelid jaws.

GLÄVES 1. Hemse Beds, Hemse Marl SE part. *Age:* Ludlow, Leintwardinian.
References: Laufeld 1974b*; Jeppsson 1974:7, 1983; Larsson 1979; Fredholm 1988.
Samples/jawed annelids: 75-30LJ, low exposure in the ditch, *Kettnerites (K.) martinssonii, K. (K.)* cf. *polonensis*, annelid jaws.

GNISVÄRD 1. Upper Visby Beds. *Age:* Early Wenlock.
References: Laufeld 1974a, b*, Larsson 1979.
Samples/jawed annelids: 67-27LJ, loose slab from the excavation of the harbour, fragments; 75-92CB, loose slabs from the excavation of the harbour, *Lanceolatites gracilis, Kettnerites (K.) martinssonii, K. (K.) abraham abraham, K. (A.) sisyphi*, annelid jaws.

GNISVÄRD 2. 637763 163833 (CJ 2707 7724), ca 3950 m SW of Tofta church. Topographical map sheet 6 I Visby NO. Geological map sheet Aa 160 Klintehamn.
Shore exposure in the innermost part of the harbour, accessible at low water.
Upper Visby Beds. Age: Early Wenlock.
Sample/jawed annelids: 79-45LJ, shallow shore exposure, *Lanceolatites gracilis, Kettnerites (K.) martinssonii, K. (K.) abraham abraham*, annelid jaws.

GODRINGS 1. Halla and Klinteberg Beds, unit a. *Age:* Late Wenlock, probably Homerian.
References: Laufeld 1974a, b*; Frykman 1989.
Samples/jawed annelids: Halla Beds: 71-86LJ, *Kettnerites (K.) polonensis, K. (A.) sisyphi, Hindenites* sp., annelid jaws; Klinteberg Beds: 77-8CB, loose slab from the ditch excavation, no annelid jaws.

GODRINGS 2. Halla Beds, uppermost part. *Age:* Late Wenlock, probably Homerian.
References: Laufeld 1974a, b*; Larsson 1979; Frykman 1989.
Samples/jawed annelids: 71-85LJ, *Kettnerites (K.)* cf. *polonensis*, fragments.

GOGS 1. Hemse Beds, Hemse Marl SE part. *Age:* Ludlow, Leintwardinian.
References: Laufeld 1974a, b*; Jeppsson 1972, 1974, 1976, 1982*, 1983; Janvier 1978; Larsson 1979; Brood 1982; Fredholm 1988.
Samples/jawed annelids: 75-64CB, 0.20 m below the uppermost bed, *Lanceolatites gracilis, Kettnerites (K.) martinssonii, K. (K.) huberti, K. (K.) burgensis*, annelid jaw fragments; 71-23LJ, excavated material, no annelid jaws.

GOTHEMSHAMMAR 1. Halla Beds, unit c, and Klinteberg Beds, unit a. *Age:* Late Wenlock.
References: Laufeld 1974a, b*; Larsson 1979; Jeppsson 1983; Frykman 1989.
Sample/jawed annelids: 75-36CB, middle part of the section, the most protruding, competent, calcarenitic bed, about 0.5 m above the base of the vertical section, *Kettnerites (K.) martinssonii, K. (K.) polonensis, Hindenites* sp., annelid jaws.

GOTHEMSHAMMAR 2. Halla Beds, unit c, and Klinteberg Beds, unit a. *Age:* Late Wenlock.
References: Laufeld 1974a, b*; Claesson 1979; Larsson 1979; Jeppsson 1983; Frykman 1989.
Samples/jawed annelids: Halla Beds, unit c: 71-60LJ, the lowest exposed bed in the section, at water level, *Kettnerites (K.)* cf. *bankvaetensis, K. (A.) microdentatus*, annelid jaws; 71-61LJ, 1 m above 71-60, *Kettnerites (K.) bankvaetensis, K. (K.) martinssonii, K. (K.) polonensis*, annelid jaws; 75-35CB, 0.30 m below the Halla–Klinteberg boundary, *Kettnerites (K.) bankvaetensis*, annelid jaws. Klinteberg Beds, unit a: 75-34CB, 0.60 m above the Halla–Klinteberg boundary, *Kettnerites (K.) bankvaetensis*, annelid jaws.

GOTHEMSHAMMAR 3. Halla Beds, unit c, and Klinteberg Beds, unit a. *Age:* Late Wenlock.
References: Laufeld 1974b*; Claesson 1979; Jeppsson 1983; Frykman 1989.
Samples/jawed annelids: Halla Beds, unit c: 73-23LJ, 1.80–1.85 m below the reference level, *Kettnerites (K.) bankvaetensis, K. (K.)* cf. *polonensis*, annelid jaws; 73-24LJ, 0.80–0.85 m below the reference level, *Kettnerites (K.) bankvaetensis*, other unidentified paulinitid taxa, annelid jaws 73-25LJ, 0.00–0.04 m below the reference level, *Kettnerites (K.) bankvaetensis*, annelid jaws. Klinteberg Beds, unit a: 73-26LJ, about 10 mm thick and some 10–20 mm above, but on the uppermost parts of, the undulating, corroded reference level, *Kettnerites (K.) bankvaetensis, K. (K.)* cf. *martinssonii*, annelid jaws; 73-27LJ, 0.15–0.25 m above the reference level, representing the uppermost limestone bed in the lowest marly beds, *Kettnerites (K.) martinssonii, K. (A.) sisyphi sisyphi*, annelid jaws; 73-28LJ, 0.80–0.85 m above the reference level, *Kettnerites (K.) bankvaetensis, K. (K.) martinssonii, K. (A.) sisyphi sisyphi*, annelid jaws; 73-29LJ, 1.65 m above the reference level and about the topmost part of the upper marly beds (Hede 1928)

Kettnerites (*K.*) *martinssonii*, annelid jaws; 73-30LJ, 1.65 m above the reference level, *Kettnerites* (*K.*) *bankvaetensis*, *K.* (*K.*) *martinssonii*, annelid jaws.

GOTHEMSHAMMAR 7. Halla Beds, unit c. *Age:* Late Wenlock.
References: Laufeld 1974a, b*; Laufeld & Jeppsson 1976.
Samples/jawed annelids: 77-45PSSFG, *Kettnerites* (*K.*) *martinssonii*, *Kettnerites* (*K.*) *polonensis* var. gandarve, *K.* (*A.*) *sisyphi sisyphi*, annelid jaws.

GOTHEMSHAMMAR 8. Halla Beds, unit c. *Age:* Late Wenlock.
References: Laufeld 1974a, b*.
Samples/jawed annelids: 75-37CB, approximately 1.75 m below the Halla–Klinteberg boundary, *Kettnerites* (*K.*) *polonensis*, *K.* (*A.*) *sisyphi*, annelid jaws.

GROGARNSHUVUD 1. Hemse Beds, unit c. *Age:* Early Ludlow.
References: Laufeld 1974a, b*; Larsson 1979; Claesson 1979; Laufeld & Martinsson 1981; Jeppsson 1982, 1983; Sundquist 1982b; Fredholm 1988.
Samples/jawed annelids: 75-51CB, at water level about 0.3–0.4 m above the bed rich in oriented orthocones, *Kettnerites* (*K.*) *huberti*, *K.* (*K.*) *polonensis*, annelid jaws; The samples from 1981, 81-40LJ–81-43LJ, represent a sequence of about 1 m accessible at low water, with 40 as the lowest and 43 as the highest sample: 81-40LJ, from an anticlinal structure beyond the shore, *Kettnerites* (*K.*) *huberti*, *K.* (*K.*) *polonensis*, annelid jaws; 81-41LJ, *Lanceolatites gracilis*, *Kettnerites* (*K.*) cf. *martinssonii*, *K.* (*K.*) *huberti*, *K.* (*K.*) *polonensis*, annelid jaws; 81-42LJ, *Kettnerites* (*K.*) *huberti*, annelid jaws; 81-43LJ, *Lanceolatites gracilis*, *Kettnerites* (*K.*) *huberti*, *K.* (*K.*) *polonensis*, annelid jaws.

GRÖNDALEN 1. Klinteberg Beds, lower part. *Age:* Wenlock–Ludlow.
References: Laufeld 1974a, b*.
Samples/jawed annelids: 71-51LJ, about half-way down from the flat upper surface of the island, sample from a 10 m thick, thin-bedded unit, about 1.5–1.6 m above its lower contact (a unit with up to 1 m thick beds), *Kettnerites* (*K.*) *polonensis*, *K.* (*A.*) *sisyphi sisyphi*, annelid jaws.

GRUNDÅRD 2. 633523 164558 (CJ 3107 3444), ca 3420 m WNW of Näs church. Topographical map sheet 5 I Hoburgen NO & 5 J Hemse NV. Geological map sheet Aa 152 Burgsvik.
Shore exposure 40 m south of the road and 40 m east of the easternmost part of the small bay.
Hemse Beds, Hemse Marl, NW part. *Age:* Ludlow.
Samples/jawed annelids: 71-146LJ, shallow shore exposure, *Kettnerites* (*K.*) *huberti*, *K.* (*A.*) *sisyphi sisyphi*, annelid jaws.

GRYMLINGS 1. Klinteberg Beds, lower–middle part. *Age:* Late Wenlock.
Reference: Jeppsson 1983*; Frykman 1989.
Samples/jawed annelids: 79-1LJ, bottom of the road-side ditch exposure, *Kettnerites* (*K.*) cf. *martinssonii*, *K.* (*A.*) *sisyphi sisyphi*, annelid jaws.

GRYNGE 1. Hemse Beds, unit d (e?). *Age:* Ludlow, Leintwardinian.
Reference: Fredholm 1988*.
Samples/annelid jaws: 82-16LJ, 5 m W of the navigation mark and about 0.5 m below the contact between the stromatoporoid limestone and the overlying, dominantly crinoidal limestone, no jawed annelids; 82-17LJ, 5 m S of the navigation mark, from above the contact between the two lithologies, annelid jaws.

GUSTAVSVIK 1. Lower Visby Beds, unit b. *Age:* Late Llandovery.
References: Laufeld 1974b*; Larsson 1979.
Samples/jawed annelids: 67-1LJ, at water level, *K.* (*K.*) cf. *versabilis*, *Kettnerites* (*A.*) *siaelsoeensis*, *K.* (*K.*) *abraham abraham*, annelid jaws; 79-20LJ, at water level, *Kettnerites* (*A.*) *siaelsoeensis*, *K.* (*K.*) *abraham abraham*, annelid jaws.

GUSTAVSVIK 2. Lower Visby Beds, unit b(?). *Age:* Late Llandovery.
References: Laufeld 1974b*.
Samples/jawed annelids: 75-102CB, uppermost bed, *Kettnerites* (*K.*) *martinssonii*, *K.* (*A.*) *siaelsoeensis*, *K.* (*K.*) *abraham abraham*, annelid jaws.

GUSTAVSVIK 3. Lower Visby Beds, unit b. *Age:* Late Llandovery.
Reference: Laufeld 1974b*.
Samples/jawed annelids: 79-21LJ, low section on the beach, *Kettnerites* (*A.*) *siaelsoeensis*, *K.* (*K.*) *abraham abraham*, annelid jaws.

GUTENVIKS 1. Hemse Beds, unit (c or) d. *Age:* Ludlow.
References: Laufeld 1974a, b*; Laufeld & Jeppsson 1976; Fredholm 1988.
Samples/jawed annelids: 71-97LJ, calcilutite, 0.50 m above the bottom of the ditch, *Kettnerites* (*K.*) *bankvaetensis*, *K.* (*K.*) *martinssonii*, *K.* (*K.*) *burgensis*, annelid jaws.

GUTEVÄGEN 2. Högklint Beds or Tofta Beds. *Age:* Early Wenlock.
References: Laufeld 1974b*.
Samples/jawed annelids: 69-54LJ, uppermost, stratified bed, rich in algal balls, annelid fragments.

GYLE 1. Hemse Beds, unit b. *Age:* Early Ludlow, Eltonian.
References: Laufeld 1974a, b*; Larsson 1979; Fredholm 1988; Frykman 1989.
Samples/jawed annelids: 75-39CB, 0.25–0.35 m from the base of the section, *Kettnerites* (*K.*) cf. *martinssonii*, annelid jaws; 75-40CB, 0.40–0.50 m from base of the section, *Kettnerites* (*K.*) *martinssonii*, *K.* (*K.*) *polonensis*, annelid jaws.

GYLE 2. 636964 166890 (CJ 5694 6696), ca 1100 m WWNW of Ala church. Topographical map sheet 6 J Roma SV. Geological map sheet Aa 170 Katthammarsvik.
Ditch section N and S of the intersection of the path and the ditch. Gyle 2 is located just W of the bridge on the NE side of the ditch.
Reference level: The boundary between the bluish to brownish, light grey limestone and the light brownish to light grey limestone (Hede 1929:29); the boundary is marked by a surface with corroded pits 10–20 mm deep and 30–40 mm wide.
Hemse Beds, unit b(?). *Age:* Early Ludlow.
References: Hede 1929:29-30 lines 23–31 and 1–6.
Sample/jawed annelids: 71-208LJ, base of the section, 0.8–1.05 m below the reference level, *Kettnerites* (*K.*) *huberti*, *K.* (*K.*) *polonensis*, annelid jaws; 71-211LJ, 0.5 m above the reference level and 5 m E of the bridge, *Kettnerites* (*K.*) *martinssonii*, annelid jaws.

HÄFTINGSKLINT 1. Upper Visby Beds and Högklint Beds, unit a. *Age:* Early Wenlock.
References: Laufeld 1974b*, Larsson 1979.
Samples/jawed annelids: Upper Visby Beds: 76-8CB about 1.0 m a.s.l., *Kettnerites* (*K.*) *abraham abraham*, annelid jaws; 76-9CB about 5 m a.s.l., *Lanceolatites gracilis* var. visby, *Kettnerites* (*K.*) cf. *martinssonii*, *K.* (*A.*) *sisyphi sisyphi*, *K.* (*K.*) *abraham abraham*, annelid jaws; 76-10CB, approximately the same level as 76-9CB but about 100 m north of the bioherm, *K.* (*K.*) *versabilis*, *Kettnerites* (*A.*) *sisyphi sisyphi*, *K.* (*K.*) *abraham abraham*, annelid jaws; 80-3CB, 0.75–0.78 m below the reference level, *Kettnerites* (*K.*) *martinssonii*, *K.* (*K.*) *abraham abraham*, annelid jaws.

HÄFTINGSKLINT 4. Lower Visby Beds, Upper Visby Beds, and Högklint Beds, unit a. *Age:* Late Llandovery to Early Wenlock.
Reference: Hede (1933; 1 to 1.5 km S of Häftingsklint).
Samples/jawed annelids: Lower Visby Beds, unit e(?), or Upper Visby Beds: 84-38LJ, 0.90 m below sea level, annelid jaws; 84-42LJ about 4.25 m a.s.l., *Lanceolatites gracilis*, *Kettnerites* (*K.*) *martinssonii*, *K.* (*A.*) *microdentatus*, *K.* (*K.*) *abraham abraham*, annelid jaws.

HAGANÄS 1. Slite Beds, Marl and unit g (Ryssnäs Limestone). *Age:* Early Wenlock.
References: Laufeld 1974a, b*; Larsson 1979; Bergman 1981, 1984*; Sundquist 1981, 1982a*; Ramsköld 1983; Jeppsson 1983.
Samples/jawed annelids: Slite Marl: 77-28CB, 4.5 m below reference level *Kettnerites* (*K.*) *martinssonii*, *K.* (*K.*) cf. *jacobi*, *K.* (*A.*) *sisyphi sisyphi*, annelid jaws; 69-10LJ, 10–20 mm thick bed at water level, no annelid jaws; 69-11LJ, immediately above 69-10, *Kettnerites* (*K.*) cf. *bankvaetensis*, *K.* (*A.*) *sisyphi* var. valle, annelid jaws; Slite Beds, unit g: 77-29CB, 0.1 m above the reference level (boundary between the Slite Marl and the Ryssnäs Limestone), no annelid jaws; 69-13LJ, 1.0 m below the top of the section, *Kettnerites* (*A.*) *sisyphi*, annelid jaws.

HÄGUR 1. Mulde Beds. *Age:* Late Wenlock.
Reference: Larsson 1979*.
Samples/jawed annelids: 81-59LJ, shallow ditch section, *Kettnerites* (*K.*) *martinssonii* var. mulde, *K.* (*A.*) *sisyphi sisyphi*, annelid jaws.

HÄGVIDE 1. Hemse Beds, Hemse Marl, SE part. *Age:* Ludlow, Leintwardinian.
References: Laufeld 1974a, b*; Larsson 1979; Jeppsson 1983*; Fredholm 1988.
Samples/jawed annelids: 71-129LJ, loose boulder, *Lanceolatites gracilis*, annelid jaws.

HÄLLAGRUND 1. Högklint Beds, unit c, and Slite Beds, unit c. *Age:* Early Wenlock.
References: Laufeld 1974a, b*; Jeppsson 1983.
Samples/jawed annelids: Högklint Beds: 77-19CB, close to the boundary, no annelid jaws. Slite Beds: 79-208LJ, uppermost bed on the southeastern side of the Hällagrundet point, *Kettnerites* (*K.*) *martinssonii*, *Hindenites gladiatus*, annelid jaws. Slite Beds: 77-18CB, 0.20–0.30 m above the boundary with the Högklint Beds, no annelid jaws.

HALLBJÄNNE 1. Eke Beds, lower part. *Age:* Ludlow, Late Leintwardinian.
References: Laufeld 1974a, b*.
Samples/jawed annelids: 72-17LJ, 0.1–0.2 m below ground level, fragments.

HALLBRO SLOTT 6. Slite Beds, unit c, e, and f. *Age:* Early Wenlock.
References: Jeppsson 1983*.
Samples/jawed annelids: Slite Beds, unit c: 73-12LJ, 2.55–2.45 m below the reference level, no annelid jaws; Slite Beds, unit e: 73-13LJ, 0.05–0.15 m above the reference level, no annelid jaws; Slite Beds, unit f or g(?): 73-14LJ, 0.80–0.90 m above the reference level, no annelid jaws; 73-15LJ, 2.50 m above the reference level, small fragments.

HÄLLINGE 1. Klinteberg Beds, unit b. *Age:* Late Wenlock.
References: Laufeld 1974a, b*; Frykman 1989.
Samples/jawed annelids: 71-93LJ, top of the section, one annelid jaw.

HÄLLINGE 2. Klinteberg Beds, unit a. *Age:* Late Wenlock.
References: Laufeld 1974a, b*; Jeppsson 1983; Aldridge & Jeppsson 1984; Frykman 1989.
Samples/jawed annelids: 71-94LJ, ditch section, fragments.

HALLSARVE 1. Hemse Beds, Hemse Marl, uppermost part, and Eke Beds, lowest part. *Age:* Ludlow, Late Leintwardinian.
References: Laufeld 1974a, b*; Jeppsson 1974, p. 10, 1983; Larsson 1979; Cherns 1982*, 1983*; Fredholm 1988.
Samples/jawed annelids: 69-26LJ, 0.25–0.15 m below the reference level, *Kettnerites* (*K.*) *polonensis*, *K.* (*K.*) *huberti*, annelid jaws; 69-27LJ, 0.05–0.10 m below the reference level and immediately below the bed containing *Shaleria impressa*, no annelid jaws; 69-28LJ, bed containing abundant *S. impressa*, *Kettnerites* (*A.*) *sisyphi klasaardensis*, *K.* (*K.*) *huberti*, annelid jaws; 69-29LJ, 0.3–0.4 m above the reference level, *Kettnerites* (*K.*) *huberti*, other paulinitid taxon, annelid jaws.

HALLS HUK 1. Upper Visby Beds and Högklint Beds, unit a. *Age:* Early Wenlock.
References: Laufeld 1974a, b*, Larsson 1979.
Samples/annelid jaws: Upper Visby Beds: 77-9CB 1.70 m above the reference level, *Kettnerites* (*K.*) *martinssonii*, *K.* (*A.*) *sisyphi sisyphi*, *K.* (*K.*) *abraham abraham*, annelid jaws.

HÄLLUDDEN 1. 642643 168420 (CK 7652 2238), ca 6440 m NNE of Fleringe church. Topographical map sheet 7 J Fårösund SO & NO. Geological map sheet Aa 171 Kappelshamn.
Fairly low shore exposure, comprising the tip of the land, north of the road.
Högklint Beds, upper part. *Age:* Early Wenlock.
Sample/jawed annelids: 82-48CB, 0.15 m below the top of the outermost part of the shallow shore exposures, annelid fragments.

HAUGKLINTAR 2. Klinteberg Beds, lower part. *Age:* Late Wenlock.
References: Laufeld 1974a, b*; Jeppsson 1983; Frykman 1989.
Samples/jawed annelids: 70-31LJ, bedded calcarenite, paulinitid jaw.

HERRVIK 2. 637048 168678 (CJ 7483 6642), ca 3350 m E of Östergarn church. Topographical map sheet 6 J Roma SO. Geological map sheet Aa 170 Katthammarsvik.
During the excavation of the middle part of the eastern inner harbour basin in 1983, the water was pumped out and the rock sample was taken in the section just outside the newly constructed concrete quay.
Hemse Beds, unit d(?). *Age:* Ludlow.
Sample/jawed annelids: 83-2CB, about 3 m below sea level, Bluish calcilutite rich in stromatoporoids with low relief, *Kettnerites* (*K.*) *polonensis*, *K.* (*A.*) cf. *fjaelensis*, *Hindenites* sp., annelid jaws.

HIDE 1. Slite Beds, Slite Marl & Slite g. *Age:* Wenlock.
References: Laufeld 1974a, b*; Larsson 1979.
Samples/jawed annelids: 73-2LJ, the lowest accessible bed, 1.95–2 m below the reference level, *Hindenites gladiatus*, *Kettnerites* (*K.*) *martinssonii*, *K.* (*K.*) *huberti*, *K.* (*A.*) *microdentatus*, *K.* (*A.*) *sisyphi sisyphi*, *Gotlandites slitensis*, annelid jaws; 73-3LJ, 1.37–1.42 m below the reference level, *Kettnerites* (*K.*) *martinssonii*, *Gotlandites slitensis*, annelid jaws; 73-4LJ, 0.0–0.15 m below the reference level, *Kettnerites* (*K.*) cf. *bankvaetensis*, *K.* (*K.*) *polonensis*, *Gotlandites slitensis*, annelid jaws. 73-5LJ, 0.0–0.05 m above the reference level, *Gotlandites slitensis*, annelid jaws; 73-7LJ, 1.1–1.6 m below the top of the section and about 4.5 m above the base of the section, no annelid jaws; 77-12CB, about 1 m above the base of the section, no annelid jaws; 77-35PSSFG, paulinitid fragments, annelid jaws.

HIDE FISKELÄGE 1. Slite Beds, Slite Marl, top and unit g. *Age:* Wenlock.
References: Laufeld 1974a, b*; Larsson 1979.
Samples/jawed annelids: Slite Marl: 77-14CB, 1.1 m below the reference level, *Gotlandites slitensis*, *Kettnerites* (*K.*) *martinssonii*, 77-15CB, about 0.30 m above the reference level, coarse fragmented limestone, no annelid jaws.

HOBURGEN 2. Burgsviks Beds, top part, and Hamra Beds, unit a. *Age:* Ludlow, Whitcliffian.
References: Laufeld 1974a, b*; Jeppsson 1974; Claesson 1979; Larsson 1979; Laufeld & Martinsson 1981; Jeppsson 1982, 1983.
Samples/jawed annelids: Burgsvik Beds: 69-39LJ, 1.20–1.05 m below the reference level, calcarenite (oolite) within the sandstone, annelid jaw; 69-40LJ, 0.80–0.60 m below the reference level, *Kettnerites* (*K.*) cf. *martinssonii*, *K.* (*K.*) *huberti*, annelid jaws; 69-41LJ, calcarenite (oolite), 0.35–0.0 m below the reference level, no annelid jaws; Hamra Beds: 69-43LJ, 2.25–2.40 m above the reference level, *Kettnerites* (*K.*) *polonensis*, annelid jaws.

HOBURGEN 3. Burgsvik Beds, Hamra Beds, unit b, c, and Sundre Beds. *Age:* Late Ludlow.
References: Laufeld 1974a, b*; Larsson 1979.
Samples/jawed annelids: Hamra Beds: 71-21LJ, sample from the roof of the lowest cave, paulinitid jaws, annelid jaws.

HOLMHÄLLAR 1. Sundre Beds, middle upper part. *Age:* Latest Ludlow.
References: Laufeld 1974a, b*; Larsson 1979; Ramsköld 1983.
Samples/jawed annelids: 75-27LJ, crinoidal calcirudite, no annelid jaws; MS905AM, fissure filling of green and greenish-grey calcilutite, *Kettnerites* (*K.*) *polonensis*, *Langeites glaber*, annelid jaws; 75-28LJ, fissure filling, no annelid jaws; 75-29LJ, fissure filling, roughly Manten's (1971) point 185, no annelid jaws.

HOLMHÄLLAR 2. 631472 165117 (CJ 3507 1355), ca 5300 m SE of Vamlingbo church. Topographical map sheet 5 I Hoburgen SO & 5 J Hemse SV. Geological map sheet Aa 152 Burgsvik.
Temporary well excavation, in the small wood just SSE of the Holmhällar boarding-house, 5 m west of the gate across the small road.
Sundre Beds. Age: Latest Ludlow.
Samples/jawed annelids: 84-111CB, excavated slabs of calcilutite, *Lanceolatites* sp., fragments.

HÖRSNE 3. Halla Beds, unit b. *Age:* Late(?) Wenlock.
Reference: Laufeld 1974b*; Odin et al. 1986.
Samples/jawed annelids: 75-36KL, annelid jaws; 82-8CB, the uppermost 0.05–0.07 m of the section, immediately above the small bioherm, no annelid jaws; 82-9CB, about 4 m E of 82-8 in a calcilutitic pocket within the middle part of the reef, about 0.7 m

above the water in the drainage ditch, annelid jaws; 82-10CB, close to 82-9 in the same pocket, *Kettnerites* (*K.*) *martinssonii*, annelid jaws; 82-11CB, very soft calcilutite in the same pocket, annelid jaws; 82-12CB, competent calcilutite, *Kettnerites* (*K.*) *martinssonii*, annelid jaws.

HÖRSNE 5. Halla Beds, unit b. *Age:* Late(?) Wenlock.
References: Laufeld 1974a, b*.
Samples/jawed annelids: 75-52CB, excavated material, *Kettnerites* (*K.*) *martinssonii*, *K.* (*K.*) aff. *jacobi*, *K.* (*A.*) *sisyphi sisyphi*, annelid jaws; 75-53CB, excavated material, *Kettnerites* (*K.*) cf. *martinssonii*, *K.* (*K.*) aff. *abraham isaac*, annelid jaws; 75-55CB, excavated material, annelid jaws.

HÖRSNE 6. Halla Beds, unit b. *Age:* Late(?) Wenlock.
Reference: Larsson 1979; Ramsköld 1985b.
Samples/jawed annelids: 74-37KL, annelid jaws; 74-39KL, *Kettnerites* sp., annelid jaws; 75-54KL, annelid jaws.

HUMMELBOSHOLM 1. Eke Beds, lower part. *Age:* Ludlow.
References: Laufeld 1974a, b*.
Sample/jawed annelids: 76-11CB, shallow ditch section, *Lanceolatites gracilis*, *Kettnerites* (*K.*) *martinssonii*, *K.* (*K.*) *huberti*, annelid jaws.

HUNNINGE 1. Klinteberg Beds, lower–middle part. *Age:* Late Wenlock.
References: Laufeld 1974a, b*; Larsson 1979; Jeppsson 1983; Frykman
1988.
Samples/jawed annelids: 71-3LJ, about 2.5–3 m above the base of the section, no annelid jaws.

HUSRYGGEN 3. Hamra Beds, unit b. *Age:* Ludlow, Whitcliffian.
Reference: Larsson 1979.
Samples/jawed annelids: 71-166LJ, lowest exposed bed, *Lanceolatites* sp., annelid jaws; 71-168LJ, 4.4 m above 71-166, paulinitid fragment, annelid jaws.

IREVIKEN 1. Lower Visby, Upper Visby, and Högklint Beds, unit a. *Age:* Late Llandovery to Early Wenlock.
References: Laufeld l974a, b*; Ramsköld 1983, 1984.
Samples/jawed annelids: Lower Visby Beds, unit e: 76-4CB, 0.5 m below the reference level, *K.* (*K.*) *versabilis*, *K.* (*K.*) *abraham abraham*, annelid jaws; Högklint Beds, unit a: 76-5CB, loose slab from the bedded, light brown calcarenite between the reef bodies, *Kettnerites* (*K.*) *abraham* cf. *abraham*, annelid jaws.

IREVIKEN 2. Lower Visby Beds, Upper Visby Beds, and Högklint Beds, unit a. *Age:* Late Llandovery(?) to Early Wenlock.
References: Laufeld 1974b*, Larsson 1979.
Samples/jawed annelids: Lower Visby Beds: 76-2CB, about 0.3 m above sea level, *Kettnerites* (*K.*) *abraham abraham*, annelid jaws; Upper Visby Beds: 76-3CB about 6 m below the reference level, *Kettnerites* (*K.*) *martinssonii*, *K.* (*K.*) *abraham abraham*, annelid jaws.

IREVIKEN 3. Lower Visby Beds, Upper Visby Beds, and Högklint Beds. *Age:* Late Llandovery to Early Wenlock.
References: Laufeld 1974a, b*; Larsson 1979; Odin et al. 1984*, 1986.
Samples/jawed annelids: Lower Visby Beds unit b: 79-15LJ, about 4 m below the reference level, *Kettnerites* (*K.*) *versabilis*, *K.* (*A.*) *siaelsoeensis*, *K.* (*K.*) *abraham abraham*, annelid jaws; 79-16LJ, 0.1–0.2 m below the reference level, *Kettnerites* (*K.*) *abraham abraham*, annelid jaws; 82-2LJ, 0.02–0.15 m above the reference level, *Kettnerites* (*A.*) *siaelsoeensis*, *K.* (*K.*) *abraham abraham*, annelid jaws; 82-3LJ, 0.85–0.90 m a.s.l., *Kettnerites* (*A.*) *siaelsoeensis*, annelid jaws; Lower Visby Beds unit c: 81-6LJ, 2.36–2.46 m above the reference level, *Kettnerites* (*K.*) cf. *versabilis*, *Lanceolatites* cf. *gracilis*, *Kettnerites* (*K.*) *abraham abraham*, annelid jaws; 82-5LJ, 1.91–2.06 m above the reference level, no annelid jaws; 82-6LJ, 2.16–2.26 m above the reference level, *Kettnerites* (*K.*) *martinssonii*, *K.* (*K.*) *abraham abraham*, annelid jaws; 82-8LJ, 2.91–3.01 m above reference level, *Lanceolatites gracilis*, *Kettnerites* (*K.*) *martinssonii*, *K. abraham abraham*, annelid jaws; Lower Visby Beds unit d: 81-7LJ, 3.96 m above the reference level, *Kettnerites* (*K.*) *versabilis*, *K.* (*K.*) *martinssonii*, *K.* (*K.*) *abraham abraham*, annelid jaws; Lower Visby Beds unit e: 82-9LJ, 4.21–4.26 m above the reference level, fragments; 82-10LJ, 5.1–5.3 m above the reference level, *Kettnerites* (*K.*) *abraham abraham*, annelid jaws; 83-25LJ, 5.32–5.44 m above the reference level, *Kettnerites* (*K.*) *abraham abraham*, annelid jaws; Upper Visby Beds: 79-18LJ, 9.25 m above the reference level, *Lanceolatites gracilis*, *Kettnerites* (*K.*) *abraham abraham*, annelid jaws; 82-11LJ, either 5.5 or 6.2 m above the reference level, *Lanceolatites gracilis*, *Kettnerites* (*K.*) *martinssonii*, *K.* (*K.*) *abraham abraham*, annelid jaws.

JAKOBSBERG 1. Slite Beds, unit g. *Age:* Wenlock.
References: Laufeld 1974a, b*; Larsson 1979.
Samples/jawed annelids: 72-4LJ, from the lower part of the eroded bed, same level as 72-47SL (Laufeld 1974a), 0.10–0.30 m below the 'step' (along the path) in the section, argillaceous calcilutite, fragments.

JUVES 1. Hamra Beds, unit c, and Sundre Beds, lower part. *Age:* Latest Ludlow.
References: Laufeld 1974b*; Larsson 1979.
Samples/jawed annelids: Hamra Beds: 76-25CB, slightly more than 2 m above the base of the section, annelid jaws.

JUVES 2. Hamra Beds, unit c, and Sundre Beds, lower part. *Age:* Latest Ludlow.
References: Laufeld 1974a, b*; Larsson 1979; Jeppsson 1982*, 1983.
Samples/jawed annelids: Hamra Beds: 71-20LJ, 5.17–5.12 m below the boundary between the Hamra and Sundre Beds, paulinitid fragment, fragments; 69-44LJ, 5.01–4.75 m below the boundary between the Hamra and Sundre Beds, fragments; 69-46LJ, 3.53–3.43 m below the boundary between the Hamra and Sundre Beds, fragments.

The stratigraphic levels of LJ samples from the Juves localities 2, 3, and 5 refer to the boundary between the Hamra and Sundre Beds which does not agree with the reference level at Juves 2 designated by Laufeld (1974b). For further information, see (Jeppsson 1982, p. 17, Fig. 2, the composite section for Juves with the levels in italics).

JUVES 3. Hamra Beds, unit c, and Sundre Beds, lower part. *Age:* Latest Ludlow.
References: Laufeld 1974a, b*; Larsson 1979; Claesson 1979; Jeppsson 1982*, 1983.
Samples/jawed annelids: Hamra Beds: 77-3LJ, 2.38–2.23 m below the reference level, no annelid jaws; 77-19LJ, 1.55–1.50 m below the reference level, annelid jaws; 69-47LJ, 1.03–0.90 m below the reference level, *Kettnerites* (*K.*) *huberti* the slender type, annelid jaws; 69-49LJ, 0.0–0.15 m below the reference level, paulinitid fragments, annelid jaws; Sundre Beds: 69-50LJ, about 3 m above the reference level, fragments.

JUVES 4. Sundre Beds, lower part. *Age:* Latest Ludlow.
References: Laufeld 1974a, b*; Larsson 1979; Jeppsson 1982*, 1983.
Samples/jawed annelids; 76-23CB, *Kettnerites* (*K.*) *bankvaetensis*, paulinitid sp. indet., annelid jaws.

JUVES 5. Hamra Beds, unit c. *Age:* Latest Ludlow.
References: Laufeld & Jeppsson 1976*; Jeppsson 1982*, 1983.
Samples/jawed annelids: 71-182LJ, 6.57–6.50 m below the reference level, no annelid jaws; 77-11LJ, 5.52–5.43 m below the reference level, no annelid jaws; 77-10LJ, 4.53–4.48 m below the reference level, no annelid jaws.

KAKHUSE 1. 637168 165477 (CJ 4298 7012), ca 3060 m NW of Guldrupe church. Topographical map sheet 6 J Roma SV. Geological Map sheet Aa 160 Klintehamn.

Shallow ditch section, on the SE side of the small road from Väte to Viklau, about 175 m NE of the cross-roads at point 41,68.
Klinteberg Beds, lower part. *Age:* Latest Wenlock – earliest Ludlow.
Samples/jawed annelids: 78-9CB, 0.25 m below the uppermost surface of the limestone, no annelid jaws.

KÄLLDAR 1. Hemse Beds, Hemse Marl NW part. *Age:* Ludlow.
References: Laufeld 1974a, b*.
Sample/jawed annelids: 71-34LJ, excavated material from ditch at Vissne myr, about 3 km south of Linde church, dumped at Källdar 1, *Lanceolatites* cf. *gracilis*, *Kettnerites* (*A.*) *sisyphi sisyphi*, annelid jaws; 75-75CB, excavated material from the same locality as 71-34LJ, *Kettnerites* (*A.*) *microdentatus*, annelid jaws.

KÄLLDAR 2. Hemse Beds, lower–middle part. *Age:* Ludlow, Bringewoodian or possibly earliest Leintwardinian.
References: Jeppsson 1983*; Fredholm 1988.
Samples/jawed annelids: 71-35LJ, middle part of the section, *Kettnerites* (*A.*) *sisyphi sisyphi*, annelid jaws.

KAMBS 5. 640444 165660 (CK 4741 0253), 3770 m WSW of Martebo church. Topographical map sheet 7 J Fårösund SV & NV. Geological map sheet Aa 183 Visby & Lummelunda.

Drainage ditch section on the northern side of the ditch, immediately E of the bridge.
Tofta Beds. *Age:* Early Wenlock.
Reference: Hede 1940, p. 35, lines 3–5 from below.
Samples/jawed annelids: 75-28CB, topmost bed in the section, no annelid jaws; 82-42LJ, topmost bed in the section, a fragment.

KAPELLSHAMN 1. Högklint Beds unit b. *Age:* Early Wenlock.
Reference: Laufeld 1974a, b*, Larsson 1979.
Samples/jawed annelids: 77-11CB, 0.5 m below the top of the section, *Kettnerites* (*K.*) *martinssonii*.

KAPELLUDDEN 1. 634751 167132 (CJ 5764 4472), ca 8500 m ESE of Burs church. Topographical map sheet 5 I Hoburgen NO & 5 J Hemse NV. Geological map sheet Aa 156 Ronehamn.

Shore exposure 200 m N of Kapellet in Kapelluddens fiskeläge marked on the topographical map sheet. The exposure is also marked on the geological map sheet.
Burgsvik Beds. *Age:* Ludlow.
References: Hede 1925a, p. 31.
Sample/jawed annelids: 83-16LJ, shallow shore exposure, *Kettnerites* (*K.*) *polonensis*, *K.* (*K.*) *huberti*, *Hindenites naerensis*, annelid jaws.

KÄRNE 3. Hemse Beds, Hemse Marl, uppermost part and Eke Beds, basal part. *Age:* Ludlow, Late Leintwardinian.
Reference: Jeppsson 1983*; Fredholm 1988.
Samples/jawed annelids: Hemse Beds: 71-195LJ, ditch exposure, 133 m south of the contact between the Hemse and Eke Beds, *Kettnerites* (*K.*) *burgensis*, *Langeites glaber*, annelid jaws; 71-196LJ, 8 m south of the contact, no annelid jaws; 71-197LJ, *Shaleria impressa*-rich level, top Hemse, 0.01–0.04 m below the contact, *Kettnerites* (*K.*) *huberti*, annelid jaw; Eke Beds: 71-198LJ, 0.0–0.06 m above the contact, *Kettnerites* (*A.*) *sisyphi klasaardensis*, *K.* (*K.*) *huberti*, annelid jaws; 71-199LJ, bryozoan-rich conglomerate, 0.06–0.16 m above the contact, *Kettnerites* (*A.*) *microdentatus*, annelid jaws; 71-200LJ, top of the section, no annelid jaws; 71-202LJ, 5 m N of the road, no annelid jaws; 71-203LJ, N of Kärne, no annnelid jaws.

KÄTTELVIKEN 5. Hamra Beds, unit b. *Age:* Ludlow, Late Whitcliffian or slightly younger.
References: Laufeld & Jeppsson 1976*; Larsson 1979; Jeppsson 1983.
Samples/jawed annelids: 71-178LJ, 3 m below the top of the section, *Kettnerites* (*K.*) *polonensis*, paulinitid sp. indet., annelid jaws; 71-179LJ, see Laufeld & Jeppsson 1976, annelid jaws; 77-33LJ, 3 m below top of the section, *Lanceolatites gracilis*, *Kettnerites* (*K.*) *martinssonii*, *K.* (*K.*) *huberti*, *K.* (*K.*) *polonensis*, *Hindenites* sp., annelid jaws; 71-176LJ, 1 m below the top of the section, *Kettnerites* (*K.*) *bankvaetensis*, *K.* (*K.*) *polonensis*, annelid jaws; 71-175LJ, uppermost exposed bed, paulinitid fragments, annelid jaws.

KATTHAMMARSVIK 1. Hemse Beds, unit a. *Age:* Early Ludlow.
References: Laufeld 1974a, b*; Laufeld & Martinsson 1981; Jeppsson 1983; Fredholm 1988.
Samples/jawed annelids: 67-30LJ, shore exposure, annelid jaws; 75-50CB, upper part of the shore exposure, annelid jaws.

KAUPARVE 1. Hamra Beds, lower–middle part. *Age:* Ludlow.
References: Laufeld 1974a, b*.
Sample/jawed annelids: 76-16CB, 0.2–0.3 m below ground level, *Kettnerites* (*K.*) *bankvaetensis*, *K.* (*K.*) *martinssonii*, annelid jaws.

KLASÅRD 1. Hemse Beds, Hemse Marl SE part. *Age:* Ludlow, Leintwardinian.
References: Jeppsson 1983*; Fredholm 1988.
Samples/jawed annelids: 71-150LJ, shore exposure, *Lanceolatites gracilis*, *Kettnerites* (*K.*) *burgensis*, *K.* (*A.*) *sisyphi klasaardensis*, *K.* (*K.*) *huberti*, annelid jaws.

KLINTEBERGET 1. Klinteberg Beds, lower and middle parts. *Age:* Late Wenlock.
References: Laufeld 1974a, b*; Claesson 1979; Larsson 1979; Laufeld & Martinsson 1981; Jeppsson 1982, 1983; Ramsköld 1983; Frykman 1989.
Samples/jawed annelids: 67-29LJ, about 8 m below the top of the cliff, *Kettnerites* (*K.*) *polonensis*, annelid jaws; 72-23KL, pockets of greenish- and brownish-grey calcilutite, *Kettnerites* (*K.*) *martinssonii*; 75-19CB, about 2 m below the top of the cliff section, annelid jaws; 75-20CB, about 3 m below the top of the cliff section, annelid jaws; 75-21CB, lutitic calcirudite above the thickly bedded limestone beds, close to top of the section, *Hindenites angustus*, annelid jaws.

KLINTEENKLAVEN 1. Slite Beds, Slite Siltstone. *Age:* Wenlock.
References: Laufeld 1974a, b*; Sivhed 1976; Bergman 1979b, 1980a.
Samples/jawed annelids: 75-24CB, 0.4 m from the top of the section, *Kettnerites* (*A.*) *sisyphi* cf. var. valle, annelid jaws.

KLINTHAGEN 1. Slite Beds, unit g. *Age:* Early Wenlock.
Reference: Jeppsson 1983*.
Samples/jawed annelids: 73-1LJ, immediately below the calciruditic bed (reefal detritus), no annelid jaws.

KLUVSTAJN 1. Slite Beds, unit b. *Age:* Early Wenlock.
References: Laufeld 1974a, b*; Jeppsson 1983.
Samples/jawed annelids: 73-16LJ, excavated material from a trench, paulinitid fragments.

KLUVSTAJN 2. 638527 164352 (CJ 3285 8445), ca 2820 m WSW of Västerhejde church. Topographical map sheet 6 I Visby NO. Geological map sheet Aa 183 Visby & Lummelunda.

Shallow ditch exposure, 6 m NW of the SE end of the truncation of the corner of the field and about 575 m NNW of point 54,4.
Slite Beds, unit a(?). *Age:* Early Wenlock.
Samples/jawed annelids: 73-17LJ, 0.1 m thick exposure, *Kettnerites* (*K.*) cf. *martinssonii*, annelid jaws.

KODINGS 3. Hemse Beds, Hemse Marl SE part. *Age:* Ludlow.
References: Laufeld 1974a, b*; Larsson 1979.
Samples/jawed annelids: 75-82CB, ditch exposure, annelid jaws.

KORPKLINT 1. Upper Visby Beds and Högklint Beds, unit a. *Age:* Early Wenlock.
References: Laufeld 1974a, b*, Larsson 1979.
Samples/jawed annelids: Upper Visby Beds: 67-2LJ, about 5 m below the reference level, *Kettnerites* (*K.*) cf. *versabilis*, *Lanceolatites* sp., *Kettnerites* (*K.*) *martinssonii*, *K.* (*K.*) *abraham abraham*, annelid jaws; 75-5CB, immediately below the reference level, *Kettnerites* (*K.*) *martinssonii*, *K.* (*K.*) *abraham abraham*; Högklint Beds, unit a: 75-4CB immediately above the reference level, no annelid jaws.

KRAKFOT 1. Klinteberg Beds, unit d(?). *Age:* Latest Wenlock – earliest Ludlow.
Reference: Frykman 1989*.
Samples/jawed annelids: 71-153LJ, bottom of the ditch, 100 m east of the road, *Kettnerites* (*K.*) *bankvaetensis*, annelid jaws.

KRASSE 1. Klinteberg Beds, middle part. *Age:* Latest Wenlock – earliest Ludlow.
References: Laufeld 1974a, b*; Frykman 1989.
Samples/jawed annelids: 78-10CB, uppermost bed in the section, no annelid jaws.

KROKEN 1. Burgsviks Beds and Hamra Beds. *Age:* Ludlow.
References: Laufeld 1974a, b*.
Sample/jawed annelids: Burgsvik Beds: 72-21LJ, 0.1–0.05 m below the Burgsvik–Hamra boundary, paulinitid sp., annelid jaws.

KROKEN 2. Burgsvik Beds and Hamra Beds. *Age:* Ludlow.
References: Laufeld 1974a, b*.
Sample/jawed annelids: 72-22LJ, lateral of a Hamra bioherm, *Kettnerites* (*K.*) *bankvaetensis*, *K.* (*K.*) *martinssonii*, *K.* (*K.*) *huberti*, *K.* (*K.*) cf. *polonensis*, *K.* (*A.*) *microdentatus*, annelid jaws.

KROKEN 3. 634784 167464 (CJ 6098 4480), ca 11580 m ESE of Burs church. Topographical map sheet 5 I Hoburgen NO & 5 J Hemse NV. Geological map sheet Aa 160 Klintehamn.

Shore exposure on the narrow point just east of the 'n' in 'Kroken' on the topographical map sheet.

Hamra Beds. Age: Ludlow.

Samples/jawed annelids: 83-12CB, shore exposure, no annelid jaws.

KULLANDS 1. Hemse Beds, Hemse Marl NW part. *Age:* Ludlow, Bringewoodian or Early Leintwardinian.

References: Laufeld 1974a, b*; Laufeld & Jeppsson 1976; Larsson 1979; Jeppsson 1983.

Samples/jawed annelids: 75-79CB, uppermost bed at the bottom of the water hole, covered by about 0.1–0.2 m soil, *Kettnerites* (*K.*) cf. *martinssonii*, annelid jaws.

KULLANDS 2. 635370 165152 (CJ 3838 5237), ca 1370 m SSW of Gerum church. Topographical map sheet 6 J Roma SV. Geological map sheet Aa 164 Hemse.

Ditch section, at the intersection of the private road (not marked on the topographical map sheet) and the ditch, about 140 m W of the westernmost house (on the topographical map sheet) at Kullands.

Hemse Beds, Hemse Marl NW part. *Age:* Ludlow.

Reference: Fredholm 1988.

Sample/jawed annelids: 84-311DF, *Kettnerites* (*K.*) *bankvaetensis*, *K.* (*K.*) *martinssonii*, *K.* (*K.*) *burgensis*, *K.* (*A.*) *fjaelensis*, annelid jaws; 84-312DF, *Lanceolatites gracilis*, *Kettnerites* (*K.*) *martinssonii*, *K.* (*A.*) *fjaelensis*, annelid jaws.

KUPPEN 1. Hemse Beds, unit d. *Age:* Ludlow, Leintwardinian.

References: Laufeld 1974a, b*; Larsson 1979; Jeppsson 1982, 1983; Kershaw 1987.

Samples/jawed annelids: 81-44LJ, 0.0–0.11 m above the reference level, no annelid jaws; 81-45LJ, 0.11–0.14 m above the reference level, no annelid jaws.

KUPPEN 2. Hemse Beds, unit d. *Age:* Ludlow, Leintwardinian.

References: Laufeld 1974a, b*; Kershaw & Riding 1978; Laufeld & Martinsson 1981; Kershaw 1981, 1987; Jeppsson 1982, 1983; Fredholm 1988.

Samples/jawed annelids: 81-46LJ, no annelid jaws; 81-47LJ, no annelid jaws.

KVÄNNVÅTEN 1. Slite Beds, *Pentamerus gothlandicus* Beds or slightly younger. *Age:* Wenlock.

References: Laufeld 1974a, b*; Larsson 1979.

Samples/jawed annelids: 77-1CB, about 2 m below the top of the section, no annelid jaws.

LAMBSKVIE 1. Hemse Beds, unit c. *Age:* Ludlow, probably Bringewoodian.

References: Laufeld 1974a, b*; Larsson 1979.

Samples/jawed annelids: 74-18KL, *Kettnerites* (*K.*) *polonensis*; 74-62KL, *Kettnerites* (*K.*) *polonensis*; 75-42CB, loose slab of excavated material, *Kettnerites* (*K.*) *polonensis*, annelid jaws; 75-44CB, top bed on the eastern side, *Kettnerites* (*K.*) *huberti*, annelid jaws.

LANGHAMMARSHAMMAR 1. Högklint Beds, lower–middle part. *Age:* Early Wenlock.

References: Laufeld 1974a, b*, Jeppsson 1983.

Samples/jawed annelids: 77-26CB, jawed annelids.

LANGHAMMARSVIKEN 2. 643385 169795 (CK 9079 2872), ca 7750 m N of Fårö church. Topographical map sheet 7 J Fårösund SO & NO. Geological map sheet Aa 180 Fårö.

Shallow exposure with large wave marks at the sea shore NW of the westernmost extension of Langhammarsviken.

Högklint Beds, lower–middle part. *Age:* Early Wenlock.

References: Hede 1936:16 lines 6–9; Bergman 1979b; Larsson 1979.

Sample/jawed annelids: 78-17CB, surface exposure, sample of the large wave-marked material, *Kettnerites* (*K.*) *martinssonii*, *K.* (*K.*) *abraham isaac*, annelid jaws.

LASSOR 1. Hemse Beds, upper part. *Age:* Ludlow.

References: Laufeld 1974 a, b*; Fredholm 1988.

Samples/jawed annelids: 72-8LJ, 0.50–0.60 m below the top of the section (same level as sample 72-51SL in Laufeld 1974a), no annelid jaws; 78-6CB, loose slabs, fragments.

LAU BACKAR 1. Eke Beds, *Rhizophyllum* Limestone. *Age:* Late Ludlow.

References: Laufeld 1974a, b*; Larsson 1979; Ramsköld 1983, 1984, 1985b.

Samples/jawed annelids: 72-27KL, annelid jaws; 77-30CB, above the bottom of the excavation and about 0.7 m below ground level, *Kettnerites* (*A.*) cf. *sisyphi*, annelid jaws.

LAUTER 1. Högklint Beds, lower middle part. *Age:* Early Wenlock.

References: Laufeld 1974a, b*.

Samples/jawed annelids: 77-24CB, surface exposure, the topmost 20–50 mm, *Kettnerites* (*K.*) *martinssonii*, *K.* (*A.*) *sisyphi sisyphi*, *K.* (*K.*) *abraham* cf. *isaac*, annelid jaws.

LAUTERHORNSVIK 2. Högklint Beds, lower–middle part. *Age:* Early Wenlock.

References: Laufeld 1974a, b*, Jeppsson 1983.

Samples/jawed annelids: 73-73LJ, low shore exposure, *Kettnerites* (*K.*) *martinssonii*, annelid jaws.

LAUTERHORNSVIK 3. Högklint Beds, lower–middle part. *Age:* Early Wenlock.

References: Larsson 1979*, Jeppsson 1983.

Samples/jawed annelids: 73-72LJ, shallow ditch exposure, *Kettnerites* (*K.*) *abraham* cf. *isaac*, annelid jaws.

LERBERGET 1. Slite Beds, Slite Marl, Lerberget Marl and *Pentamerus gothlandicus* Beds. *Age:* Wenlock.

References: Laufeld 1974a, b*; Larsson 1979; Jeppsson 1983.

Samples/annelid jaws: Lerberget Marl: 79-37LJ, lowest exposed bed, *Kettnerites* (*A.*) *sisyphi sisyphi*, annelid jaws; 70-37LJ, 0.30 m above the base of the section, fragments; 70-38LJ, 0.95 m above the base of the section, *Kettnerites* (*A.*) *sisyphi sisyphi*; 70-39LJ, 2.75 m above the base of the section, bed rich in *Ketophyllum*, fragments; *Pentamerus gothlandicus*?: 70-40LJ, 5.15 m above the base of the section, paulinitid sp. indet., fragments; 70-41LJ, 7.0 m above the base of the section, paulinitid sp. indet.; 70-42LJ, 9.15 m above the base of the section and about at the top of the marl, *Kettnerites* cf. *abraham*, *K.* (*A.*) cf. *sisyphi*, annelid jaws; 70-43LJ, about 5 m above the base of the succeeding limestone, no annelid jaws; 70-44LJ, about 15 m above the base of the limestone, no annelid jaws.

LICKERS 1. 641607 166265 (CK 5399 1335), ca 4840 m N of Stenkyrka church. Topographical map sheet 7 J Fårösund SV & NV. Geological map sheet Aa 183 Visby & Lummelunda.

Locality comprising the beach and adjacent cliff, the area between the bioherms.

Lower Visby Beds, unit b or c, on the shore. *Age:* Late Llandovery.

Samples/jawed annelids: 81-10LJ, 0.1 m above mean sea level, *Lanceolatites* cf. *gracilis*, *Kettnerites* (*K.*) *abraham abraham*, annelid jaws.

LICKERSHAMN 2. Upper Visby Beds and Högklint Beds, units a and b(?). *Age:* Early Wenlock.

Reference: Laufeld 1974a, b*, Larsson 1979.

Samples/jawed annelids: Upper Visby Beds: 73-54LJ, about 0.9 above the base of the section, *Kettnerites* (*K.*) *abraham abraham*, annelid jaws; 73-55LJ, 2.65 above the base of the section, a 60–80 mm thick bed, *Lanceolatites gracilis*, *Kettnerites* (*A.*) *sisyphi sisyphi*, annelid jaws; Upper Visby Beds(?); 73-56LJ, 4.35 m above the base of the section, *Kettnerites* (*K.*) *martinssonii*, *K.* (*K.*) *abraham abraham*, annelid jaws; Högklint Beds, probably unit a; 73-58LJ, 5.35 m above the base of the section, a 75 mm thick bed immediately above a bentonite(?), *Kettnerites* (*K.*) *martinssonii*, annelid jaws; 73-59LJ, 8.30 m above the base of the section, *Kettnerites* (*K.*) *martinssonii*, *K.* (*A.*) *sisyphi sisyphi*, *K.* (*K.*) *abraham abraham*, annelid jaws; 75-103CB, reef level in lowest Högklint, no annelid jaws; 75-106CB, about 4 m below the reference level, no annelid jaws; Högklint Beds, unit b(?); 73-60LJ, 9.85 m above the base of the section, *Kettnerites* (*K.*) *martinssonii*, *K.* (*A.*) *sisyphi sisyphi*, *K.* (*K.*) *abraham* cf. *isaac*, annelid jaws; 73-61LJ, 11.90 m above the base of the section, *Kettnerites* (*K.*) *martinssonii*, annelid jaws; 73-62LJ, 14.05 m above the base of the section, *Kettnerites* (*K.*) *martinssonii*, annelid jaws.

LIKMIDE 2. Hemse Beds, Hemse Marl SE. *Age:* Ludlow, Late Leintwardinian.

Reference: Fredholm 1988*.

Samples/jawed annelids: 82-28LJ, shallow ditch, *Lanceolatites gracilis*, *Kettnerites* (*K.*) *burgensis*, *K.* (*A.*) *fjaelensis*, annelid jaws.

LILLA HALLVARDS 1. Hemse Beds, Hemse Marl NW part. *Age:* Early Ludlow.

References: Laufeld 1974a, b*; Larsson 1979; Jaeger 1981; Laufeld & Martinsson 1981; Jeppsson 1983; Fredholm 1988.

Samples/jawed annelids: 71-143LJ, ditch exposure, *Lanceolatites gracilis*, *Kettnerites* (*K.*) cf. *martinssonii*, *K.* (*A.*) *sisyphi*, annelid jaws.

LILLA HALLVARDS 4. Hemse Beds, Hemse Marl NW part. *Age:* Early Ludlow.

References: Larsson 1979*; Fredholm 1988.

Samples/jawed annelids: 84-314DF, *Lanceolatites gracilis*, *Kettnerites* (*K.*) *martinssonii*, *K.* (*A.*) *microdentatus*, annelid jaws.

LINDE 1. Hemse Beds, upper part. *Age:* Ludlow.

References: Laufeld 1974a, b*; Larsson 1979; Ramsköld 1983.

Samples/jawed annelids: 71-112LJ, about 2 m above road surface, no annelid jaws; 71-114LJ, on the slope about 0.5 m above the edge of the section, no annelid jaws.

LINVIKEN 2. 633493 164598 (CJ 3146 3419), ca 2950 m WNW of Näs church. Topographical map sheet 5 I Hoburgen NO & 5 J Hemse NV. Geological map sheet Aa 152 Burgsvik.

Shore exposure 750 m SSE of the solitary house, close to the stone fence that starts by the cattle grid on the road NE of the locality.

Hemse Beds, Hemse Marl NW part. *Age:* Ludlow.

References: Hede 1919, pp. 17-19, loc. 6; Hede 1942, p. 20, loc. 4C.

Samples/jawed annelids: 71-147LJ, shore exposure, *Kettnerites* (*K.*) *huberti*, *K.* (*K.*) *burgensis*, annelid jaws.

LJUGARN 1. Hemse Beds, upper part, probably unit d. *Age:* Ludlow.

References: Laufeld 1974a, b*; Watkins 1975; Jeppsson 1983; Fredholm 1988.

Samples/jawed annelids: 79-51LJ, shore exposure, no annelid jaws.

LOGGARVE 1. Mulde Beds, uppermost part and Klinteberg Beds, lowest part. *Age:* Late Wenlock.

References: Jeppsson 1982, 1983*; Frykman 1989.

Samples/jawed annelids: Mulde Beds: 77-22LJ, 0.50 above the reference level. A comparison with Loggarve 2 has shown that only the upper 0.75 m of the strata belongs to the Klinteberg Beds: *Kettnerites* (*K.*) *polonensis*, *K.* (*A.*) *sisyphi sisyphi*, annelid jaws; Klinteberg Beds: 77-23LJ, 2.5 m above the reference level, no annelid jaws.

LOGGARVE 2. Mulde Beds and Klinteberg Beds, lowest part. *Age:* Late Wenlock.

The contact between the basal 0.1 m of soft clay and the succeeding harder beds remains the best reference level (Lennart Jeppsson, personal communication). The contact is found about 2 m below the base of the Klinteberg Beds, defined by the initial occurrence of the brachiopod *Conchidium conchidium*.

References: Jeppsson 1982*; Ramsköld 1985b; Frykman 1989.

Samples/jawed annelids: Mulde Beds: 82-21LJ, 1.98–2.05 m below the reference level, *Kettnerites* (*K.*) *martinssonii*, *K.* (*A.*) *sisyphi sisyphi*, annelid jaws; 82-22LJ, 1.80–1.72 m below the reference level, *Kettnerites* (*K.*) *martinssonii*, *K.* (*K.*) *polonensis* var. gandarve, annelid jaws; 82-23LJ, 0.87–0.80 m below the reference level, *Kettnerites* (*K.*) sp., annelid jaws; 82-24LJ, top Mulde, 0.10-0.00 m below the reference level, *Kettnerites* (*A.*) *sisyphi sisyphi*, annelid jaws; 82-25LJ, base of Klinteberg Beds, 0.02–0.05 m above the reference level, *Kettnerites* (*K.*) *martinssonii* var. mulde, *Hindenites angustus*, annelid jaws.

LUKSE 1. Hemse Beds, Hemse Marl NW part. *Age:* Ludlow.

References: Laufeld 1974a, b*; Larsson 1979; Ramsköld 1983; Jeppsson 1983; Fredholm 1988.

Samples/jawed annelids: 71-12LJ, about 0.2 m below the top of the culvert, *Lanceolatites gracilis*, *Kettnerites* (*A.*) *sisyphi sisyphi*, paulinitid sp., annelid jaws; 71-13LJ, loose slabs, *Lanceolatites gracilis*, *Kettnerites* (*A.*) *fjaelensis*, annelid jaws.

LYRUNGS 1. Hemse Beds, probably middle part. *Age:* Ludlow.

References: Laufeld 1974a, b*; Fredholm 1988.

Samples/jawed annelids: 79-215LJ, rivulet exposure, no annelid jaws.

MALMS 1. Hemse Beds, Hemse Marl, uppermost part and Eke Beds, lowest part. *Age:* Ludlow, Late Leintwardinian.

References: Laufeld 1974a, b*; Larsson 1979; Cherns 1983*.

Samples/jawed annelids: Hemse Beds: 75-85CB, the uppermost Hemse bed, paulinitid jaws, annelid jaws; Eke Beds: 75-86CB, 0.20 m above the reference level, no annelid jaws.

MARTILLE 7. Slite Beds, unit d. *Age:* Early Wenlock.

Reference: Jeppsson 1983*.

Samples/jawed annelids: 73-18LJ, 0.05 m below soil surface (not 0.5 m, personal communication, Lennart Jeppsson; compare Jeppsson 1983), fragments.

MILLKLINT 3. Hemse Beds, unit e (Millklint Limestone). *Age:* Ludlow, Late Leintwardinian.

References: Laufeld 1974a, b*; Jeppsson 1974, 1983; Laufeld & Jeppsson 1976.

Samples/jawed annelids: 69-30LJ, base of the section, slightly more than 1 m below the reference level, no annelid jaws; 69-31LJ, 0.50–0.60 m above base of section, no annelid jaws; 69-32LJ, 1.0–1.1 m above base of section, no annelid jaws; 69-35LJ, 2.50–2.65 m above base of section, no annelid jaws; 69-36LJ, annelid jaws; 69-37LJ, 3.5–3.6 m above base of section, no annelid jaws.

MOJNER 4. 639840 167882 (CJ 6904 9480), ca 2720 m ESE of Boge church. Topographical map sheet 6 J Roma NV & NO. Geological map sheet Aa 169 Slite.

Small, shore-line exposure, about 400 m NE of the fishing huts (not marked on the topographical map sheet). There is a wide area with beach ridges outside of the forest. Wheel tracks from this area join the small road that is marked on the topographical map. The exposure is beyond the beach ridges, nearly straight down to the shore from this intersection, at the southern end of the about 200 m long, reed-free shore.

Slite Beds, Slite Marl.

Samples/jawed annelids: 84-94LJ, shore exposure, annelid jaws.

MÖLLBOS 1. Halla Beds, unit b. *Age:* Late Wenlock.

References: Laufeld 1974a, b*; Laufeld & Jeppsson 1976; Claesson 1979; Larsson 1979; Laufeld & Martinsson 1981; Liljedahl 1981, 1983, 1984, 1985, 1986; Stridsberg 1981a, b, 1985; Jeppsson 1983; Ramsköld 1985b; Frykman 1989.

Samples/jawed annelids: 67-25LJ, ditch section, *Kettnerites* (*K.*) *bankvaetensis*, *K.* (*K.*) *polonensis*, annelid jaws; 67-26LJ, ditch section, *Kettnerites* (*K.*) *martinssonii*, *K.* (*K.*) *polonensis*, annelid jaws; 75-16CB, about 20 m NW of the reference point and 0.40–0.45 m below the top of the section, annelid jaws; 75-17CB, immediately above 75-16CB, 0.30–0.40 m below the top of the section, *Kettnerites* (*K.*) *bankvaetensis*, *K.* (*K.*) *martinssonii*, annelid jaws; 77-28LJ, top of the section(?), *Kettnerites* (*K.*) *bankvaetensis*, *K.* (*A.*) *sisyphi* cf. var. valle, annelid jaws.

MÖLLBOS 2. Halla Beds, unit b. *Age:* Late Wenlock.

References: Laufeld 1974a, b*; Larsson 1979.

Samples/jawed annelids: 79-81LJ, *Kettnerites* (*K.*) *polonensis*, annelid jaws.

MÖLNER 1. Mulde Beds, upper part. *Age:* Late Wenlock.

References: Laufeld 1974a, b*; Larsson 1979; Jeppsson 1983.

Samples/jawed annelids: 71-6LJ, top of the section, paulinitid fragments, annelid jaws; 71-7LJ, 1.70 m below the top of the section and 0.70 m above the base of the section, *Kettnerites* (*K.*) *martinssonii*, *K.* (*A.*) *sisyphi sisyphi*, annelid jaws.

MULDE 2. Mulde Beds, uppermost part. *Age:* Late Wenlock.

References: Laufeld 1974a, b*.

Samples/jawed annelids: 75-26CB, middle of the ditch section, *Kettnerites* (*K.*) *martinssonii* var. mulde, *K.* (*K.*) *polonensis*, *K.* (*A.*) *sisyphi sisyphi*, annelid jaws.

MULDE TEGELBRUK 1. Mulde Beds. *Age:* Late Wenlock.

References: Laufeld 1974a, b*; Larsson 1979.

Samples/jawed annelids: 67-28LJ, loose slabs, *Kettnerites* (*K.*) *martinssonii* var. mulde, *K.* (*A.*) *sisyphi sisyphi*, annelid jaws; 82-7CB, loose slabs, *Kettnerites* (*K.*) *martinssonii* var. mulde, *K.* (*K.*) *polonensis*, *K.* (*A.*) *sisyphi sisyphi*, *K.* (*A.*) *sisyphi* cf. valle, annelid jaws.

MUNKEBOS 1. Slite Beds, Slite Marl, *Pentamerus gothlandicus* Beds or slightly younger. *Age:* Wenlock.
References: Laufeld 1974a, b*.
Samples/jawed annelids: 71-84LJ, rivulet section 25 m W of the reference point, *Gotlandites slitensis, Hindenites gladiatus, Kettnerites (A.) sisyphi sisyphi*, annelid jaws.

MYRSNE 1. Slite Beds, Slite Marl. *Age:* Wenlock.
Reference: Larsson 1979*.
Samples/jawed annelids: 82-43LJ, low ditch section, *Kettnerites (K.) huberti, K. (K.) jacobi, K. (A.) sisyphi sisyphi*, annelid jaws.

NABBAN 2. Eke Beds, lower part. *Age:* Ludlow, Late Leintwardinian.
References: Laufeld 1974a, b*.
Samples/jawed annelids: 72-16LJ, shore exposure, same level as sample 72-60SL (Laufeld 1974a), *Kettnerites (K.)* cf. *bankvaetensis, K. (K.) martinssonii*, annelid jaws.

NÄR 2. Hemse Beds, Hemse Marl, uppermost part. *Age:* Ludlow, Late(?) Leintwardinian.
References: Fredholm 1988*.
Samples/jawed annelids: 84-306DF, 0.82–0.78 m below the reference level, *Lanceolatites gracilis, Kettnerites (K.) martinssonii, K. (K.) huberti, K. (K.) polonensis*, annelid jaws; 82-326DF, 0.05–0.10 m above the reference level, *Lanceolatites gracilis, Kettnerites (K.) martinssonii, K. (K.) huberti*, annelid jaws.

NÄRS FYR 1. Hamra Beds, lower part. *Age:* Late Ludlow, Whitcliffian.
References: Laufeld 1974a, b*.
Samples/jawed annelids: 83-9CB, east of Närs Fyr 1, extending the locality eastward, shore exposure in the southernmost part of Närsholmen immediately east of the small sea stack area, about 1 m above water level, close to a filled fissure, no annelid jaws; 83-8CB, fissure filling, fragments.

NÄRSHAMN 2. 634790 167246 (CJ 5883 4503), ca 9460 m ESE of Burs church. Topographical map sheet 5 I Hoburgen NO & 5 J Hemse NV. Geological map sheet Aa 156 Ronehamn.
An exposure in the water, about 1 m from the beach in the shallow harbour between the two jetties, closer to the northern than the southern jetty.
Burgsvik Beds, lower part(?).
References: Hede 1925a, the area in general.
Samples/jawed annelids: 83-12LJ, shallow exposure, *Kettnerites (K.) martinssonii, K. (A.) microdentatus, K. (K.) polonensis, Hindenites naerensis*, annelid jaws.

NÄRSHAMN 3. 634770 167218 (CJ 5852 4482), ca 9490 m ESE of Burs church. Topographical map sheet 5 I Hoburgen NO & 5 J Hemse NV. Geological map sheet Aa 156 Ronehamn.
Shore exposure with low sections with fossil wave marks, where the road is at its closest to the sea, about 200 m south of the southernmost jetty at Närshamn.
Burgsviks Beds, lower part(?).
References: Hede 1925a, p. 31, the area in general.
Samples/jawed annelids: 83-13LJ, above an uneven erosion surface, *Kettnerites (K.)* cf. *martinssonii, K. (A.) microdentatus*, annelid jaws.

NISSE 1. Hemse Beds, Hemse Marl NW part. *Age:* Ludlow.
References: Laufeld 1974a, b*; Larsson 1979.
Samples/jawed annelids: 75-98CB, in a water hole close to the ditch, nodular lutitic calcarenite in soft marl, *Kettnerites (K.) huberti*, annelid jaws.

NORRVANGE 1. Slite Beds, unit a. *Age:* Early Wenlock.
References: Laufeld 1974a, b*.
Samples/jawed annelids: 82-46CB, 0.65 m below the top of the inland cliff section, annelid jaws; 82-47CB, about 0.2 m below 82-46CB, no annelid jaws.

NORS 1. Högklint Beds, unit b upper part. *Age:* Early Wenlock.
References: Laufeld 1974a, b*.
Samples/jawed annelids: 82-50CB, uppermost exposed bed at the strand, no annelid jaws.

NORS STENBROTT 1. 642580 168514 (CK 7740 2166), ca 6230 m NNE of Fleringe church. Topographical map sheet 7 J Fårösund SO & NO. Geological map sheet Aa 171 Kappelshamn.
Quarry, situated at the end of the small road marked with red on the map. The quarry, which is not marked on the 2nd edition of the topographical map, is marked on the 3rd edition. The section is immediately SE of the entrance of the quarry.
Reference point: The easternmost point of the quarry close to and south of the entrance.
Reference level: The contact between the lower brownish-grey, thin-bedded calcarenite and the light-coloured, more thickly bedded calcarenite, about 1.85 m above the base of the quarry.
Högklint Beds, unit b(?). *Age:* Wenlock, Sheinwoodian.
Samples/jawed annelids: 82-49CB, about 1.6 m above the base of the section and 0.2–0.25 m below the very distinct contact between the light and the dark calcarenite, *Kettnerites (K.) abraham isaac, K. (K.)* aff. *bankvaetensis*, annelid jaws.

NYAN 1. Eke Beds, lower part. *Age:* Ludlow, Late Leintwardinian.
References: Laufeld 1974a, b*; Larsson 1979.
Sample/jawed annelids: 72-9LJ, excavated material, no annelid jaws; 72-10LJ, excavated material, *Kettnerites (K.)* cf. *bankvaetensis*, annelid jaws; 75-63CB, excavated material, paulinitid jaw.

NYAN 2. Hemse Beds, uppermost part and Eke Beds, lowest part. *Age:* Ludlow, Late Leintwardinian.
References: Laufeld 1974a, b*; Larsson 1979; Cherns 1983*; Fredholm 1988.
Samples/jawed annelids: Hemse Beds: 72-14LJ, 0.78–0.81 m below the reference level, fragments; 72-13LJ, 0.45–0.51 m below the reference level, *Kettnerites (K.) huberti*, annelid jaws; 72-12LJ, 0–0.10 m below the reference level, *Kettnerites (K.) huberti, K. (K.)* cf. *polonensis*, annelid jaws; Eke Beds: 72-11LJ, 0–0.21 m above the reference level, paulinitid jaws, annelid jaws; 75-62CB, uppermost bed in the section, no annelid jaws.

NYGÅRDS 1. Slite Beds, Slite Marl, *Pentamerus gothlandicus* Beds. *Age:* Wenlock.
References: Laufeld 1974a, b*; Larsson 1979; Ramsköld 1983.
Samples/jawed annelids: 75-14CB, about 0.75 m below ground level, *Kettnerites (K.) martinssonii, K. (A.) sisyphi sisyphi, Gotlandites slitensis*, annelid jaws.

NYGÅRDS 2. Halla Beds, unit b. *Age:* Late(?) Wenlock.
References: Laufeld 1974a, b*.
Samples/jawed annelids: 75-15CB, loose slabs, *Kettnerites (K.) bankvaetensis, K. (A.) sisyphi sisyphi*, annelid jaws; 81-75CB, shallow ditch, *Kettnerites (K.) bankvaetensis, K. (K.) martinssonii, K. (K.) polonensis, Hindenites gladiatus*, annelid jaws.

NYGÅRDSBÄCKPROFILEN 1. Lower and Upper Visby Beds. *Age:* Early Wenlock.
Reference: Jeppsson 1983*.
Samples/jawed annelids: Lower Visby Beds, unit e: 81-2LJ, 0.0 m a.s.l., *Kettnerites (K.) martinssonii, K. (K.) abraham abraham, K. (K.) versabilis*, annelid jaws; 79-42LJ, about 2.25 m a.s.l., *K. (K.) versabilis, K. (K.) versabilis* C form, *Lanceolatites gracilis, Kettnerites (K.) abraham abraham*, annelid jaws; Upper Visby Beds: 79-43LJ, about 4.6 m a.s.l., *Kettnerites (K.) abraham abraham*, annelid jaws.

NYHAMN 1. Lower Visby Beds, unit b. *Age:* Late Llandovery to Early Wenlock.
Reference: Laufeld l974a, b*, Larsson 1979.
Samples/jawed annelids: 79-217LJ, about 2.5 m a.s.l., *Kettnerites (K.) versabilis, K. (K.) abraham abraham*, annelid jaws; 81-28LJ, about 1.5 m a.s.l., jaw fragments.

NYHAMN 2. Lower Visby Beds unit b. *Age:* Late Llandovery to Early Wenlock.
References: Laufeld 1974b*, Larsson l979.
Samples/jawed annelids: 75-2CB, shore exposure, *Kettnerites (K.) abraham abraham*, annelid jaws.

NYHAMN 4. 640700 165521 (CK 4615 0523), ca 2720 m WSW of Lummelunda church. Topographical map sheet 7 J Fårösund SV & NV. Geological map sheet Aa 183 Visby & Lummelunda.
Shore exposure 40–50 m N of the bay and 5 m N of the small clearance in the beach cobble made to permit small boats to be dragged up on land. The locality is between the westernmost, medium-sized lichen-covered boulders on the shore.
Lower Visby Beds, unit b.
Samples/jawed annelids: 79-14LJ, surface exposure, *Kettnerites (K.) versabilis, K. (K.) martinssonii, K. (K.) abraham abraham, Kettnerites* sp., annelid jaws.

NYHAMN 5. 640620 165510 (CK 4592 0445), ca 3240 m SW of Lummelunda church. Topographical map sheet 7 J Fårösund SV & NV. Geological map sheet Aa 183 Visby & Lummelunda.
Small shore exposure between erratic boulders. The locality is marked on the geological map about 1000 m S of Nyhamn.
Lower Visby Beds, unit b.
Samples/jawed annelids: 84-11LJ, shore exposure, *Kettnerites (A.) microdentatus, Kettnerites* sp., annelid jaws.

NYMÅNETORP 1. Högklint Beds, unit b. *Age:* Early Wenlock.
References: Laufeld 1974a, b*; Laufeld & Martinsson 1981.
Samples/jawed annelids: 81-64LJ, 2.0 m below the reference level, no annelid jaws.

OIVIDE 1. Slite Beds, unit f (*Rhipidium tenuistriatum* Beds). *Age:* Early Wenlock.
References: Laufeld 1974a, b*; Larsson 1979; Jeppsson 1982, 1983.
Samples/jawed annelids: 79-44LJ, 0.5 m below the top of the section, *Kettnerites* fragment, annelid jaws.

OLLAJVS 1. Hamra Beds, unit c. *Age:* Late Ludlow.
References: Laufeld 1974a, b*.
Samples/jawed annelids: 76-17CB, ditch section, about 0.5 m above the bottom of the ditch, *Kettnerites (K.)* cf. *bankvaetensis*, annelid jaws.

OLSVENNE 3. Eke Beds, lowest part. *Age:* Ludlow, Leintwardinian.
References: Laufeld 1974a, b*.
Samples/jawed annelids: MS901AM, paulinitid sp., annelid jaws; 74-32KL, *Kettnerites* sp., annelid jaws.

ÖNDARVE 2. Hemse Beds, Hemse Marl, uppermost part, Eke, base. *Age:* Ludlow, Late Leintwardinian.
Reference: Fredholm 1988.
Samples/jawed annelids: Hemse Beds: 83-5LJ, ditch exposure, *Kettnerites (K.) martinssonii, K. (K.) huberti*, annelid jaws.

ÖSTERGARNSHOLM 1. Hemse Beds, unit e(?). *Age:* Ludlow, Leintwardinian.
References: Laufeld 1974a, b*; Larsson 1979; Jeppsson 1983.
Samples/jawed annelids: 78-19CB, about 0.50 m above the base of the section which is covered by loose gravel, no annelid jaws.

ÖSTERGARNSHOLM 2. Hemse Beds, unit e(?). *Age:* Ludlow, Leintwardinian.
References: Laufeld 1974a, b*; Larsson 1979; Jeppsson 1982; Fredholm 1988.
Sample/jawed annelids: 78-18CB, 0.6 m above the flat surface, just above sea level, no annelid jaws.

ÖSTERGARNSHOLM 3. Hemse Beds, unit e(?). *Age:* Ludlow, Leintwardinian.
Reference: Jeppsson 1983; Fredholm 1988*.
Sample/jawed annelids: 78-20CB, middle part of the shallow shore exposure, no annelid jaws.

OTES 1. Sundre Beds, middle part. *Age:* Latest Ludlow.
References: Laufeld 1974a, b*.
Samples/jawed annelids: 75-16LJ, about 200 m S of the monument, no annelid jaws.

PETSARVE 2. Eke Beds, middle–upper part. *Age:* Ludlow.
References: Laufeld 1974a, b*.
Samples/jawed annelids: 77-34CB, excavated material from ditch, *Lanceolatites gracilis, Kettnerites (K.) martinssonii, K. (K.) huberti*, annelid jaws.

PETSARVE 15. Eke Beds, middle–upper part. *Age:* Late Ludlow(?).
References: Laufeld 1974a, b*; Larsson 1979.
Samples/jawed annelids: 77-33CB, bottom of the ditch, *Kettnerites (K.) martinssonii, K. (K.) huberti*, annelid jaws.

PRÄSTBÅTELS 2. Klinteberg Beds, lower part. *Age:* Latest Wenlock.
References: Laufeld 1974a, b*, Frykman 1989.
Samples/jawed annelids: 75-93CB, 1.20 m below the top of the section and about 0.3–0.5 m above the base of the section, no annelid jaws; 75-94CB, 0.85 m above 75-93CB, no annelid jaws; 75-95CB, loose slab, no annelid jaws.

RÅGÅKRE 1. Slite Beds, Slite Siltstone, Halla Beds and Klinteberg Beds, lower part. *Age:* Late Wenlock.
References: Laufeld 1974a, b*; Sivhed 1976 (Dans 2); Bergman 1979b, 1980a; Frykman 1989.
Samples/jawed annelids: 76-31CB, from the limestone with wave marks, rich in bryozoans, at the base of the quarry, *Kettnerites (K.) martinssonii*, annelid jaws; 76-28CB, soft calcilutite above the wave marks in the floor of the quarry, *Kettnerites (K.)* cf. *martinssonii, K. (K.)* cf. *polonensis*, annelid jaws.

RANGSARVE 1. Hemse Beds, upper part. *Age:* Ludlow, Leintwardinian.
References: Laufeld 1974a, b*; Laufeld & Jeppsson 1976; Claesson 1979; Larsson 1979; Laufeld & Martinsson 1981; Ramsköld 1983; Jeppsson 1983; Fredholm 1988.
Samples/jawed annelids: 71-106LJ, base of the section, a more than 0.35 m thick bed, fragments; 71-107LJ, 0.35–0.40 m above the basal bed, fragments; 71-108LJ, 1.00–1.08 m above the base of the section, *Kettnerites (K.)* cf. *martinssonii, K. (A.) sisyphi*, annelid jaws; 71-109LJ, 2.28 m above the base of the section, annelid jaws; 71-110LJ, 3.18 m above the base of the section, fragmented paulinitid jaws, annelid jaws; 75-59CB, 0.5 m above road surface, *Kettnerites (A.)* cf. *sisyphi*, annelid jaws; 75-58CB, about 1.75 m above road surface, no annelid jaws; 75-57CB, about 3.0 m above road surface, no annelid jaws; 75-56CB, top of the section, 4.5 m above road surface, no annelid jaws.

ROBBJÄNS KVARN 2. Slite Beds, Slite Siltstone, and Halla Beds. *Age:* Wenlock.
References: Laufeld 1974a, b*; Sivhed 1976; Bergman 1979b, 1980a.
Samples/jawed annelids: Halla Beds: 71-139LJ, at the east–south-easternmost extension of the locality, annelid jaws.

ROBBJÄNS KVARN 3. 636361 164527 (CJ 3294 6274), ca 820 m SW of Klinte church. Topographical map sheet 6 I Visby SO. Geological map sheet Aa 160 Klintehamn.
Ditch exposure ca 100 m E of the house at Robbjäns Kvarn 2.
Mulde Beds, lower part. *Age:* Late Wenlock.
Samples/jawed annelids: 71-138LJ, ditch exposure, *Kettnerites (A.) sisyphi sisyphi*, annelid jaws.

RONEHAMN 1. Eke Beds and Burgsvik Beds. *Age:* Late Ludlow.
References: Laufeld 1974a, b*.
Samples/jawed annelids: Eke Beds: 75-84CB, loose slabs of excavated material about 15 m N of the northernmost oil storage tank, annelid jaws.

RONEHAMN 2. Eke Beds and Burgsvik Beds. *Age:* Late Ludlow.
Reference: Larsson 1979.
Samples/jawed annelids: Eke Beds: 75-24LJ, exposure on the shore, *Kettnerites (K.) polonensis*, annelid jaws; 75-25LJ, loose material, no annelid jaws.

RONNINGS 1. Eke Beds, upper part. *Age:* Ludlow, Whitcliffian.
References: Laufeld 1974a, b*; Larsson 1979; Jeppsson 1983.
Samples/jawed annelids: 71-190LJ, uppermost exposed bed, *Kettnerites (K.) martinssonii, K. (K.) huberti*, annelid jaws; 75-26LJ, 0.2 m below 71-190, *Kettnerites (K.) martinssonii*, annelid jaws.

RÖNNKLINT 1. Lower Visby Beds, Upper Visby Beds, and Högklint Beds. *Age:* Late Llandovery to Early Wenlock.
References: Hede 1940, p. 13 lines 16–17; Brood 1982, 1985; Jeppsson 1983*; Ramsköld 1984, 1985b.
Samples/jawed annelids: Lower Visby Beds, unit b: 81-11LJ, about 7.6 m below the reference level at the sea level, *Lanceolatites gracilis, Kettnerites (K.) abraham abraham, K. (A.) siaelsoeensis*, annelid jaws; 79-211LJ, about 6.10–6.30 m below the reference level, *Kettnerites (K.) martinssonii, K. (K.) abraham abraham*, annelid jaws; 79-212LJ, 4.05–4.00 m below the reference level, *Kettnerites (K.) martinssonii, K. (K.) abraham abraham*, annelid jaws; 79-213LJ, 2.05–1.95 m below the reference level, no annelid jaws; 79-214LJ, 0.05–0.08 m below the reference level, *Kettnerites*

(*A.*) *sisyphi sisyphi, K.* (*A.*) *siaelsoeensis, K.* (*K.*) *abraham abraham,* annelid jaws; 81-24LJ, 0.25–0.30 m above the reference level, *Kettnerites* (*K.*) *abraham abraham,* annelid jaws.

RUDVIER 1. Hemse Beds, unit c(?) and d. *Age:* Ludlow.
References: Mori 1970* loc. 116; Ramsköld 1986*; Fredholm 1988.
Samples/jawed annelids: Hemse Beds, unit c(?): 84-310DF, from a lens of dense marly limestone within the shale, *Kettnerites* (*K.*) *polonensis,* annelid jaws.

SAXRIV 1. Högklint Beds unit b. *Age:* Early Wenlock.
Reference: Hede 1933*:23 lines 22 to 30, page 24 lines 1–15; Larsson 1979*.
Reference point: The boulders leaning against the section seen on the photograph (Hede 1933, p. 25), the boundary of Hede's local units a–b.
Reference level: The deepest eroded bed about 0.5 m above water level.
Samples/jawed annelids: 84-77LJ, 0.45–0.40 m below the reference level, *Kettnerites* (*K.*) *martinssonii, K.* (*K.*) cf. *abraham, K.* (*A.*) *sisyphi sisyphi, Hindenites gladiatus,* annelid jaws; 84-79LJ, 3.8 m above the reference level, Hede's local unit c, fragments.

SIBBJÄNS 2. Hamra Beds, unit b. *Age:* Latest Ludlow.
References: Laufeld 1974a, b*.
Samples/jawed annelids: 82-32LJ, bottom of the ditch, *Kettnerites* (*K.*) *polonensis,* annelid jaws.

SIGVALDE 2. Hemse Beds, lower–middle (probably unit c, perhaps also unit d). *Age:* Early(?) Ludlow.
Reference: Jeppsson 1983*; Fredholm 1988.
Reference level: The upper surface of the *Ilionia prisca* bed (Hede 1925a:17).
Reference point: The fissure about 30 mm from the left side of the photograph in Munthe (1910) and Hede (1925a, p. 17).
Samples/jawed annelids: 71-116LJ, 0.95–0.90 m below the reference level, *Kettnerites* (*K.*) *martinssonii, K.* (*K.*) cf. *polonensis,* annelid jaws; 71-115LJ, the *Ilionia prisca* bed, *Kettnerites* (*K.*) *martinssonii, K.* (*K.*) *polonensis, K.* (*A.*) *sisyphi sisyphi,* annelid jaws; 71-117LJ, 0.70–0.80 m above the reference level and adjacent to the reference point, fragments.

SION 1. Slite Beds, unit f (*Rhipidium tenuistriatum* Beds). *Age:* Early Wenlock.
References: Laufeld 1974a, b*; Jeppsson 1983.
Samples/jawed annelids: 72-3LJ, shallow exposure, same sample level as 72-46SL (Laufeld 1974a), no annelid jaws.

SJÄLSÖ 1. Lower Visby Beds, unit b. *Age:* Late Llandovery.
Reference: Larsson 1979*.
Samples/jawed annelids: 79-12LJ, surface exposure, *Kettnerites* (*K.*) *abraham abraham, K.* (*A.*) *siaelsoeensis,* annelid jaws.

SJÄLSÖ 3. 639920 165190 (CJ 4227 9770), 4200 m W of Väskinde church. Topographical map sheet 6 J Roma NV & NO. Geological map sheet Aa 183 Visby & Lummelunda.
Bottom of the shallow harbour beside the jetty marked with the year 1970.
Lower Visby Beds, unit b (?). *Age:* Late Llandovery.
Samples/jawed annelids: 81-55LJ, topmost bed at the outer end of the jetty, annelid jaws.

SJAUSTREHAMMAR 1. 636453 168056 (CJ 6815 6097), ca. 3770 m S of Gammelgarn church. Topographical map sheet 6 J Roma SO. Geological map sheet Aa 170 Katthammarsvik.
Cliff section on the southern part of the peninsula Sjaustrehammar, SW of the triangulation point. The locality is shown on the photograph in Hede 1929, p. 45.
Hemse Beds, unit d. *Age:* Ludlow.
Reference: Hede 1929, pp. 44–45.
Samples/jawed annelids: 82-18LJ, 0.5 m below the top of the distinct, thinly bedded calcilutite, no annelid jaws; 82-19LJ, 0.5 m above the thinly bedded calcilutitic bed, *Kettnerites* (*K.*) *polonensis* var. sjaustre, *Kettnerites* (*K.*) sp., annelid jaws; 82-20LJ, the *Megalomus* limestone below the thinly bedded calcilutitic bed, no annelid jaws.

SKALASANDSVIK 2. 643068 170803 (DK 0045 2479), ca 11680 m ENE of Fårö church. Topographical map sheet 7 K Ullahau NV. Geological map sheet Aa 180 Fårö.
Shore section about 1340 m E of the triangulation point 22,3 at Lansa nässkifte, the area with ripple marks and west of them.
Slite Beds, Slite Marl. *Age:* Wenlock.
Reference level: The upper surface of the wave marks.
References: Hede 1936, p. 33; Bergman 1979b, p. 221, loc. 6 and 7.
Samples/jawed annelids: 80-5CB, 0.5 m above the wave mark level and 40 m W of the westernmost wave marks, fragments.

SKENALDEN 1. 640004 168599 (CJ 7628 9595), ca 9950 m SSE of Hellvi church. Topographical map sheet 7 J Fårösund SO & NO. Geological map sheet Aa 169 Slite.
Shore exposure on the southern point of the island Skenalden about 1 m above sea level.
Slite Beds, unit g. *Age:* Wenlock.
Samples/jawed annelids: 75-6CB, argillaceous calcarenite, rich in stromatoporoids, annelid jaws.

SKENALDEN 2. 640016 168603 (CJ 7633 9607), ca 9850 m SSE of Hellvi church. Topographical map sheet 7 J Fårösund SO & NO. Geological map sheet Aa 169 Slite.
Cliff section on the eastern shore of the island of Skenalden.
Reference point: The ca 1 m wide, dike-like structure composed of argillaceous material darker than the surrounding limestone.
Slite Beds, unit g. *Age:* Wenlock.
Samples/jawed annelids: 75-7CB, from the lower part of the structure, no annelid jaws; 75-8CB, about 1 m above previous sample and from the upper part of the section, paulinitid jaws, annelid jaws.

SLÄTHÄLLAR 3. Högklint Beds, middle part. *Age:* Early Wenlock.
Reference: Larsson 1979.
Samples/jawed annelids: 75-42KL, annelid jaws.

SLITEBROTTET 1 and 2. Slite Beds, Slite Marl and Slite g. *Age:* Wenlock.
References: Laufeld 1974a, b*; Walmsley & Boucot 1975, p. 65; Eisenack 1975, Fig. 18; Larsson 1979; Jeppsson 1983*; Odin et al. 1984, 1986.
Samples/jawed annelids: SLITEBROTTET 1 (all levels are given in relation to the top of the section, except those referring to marker levels): 73-32LJ, 0.15 m, *Kettnerites* (*K.*) *martinssonii, K.* (*A.*) *sisyphi sisyphi, Gotlandites slitensis, Hindenites* sp., annelid jaws; 73-33LJ, 2.0 m, *Gotlandites sp.,* annelid jaws; 73-34LJ, 4.40 m below the top of the section and 1.20 m above a 0.37 m thick, deeply eroded marly, level which can be followed around the quarry, *Kettnerites* (*K.*) *martinssonii, K.* (*A.*) *sisyphi sisyphi, K.* (*K.*) cf. *jacobi, Gotlandites slitensis, Hindenites* sp., annelid jaws; 73-36LJ, 8.55 m below the top of the section, immediately below a 0.20 m, deeply eroded level and 0.50 m above another deeply eroded level (0.23 m thick), no annelid jaws; 73-37LJ, 10.45 m, *Kettnerites* (*K.*) *bankvaetensis,* annelid jaws; 73-38LJ, 11.45 m below the top of the section and 0.45 m below a drilled drainage hole in the section, *Kettnerites* (*K.*) *martinssonii, K.* (*A.*) *sisyphi sisyphi,* annelid jaws; 73-39LJ, 13.60 m fragments; 73-40LJ, 15.70 m, *Kettnerites* (*K.*) *martinssonii, K.* (*A.*) *sisyphi sisyphi, Gotlandites slitensis,* annelid jaws; 73-41LJ, 18.60 m, *Kettnerites* (*K.*) *bankvaetensis, K.* (*K.*) *martinssonii, K.* (*A.*) *sisyphi sisyphi,* annelid jaws; 73-42LJ, 20.90 m, *Kettnerites* (*K.*) *martinssonii, K.* (*A.*) *sisyphi sisyphi, Hindenites* sp., annelid jaws; 73-44LJ, 25.60 m, no annelid jaws. SLITEBROTTET 2 (levels are related to the top of the section measured from Slitebrottet 1 and thus forming a continuous section): 83-31LJ, 35.90 m below the top of the section and 0.00–0.10 m above the lower reference level, *Kettnerites* (*K.*) *martinssonii, K.* (*K.*) *jacobi, K.* (*K.*) *polonensis, K.* (*A.*) cf. *microdentatus, Hindenites* cf. *gladiatus,* annelid jaws; 83-32LJ, 36.10 m below the top of the section and 0.00–0.20 m below the lower reference level, annelid jaws; 83-33LJ, 37.70 m below the top of the section and 1.67–1.70 m below the lower reference level, *Kettnerites* (*K.*) *martinssonii, K.* (*K.*) *jacobi, K.* (*A.*) cf. *sisyphi,* annelid jaws; 83–34, 40.60 m, *Kettnerites* (*K.*) *martinssonii, K.* (*K.*) *jacobi, K.* (*A.*) cf. *microdentatus,* annelid jaws; 73-48LJ, 40.80 m, *Kettnerites* (*K.*) *martinssonii, K.* (*K.*) *jacobi, K.* (*A.*) *sisyphi sisyphi, K.* (*A.*) *microdentatus,* annelid jaws; 83-35LJ, 41.70 m, *Kettnerites* (*K.*) *martinssonii, K.* (*K.*) *jacobi, K.* (*A.*) *sisyphi sisyphi* annelid jaws; 73-49LJ, 42.90 m, paulinitid fragments, annelid jaws; 83-36LJ, 43.95 m, *Kettnerites* (*K.*) *martinssonii, K.* (*K.*) *jacobi, K.* (*A.*) *sisyphi sisyphi, Hindenites gladiatus,* annelid jaws; 84-1LJ, 49.60 m, *Kettnerites* (*K.*) *jacobi, K.* (*A.*) *sisyphi sisyphi,* annelid jaws.

SMISS 1. Klinteberg Beds, Klinteberg Marl, top. *Age:* Ludlow.
References: Laufeld 1974a, b*; Larsson 1979; Frykman 1989.
Samples/jawed annelids: 75-76CB, loose slabs from the temporary excavation, *Kettnerites* (*A.*) *microdentatus,* annelid jaws.

SNÄCK 1. 639670 165075 (CJ4096 9529), ca 5900 m WSW of Väskinde church. Topographical map sheet 6 J Roma NV & NO. Geological map sheet Aa 183 Visby & Lummelunda.
Temporary excavation, about 100 m NW of the northwesternmost corner of the main building of Snäck, not marked on the topographical map.
Lower Visby Beds. Age: Late Llandovery.
Samples/jawed annelids: 83-1CB, about 1.5 m below the road surface, *Kettnerites* (*A.*) cf. *microdentatus,* annelid jaws.

SNÄCKGÄRDSBADEN 1. Upper Visby Beds and Högklint Beds, unit a. *Age:* Early Wenlock.
References: Laufeld 1974a, b*, Larsson 1979.
Samples/jawed annelids: Upper Visby Beds: 79-22LJ, 0.5 m above road surface, lowest part of the section, no annelid jaws; 79-25LJ, 6.3 m above sample 79-22, probably about 1.0 m below the reference level, *Lanceolatites gracilis, Kettnerites* (*K.*) *martinssonii, K.* (*K.*) *abraham abraham,* annelid jaws; 79-26LJ, about 2 m below the reference level, *Kettnerites* (*K.*) *abraham abraham,* annelid jaws; Högklint Beds, unit a: 79-27LJ, lowest Högklint Beds, unit a, the reference level, *Kettnerites* (*K.*) *martinssonii,* annelid jaws; 79-28LJ, about 2 m above the reference level, *Kettnerites* (*K.*) *martinssonii, K.* (*A.*) *sisyphi,* annelid jaws.

SNAUVALDS 1. Hemse Beds, Hemse Marl NW part. *Age:* Ludlow, Bringewoodian or more probably Early Leintwardinian.
References: Laufeld 1974a, b*; Larsson 1979; Jeppsson 1983; Fredholm 1988.
Samples/jawed annelids: 71-14LJ, ditch section, *Kettnerites* (*K.*) *huberti, K.* (*A.*) *sisyphi sisyphi,* annelid jaws.

SNODER 2. Hemse Beds, Hemse Marl NW part. *Age:* Ludlow.
References: Laufeld 1974a, b*; Larsson 1979; Ramsköld 1983, 1984, 1985b.
Samples/jawed annelids: 82-14CB, loose slab from the drainage ditch, *Lanceolatites gracilis, Kettnerites* (*K.*) *martinssonii* var. mulde, *K.* (*A.*) *sisyphi sisyphi, K.* (*A.*) *fjaelensis,* annelid jaws.

SNODER 3. Hemse Beds, Hemse Marl NW part. *Age:* Ludlow, Bringewoodian or Early Leintwardinian.
Reference: Jeppsson 1983*; Fredholm 1988.
Samples/jawed annelids: 71-10LJ, bottom of the small ditch, *Kettnerites* (*K.*) *martinssonii* var. mulde, *K.* (*A.*) cf. *sisyphi,* annelid jaws.

SÖDRA TORG 1. 639279 164840 (CJ 3828 9160), ca 800 SW of Visby cathedral. Topographical map sheet 6 I Visby NO. Geological map sheet Aa 183 Visby & Lummelunda.
Temporary excavation along the western side of the market place (Södra Torg), about 28–29 m a.s.l.
Högklint Beds, unit c. *Age:* Early Wenlock.
Samples/jawed annelids: 75-6LJ, about 0.75 m below the top of the section, no annelid jaws; 75-7LJ, about 0.7 m above 75-6, no annelid jaws.

SOJVIDE 1. Slite Beds, unit f (*Rhipidium tenuistriatum* Beds). *Age:* Early Wenlock.
References: Larsson 1979*; Jeppsson 1983.
Samples/jawed annelids: 81-58LJ, very small ditch exposure, *Kettnerites* (*A.*) *sisyphi sisyphi,* paulinitid sp., annelid jaws.

SPROGE 3. 635024 164538 (CJ 3203 4960), ca 280 m NE of Sproge church. Topographical map sheet 6I Visby SO. Geological map sheet Aa 164 Hemse.
Ditch exposure along and SE of the road between Levide and Sproge, 25 m NE of the minor road towards the SE and about 225 m E of the main road (140).
Hemse Beds, Hemse Marl NW part. *Age:* Ludlow.
Samples/jawed annelids: 79-9LJ, bottom of the ditch, no annelid jaws.

SPROGE 4. 635016 164434 (CJ 3097 4940), ca 840 m W of Sproge church. Topographical map sheet 6I Visby SO. Geological map sheet Aa 164 Hemse.
Ditch exposure along and south of the road, about 100 m W of the farm house at Alvegårde, just E of the entrance to the field.
Hemse Beds, Hemse Marl NW part. *Age:* Ludlow.
Samples/jawed annelids: 84-80LJ, bottom of the ditch, *Kettnerites* (*K.*) *martinssonii* var. mulde, *K.* (*A.*) *sisyphi sisyphi, K.* (*A.*) *microdentatus, K.* (*A.*) *fjaelensis,* annelid jaws.

SPROGE 5. 634984 164382 (CJ 3045 4910), ca 1380 m WSW of Sproge church. Topographical map sheet 5 I Hoburgen NO & 5 J Hemse NV. Geological map sheet Aa 164 Hemse.
Road-side ditch exposure E of the road WSW of Alvegårde, about 200 m N of the small road leading to the farm SSW of Alvegårde. The locality is where a small ditch runs from a bushy area east of the road and meets the main ditch beside the road.
Hemse Beds: Hemse Marl NW part. *Age:* Ludlow.
Samples/jawed annelids: 84-81LJ, 0.5 m above the bottom of the ditch, annelid jaws.

STAVE 1. Slite Beds, Slite Marl, central part. *Age:* Early Wenlock.
References: Laufeld 1974a, b*; Larsson 1979.
Samples/jawed annelids: 75-11CB, 0.5 m below ground surface, *Kettnerites* (*K.*) *martinssonii, K.* (*K.*) cf. *jacobi, K.* (*A.*) *microdentatus,* annelid jaws.

STENSTU 2. 637539 167433 (CJ 6278 7224), ca 1150 m SSE of Anga church. Topographical map sheet 6 J Roma NV & NO. Geological map sheet Aa 170 Katthammarsvik.
Ditch exposure W of the old road at point 12,8; note that the road has now been rebuilt.
Klinteberg Beds, unit f. *Age:* Early Ludlow.
Samples/jawed annelids: 75-39LJ, ditch exposure, fragments.

STENSTUGÅRDS 1. Klinteberg Beds, lower part. *Age:* Latest Wenlock.
References: Laufeld 1974a, b*.
Samples/jawed annelids: 78-7CB, 0.15 m below the top of the section in the SE corner of the quarry, no annelid jaws.

STORA BANNE 2. Slite Beds, unit g, lower part. *Age:* Wenlock.
Reference: Laufeld 1974b*.
Samples/jawed annelids: 69-1LJ, 0.40–0.50 m above the base of the section, no annelid jaws.

STORA BANNE 3. Slite Beds, unit g, lower part. *Age:* Wenlock.
Reference: Laufeld 1974b*.
Samples/jawed annelids: 69-2LJ, 1.40 m below the top of the section, no annelid jaws; 69-3LJ, 3.55 m below the top of the section, no annelid jaws.

STORA KRUSE 1. 634487 165311 (CJ 3932 4343), ca 1380 m WSW of Alva church. Topographical map sheet 5 I Hoburgen NO & 5 J Hemse NV. Geological map sheet Aa 164 Hemse.
Ditch, 25 m E of the easternmost house (not marked on the topographical map sheet) at Store Kruse, a ditch running N–S along the boundary between the two fields, north of and by the road.
Hemse Beds, Hemse Marl SE part. *Age:* Ludlow.
Samples/jawed annelids: 82-29LJ, loose slabs within the ditch, *Lanceolatites gracilis, Langeites glaber, Kettnerites* (*A.*) *sisyphi klasaardensis, K.* (*K.*) *huberti,* annelid jaws.

STORA MAFRIDS 2. 637287 164225 (CJ 3060 7221), ca 2500 m NW of Västergarn church. Topographical map sheet 6 I Visby SO. Geological map sheet Aa 160 Klintehamn.
Ditch exposure, 15 to 20 m NW of the road from the small ditch that runs NW to Idån.
Slite Beds, unit f, *Rhipidium tenuistriatum* zone or slightly younger.
Samples/jawed annelids: 84-67LJ, excavated material, *Kettnerites* (*K.*) *jacobi, K.* (*A.*) *microdentatus,* annelid jaws.

STORA MYRE 1. Slite Beds, unit d. *Age:* Early Wenlock.
References: Laufeld 1974a, b*; Larsson 1979; Jeppsson 1983.
Samples/jawed annelids: 71-74LJ, surface exposure, *Kettnerites* (*K.*) *martinssonii, K.* (*K.*) *bankvaetensis,* annelid jaws; 75-29CB, surface exposure, *Kettnerites* (*K.*) *martinssonii, K.* (*A.*) *sisyphi sisyphi,* annelid jaws.

STORA TUNE 1. Klinteberg Beds, lower part. *Age:* Latest Wenlock.
References: Laufeld 1974a, b*; Frykman 1989.

Samples/jawed annelids: 78-8CB, a surface exposure SE of the building and 3 m from the road, no annelid jaws.

STORA VIKARE 2. Halla Beds, unit b. *Age:* Late Wenlock.
References: Laufeld 1974a, b*.
Samples/jawed annelids: 75-18CB, shallow excavation, *Kettnerites* (*K.*) *polonensis, K.* (*A.*) *sisyphi sisyphi,* annelid jaws.

STORUGNS 1. Slite Beds, unit e (Kalbjerga Limestone). *Age:* Early Wenlock.
Reference: Laufeld 1974b*.
Samples/jawed annelids: 78-1CB, about 1.5 m above the base of the section, no annelid jaws; 78-2CB, about 6 m above the base of the section, crinoid- and stromatoporoid-rich limestone immediately above a homogeneous white limestone, no annelid jaws; 69-16LJ, no annelid jaws.

STORUGNS 1B. 641572 167915 (CK 7058 1217), ca 5100 m NNE of Lärbro church. Topographical map sheet 7 J Fårösund SO & NO. Geological map sheet Aa 171 Kappelshamn.
Part of the large quarry Storugns, ca 320 m NNE of point 28,3 at Storugns and about 160 m E of Storugns 1, where the road from Klinthagen quarry, S of Storugns, enters the quarry.
Reference level: A marker level high up on the walls can be followed around a large part of the quarry. It consists of 0.52 m argillaceous limestone with several pyrite horizons. The lower boundary consists of a bed up to 10 mm thick, with pyrite, the base of which is the reference level. There are also a number of higher pyrite levels, up to 1.10 m above the reference level. *Slite Beds,* units e and g. *Age:* Wenlock.
Samples/jawed annelids: 84-29LJ, 0.01–0.20 m above the reference level, *Kettnerites* (*K.*) *bankvaetensis,* annelid jaws; 84-28LJ, 0.32–0.52 m above the reference level, *Kettnerites* (*K.*) *bankvaetensis, K.* (*K.*) *martinssonii,* annelid jaws.

STRANDAKERSVIKEN 1. Högklint Beds, unit c and Slite Beds, unit c. *Age:* Early Wenlock.
References: Laufeld 1974a, b*.
Samples/jawed annelids: Högklint Beds, unit c: 77-21CB, about 0.4 m below the reference level, *Kettnerites* (*K.*) *martinssonii, K.* (*K.*) *polonensis, Hindenites angustus,* annelid jaws. Slite Beds: 77-22CB, 0.1–0.2 m above the reference level, no annelid jaws.

STRANDS 1. Hamra Beds, unit b. *Age:* Latest Ludlow.
References: Laufeld 1974a, b*; Larsson 1979; Jeppsson 1983.
Samples/jawed annelids: 75-14LJ, bottom of the ditch, *Lanceolatites* sp. A, *Kettnerites* (*K.*) *bankvaetensis, K.* (*K.*) *huberti, K.* (*K.*) cf. *polonensis,* annelid jaws; 76-19CB, bottom of the ditch, *Lanceolatites* sp. A, *Kettnerites* (*K.*) *bankvaetensis, K.* (*K.*) *huberti,* annelid jaws.

STUTSVIKEN 1. Högklint Beds unit c. *Age:* Early Wenlock.
References: Laufeld 1974a, b*.
Samples/jawed annelids: 77-20CB, *Kettnerites* (*K.*) *martinssonii, K.* (*K.*) *abraham isaac,* annelid jaws.

SUDERBYS 3. 638740 167425 (CJ 6356 8423), ca 1050 m WNW of Gothem church. Topographical map sheet 6 J Roma NV & NO. Geological map sheet Aa 169 Slite.
Ditch exposure, directly north of Suderbys in the ditch along the road and about 1 m W of the N–S ditch.
Klinteberg Beds, unit b or c. *Age:* Late Wenlock.
Reference: Hede 1928, p. 58, lines 28–31.
Samples/jawed annelids: 75-35LJ, ditch exposure, *Kettnerites* (*K.*) *bankvaetensis,* annelid jaw fragments.

SUDERS 1. Sundre Beds, middle part. *Age:* Latest Ludlow.
References: Laufeld 1974a, b*.
Samples/jawed annelids: 75-18LJ, bottom of the ditch, annelid jaws; 76-27CB, about 0.2 m below the uppermost bed, annelid jaws.

SUTARVE 2. Hemse Beds, lower part or Klinteberg Beds, unit f, top. *Age:* Early Ludlow.
References: Laufeld 1974a, b*; Laufeld & Jeppsson 1976; Jeppsson 1983; Frykman 1989.
Samples/jawed annelids: 71-102LJ, ditch exposure, annelid jaws.

SUTARVE 3. Klinteberg Beds, unit f, upper part. *Age:* Early Ludlow.
References: Laufeld 1974a, b*; Frykman 1989.
Samples/jawed annelids: 75-49CB, the sample belongs somewhere within the uppermost 4 m of unit f, no annelid jaws.

SVARVARE 1. Slite Beds, Slite Marl, *Pentamerus gothlandicus* Beds. *Age:* Wenlock.
References: Laufeld 1974a, b*; Larsson 1979; Sivhed 1976; Bergman 1980a; Ramsköld 1985b.
Samples/jawed annelids: 75-6KL, *Kettnerites* (*A.*) *sisyphi* var. valle, annelid jaws; 75-7KL, *Kettnerites* (*K.*) *martinssonii, K.* (*A.*) *sisyphi* var. valle, annelid jaws.

SVARVARE 3. Slite Beds, Slite Marl, *Pentamerus gothlandicus* Beds. *Age:* Wenlock.
Reference: Larsson 1979*.
Samples/jawed annelids: 81-57LJ, ditch, *Kettnerites* (*A.*) *sisyphi* var. valle, annelid jaws.

SVARVEN 1. Högklint Beds unit c. *Age:* Early Wenlock.
References: Laufeld 1974a, b*, Larsson 1979, Jeppsson 1983.
Samples/jawed annelids: 78-3CB, 0.30 m above the reference level, *Kettnerites* (*K.*) *abraham* cf. *isaac,* annelid jaws; 84-34LJ, 2.40–2.20 m below the reference level, *Hindenites gladiatus, Kettnerites* (*K.*) *martinssonii, K.* (*K.*) *abraham isaac, K.* (*A.*) *sisyphi sisyphi,* annelid jaws.

SYSNE 1. Hemse Beds, unit d(?). *Age:* Ludlow, Late Leintwardinian.
References: Larsson 1979*; Jeppsson 1983; Fredholm 1988.
Samples/jawed annelids: 81-48LJ, at the same level as the road surface, annelid jaws; 81-50LJ, 4.40 m above 81-48, fragments.

SYSNEUDD 1. Hemse Beds, unit d. *Age:* Ludlow.
Reference: Larsson 1979*.
Samples/jawed annelids: 78-5CB, from the base of the bioherm, no annelid jaws; 78-22CB, 0.30 m above 78-5CB, no annelid jaws.

TALINGS 1. Slite Beds, unit g. *Age:* Wenlock.
References: Laufeld 1974a, b*; Larsson 1979.
Samples/jawed annelids: 80-2CB, at the same level as the road surface, *Kettnerites* (*K.*) cf. *martinssonii, K.* (*A.*) *sisyphi sisyphi, Gotlandites slitensis,* annelid jaws.

TÄNGLINGS 2. Hemse Beds, lower–middle part. *Age:* Ludlow.
References: Laufeld 1974a, b*; Larsson 1979; Fredholm 1988.
Samples/jawed annelids: 75-89CB, 0.50 m below the surface of the road, annelid jaws; 75-90CB, 0.20 m above the road surface, annelid jaws.

TÄNGLINGS KVARN 1. Hemse Beds, probably middle or upper part. *Age:* Ludlow.
Reference: Stridsberg 1985*; Fredholm 1988.
Sample/jawed annelids: 83-1SS, from the upper part of the southern wall of the section, no annelid jaws.

TINGS 1. Klinteberg Beds, unit f. *Age:* Early Ludlow.
References: Laufeld 1974a, b*; Larsson 1979; Frykman 1989.
Samples/jawed annelids: 71-214LJ, 0.45-0.55 m below the top of the section, no annelid jaws.

TINGSTÄDE 1. 640453 166689 (CK 5760 0184), ca 510 m SW of Tingstäde church. Topographical map sheet 7 J Fårösund SV & NV. Geological map sheet Aa 169 Slite.
Temporary excavation in connection with the construction of houses, about 20 m NW of the bend in the main road. The sample was collected near the southeasternmost point of the row of small houses. The bluish marl is very rich in stromatoporoids.
Slite Beds, unit g(?). *Age:* Wenlock.
Sample/jawed annelid: 80-6CB, excavated material from about 1–2 m below ground surface, no annelid jaws.

TJÄNGDARVE 1. Hemse Beds, Hemse Marl SE. *Age:* Ludlow, Leintwardinian.
References: Laufeld 1974a, b*; Larsson 1979; Fredholm 1988.
Samples/jawed annelids: 71-188LJ, loose slabs on the field, *Kettnerites* (*K.*) *huberti,* annelid jaws.

TJAUTET 1. Slite Beds, unit e and lower part of unit g. *Age:* Wenlock.
References: Laufeld 1974a, b*; Larsson 1979.
Samples/jawed annelids: Slite Beds, unit e: 72-6LJ, 0.65–0.69 m below the reference level, no annelid jaws; 72-7LJ, 0.00–0.05 m below the reference level, no annelid jaws.

TJELDERSHOLM 1. Slite Beds, *Pentamerus gothlandicus* Beds, and beds immediately younger. *Age:* Late(?) Wenlock.
References: Laufeld 1974a, b*; Larsson 1979; Jeppsson 1983*; Ramsköld 1985b.
Samples/jawed annelids: 73-8LJ, 0.35–0.40 m below the reference level, *Kettnerites* (*K.*) *martinssonii,* annelid jaws; 73-9LJ, 0.0–0.07 m below the reference level, *Hindenites* sp., *Kettnerites* (*K.*) *martinssonii, K.* (*K.*) *jacobi, K.* (*A.*) *sisyphi sisyphi, Gotlandites slitensis,* annelid jaws; 75-31CB, about 0.35 m above the reference level, *Kettnerites* (*A.*) *sisyphi sisyphi, Gotlandites slitensis,* annelid jaws; 75-33CB, about 0.35 m above the reference level and 30 m NW of the reference point, *Kettnerites* (*K.*) *martinssonii, K.* (*A.*) *sisyphi sisyphi, Gotlandites slitensis,* annelid jaws; 75-32CB, 0.40 m above the reference level, *Gotlandites slitensis,* annelid jaws; 73-10LJ, 0.50–0.55 m above the reference level, *Kettnerites* (*K.*) *martinssonii, K.* (*A.*) *sisyphi sisyphi, Gotlandites slitensis,* annelid jaws; 73-11LJ, about 3.70 m above the reference level, fragments.

TJULS 3. Slite Beds, Slite Marl. *Age:* Wenlock.
Reference: Larsson 1979*.
Samples/annelid jaws: 84-68LJ, immediately SE of the bridge, ditch exposure, fragments.

TOMASE 1. Slite Beds, unit g. *Age:* Wenlock.
Reference: Larsson 1979*; Ramsköld 1983.
Samples/jawed annelids: 82-39LJ, south of the road, ditch section, 0.40–0.60 m below the plastic clay layer mentioned by Hede, no annelids; 82-41LJ, 0.60–0.70 m above the plastic clay layer, no annelids.

TOMTBODARNE 1. 634368 166522 (CJ 5128 4136), ca 6250 m ESE of Rone church. Topographical map sheet 5 I Hoburgen NO & 5 J Hemse NV. Geological map sheet Aa 156 Ronehamn.
Shallow shore exposure marked on the geological map, about 350 m E of the bend in the road close to the fishing huts marked on the topographical map sheet.
Eke Beds. Age: Ludlow, possibly Late Leintwardinian.
Samples/jawed annelids: 84-110CB, uppermost bed, *Kettnerites* (*K.*) *martinssonii, K.* (*K.*) cf. *huberti, K.* (*A.*) cf. *fjaelensis,* annelid jaws.

TRÄDGÅRDEN 1. Slite Beds and Halla Beds. *Age:* Late Wenlock.
References: Laufeld 1974a, b*; Larsson 1979.
Samples/jawed annelids: Slite Beds, Slite Marl, *Pentamerus gothlandicus* Beds: 71-57LJ, 3.40 m above the base of the section and about 1.60 below the reference level, *Kettnerites* (*K.*) *polonensis, K.* (*A.*) *sisyphi sisyphi,* annelid jaws.

TRÄSKE 1. Hemse Beds, unit b. *Age:* Ludlow.
References: Laufeld 1974a, b*; Larsson 1979; Fredholm 1988.
Samples/jawed annelids: 71-103LJ, low ditch exposure, *Kettnerites* (*K.*) *martinssonii, K.* (*K.*) *polonensis, K.* (*K.*) cf. *huberti,* annelid jaws.

TULE 1. KLinteberg Beds, unit e(?) *Age:* Latest Wenlock or earliest Ludlow.
References: Laufeld 1974a, b*; Frykman 1989.
Samples/jawed annelids: 71-95LJ, bed accessible less than 0.1–0.2 m below soil surface, paulinitid fragments, annelid jaws.

VAKTÅRD 2. Hemse Beds, Hemse Marl SE part. *Age:* Ludlow, Leintwardinian.
References: Larsson 1979*.
Samples/jawed annelids: 81-36LJ, shore exposure, *Lanceolatites gracilis, Kettnerites* (*A.*) *sisyphi klasaardensis, K.* (*K.*) *burgensis,* annelid jaws.

VAKTÅRD 3. Hemse Beds, Hemse Marl, SE part. *Age:* Ludlow, Leintwardinian.
Reference: Jeppsson 1983; Fredholm 1988.
Samples/jawed annelids: 71-149LJ, shallow shore exposure, *Kettnerites* (*K.*) *burgensis,* annelid jaws.

VAKTÅRD 4. 633359 164611 (CJ 3168 3275), ca 3100 m WSW of Näs church. Topographical map sheet 5 I Hoburgen NO & 5 J Hemse NV. Geological map sheet Aa 152 Burgsvik.
Ditch exposure south of and along the road, 200 m W of Vaktård 5.
Hemse Beds, Hemse Marl SE part. *Age:* Ludlow, Leintwardinian.
Samples/jawed annelids: 81-35LJ, shallow ditch exposure, *Lanceolatites gracilis, Kettnerites* (*K.*) *huberti, K.* (*A.*) *sisyphi klasaardensis, K.* (*K.*) *burgensis,* annelid jaws.

VAKTÅRD 5. 633364 164651 (CJ 3188 3276), ca 2400 m WSW of Näs church. Topographical map sheet 5 I Hoburgen NO & 5 J Hemse NV. Geological map sheet Aa 152 Burgsvik.
Ditch exposure south of and along the road, 150 m W of the highest point of the road. North of the road is the boundary between pasture and tilled land.
Hemse Beds, Hemse Marl SE part. *Age:* Ludlow, Leintwardinian.
Samples/jawed annelids: 81-37LJ, shallow ditch exposure, *Lanceolatites gracilis, Kettnerites* (*A.*) *sisyphi klasaardensis, K.* (*K.*) *huberti, K.* (*K.*) *burgensis,* annelid jaws.

VALBYBODAR 1. 635895 164258 (CJ 2987 5829), ca 1030 m WSW of Fröjel church. Topographical map sheet 6 I Visby SO. Geological map sheet Aa 164 Hemse.
A shore exposure, about 10 m long, just north of the small bay at Valbybodar, 175 m NW of the single house marked on the first edition of the topographical map sheet.
Slite Beds, Slite Siltstone. *Age:* Wenlock.
Reference: Hede 1942, loc 1a.
Samples/jawed annelids: 82-26LJ, 0.2 m below mean water level, *Hindenites gladiatus, Kettnerites* (*K.*) cf. *bankvaetensis, K.* (*K.*) *martinssonii, K.* (*K.*) *polonensis, K.* (*A.*) *sisyphi* cf. var. valle, annelid jaws.

VALBYTTE 1. Slite Beds, Slite Marl slightly younger than the *Rhipidium tenuistriatum* Beds. *Age:* Wenlock.
References: Laufeld 1974a, b*; Larsson 1979; Ramsköld 1983, 1984; 1985b.
Samples/jawed annelids: 76-45KL, *Kettnerites* (*K.*) *jacobi,* annelid jaws.

VALBYTTE 2. Slite Beds, Slite Marl slightly younger than the *Rhipidium tenuistriatum* Beds. *Age:* Wenlock.
Reference: Larsson 1979*.
Samples/jawed annelids: 75-30KL, annelid jaws.

VALLE 1. Slite Beds, *Pentamerus gothlandicus* Beds. *Age:* Late(?) Wenlock.
References: Laufeld 1974a, b*; Larsson 1979.
Samples/jawed annelids: 66-144SL, 0.0–0.15 m below top of the section, *Kettnerites* (*A.*) *sisyphi* var. valle, annelid jaws.

VALLE 2. Slite Beds, *Pentamerus gothlandicus* Beds. *Age:* Late(?) Wenlock.
References: Laufeld 1974a, b*; Larsson 1979.
Samples/jawed annelids: 66-145SL, 1.00–1.25 m below the top of the section, *Kettnerites* (*K.*) *martinssonii, K.* (*A.*) *microdentatus, K.* (*A.*) *sisyphi* var. valle, annelid jaws.

VALLEVIKEN 1. Slite Beds, Slite Marl. *Age:* Wenlock.
References: Laufeld 1974a, b*; Larsson 1979; Ramsköld 1985b.
Samples/jawed annelids: 70-10LJ, 0.40–0.50 m below the top of the section in the northern part, *Kettnerites* (*K.*) cf. *martinssonii,* annelid jaws; 70-11LJ, 0.10 m above water level, *Kettnerites* (*A.*) *microdentatus,* annelid jaws; 81-1CB, sample from the NE part of the quarry. The bed of which the upper surface is developed as a hardground, covers a large area some metres below the top of the ground surface, *Kettnerites* (*K.*) *bankvaetensis, K.* (*K.*) *martinssonii, K.* (*A.*) *microdentatus,* annelid jaws.

VALLMYR 1. Klinteberg Beds, unit d. *Age:* Wenlock, very close to the end.
References: Larsson 1979*; Jeppsson 1983; Frykman 1989.
Samples/jawed annelids: 75-38LJ, on the northern side of the ditch, *Kettnerites* (*K.*) *martinssonii,* annelid jaws.

VALLSTENA 2. Slite Beds, *Pentamerus gothlandicus* Beds or slightly older. *Age:* Late(?) Wenlock.
References: Laufeld 1974a, b*; Larsson 1979.
Samples/jawed annelids: 67-23LJ, excavated material, *Kettnerites* (*K.*) cf. *bankvaetensis, K.* (*K.*) cf. *martinssonii, Gotlandites slitensis,* annelid jaws; 77-2CB, 0.0–0.1 m of the top of the section, *Kettnerites* (*K.*) cf. *bankvaetensis, Gotlandites slitensis, Hindenites gladiatus,* paulinitid sp., annelid jaws; 77-3CB, 0.50 m below the top of

the section, *Kettnerites* (*K.*) *martinssonii, K.* (*K.*) *jacobi, K.* (*A.*) *sisyphi sisyphi, Gotlandites slitensis, Hindenites gladiatus*, annelid jaws.

VALVE 3. 637322 164010 (CJ 2851 7273), ca 2530 m NNW of Västergarn church. Topographical map sheet 6 I Visby SO. Geological map sheet Aa 160 Klintehamn.

Ditch exposure east of the house at the western side of the road running N–S from the cross-roads south of Valve towards Sigvards (NE of Paviken).
Slite Beds, unit a. *Age:* Wenlock.
References: Hede 1927a, p. 24, lines 34–44.
Samples/jawed annelids: 84-66LJ, ditch exposure, *Kettnerites* (*K.*) *martinssonii, K.* (*A.*) *microdentatus, K.* (*A.*) *sisyphi sisyphi*, annelid jaws.

VÄRSÄNDE 1. Mulde Beds, lowest part. *Age:* Late Wenlock.
References: Laufeld 1974a, b*.
Samples/jawed annelids: 75-23CB, shallow ditch section, *Kettnerites* (*K.*) *martinssonii, K.* (*A.*) *fjaelensis, K.* (*A.*) *sisyphi sisyphi*, annelid jaws; 82-41CB, lowest bed, rich in trilobites, *Kettnerites* (*K.*) *martinssonii, K.* (*A.*) *microdentatus*, annelid jaws; 75-22CB, about 350 m S of the northernmost house at Värsände, *Kettnerites* (*K.*) *martinssonii, K.* (*A.*) *fjaelensis, K.* (*A.*) *sisyphi sisyphi*, annelid jaws; 82-42CB, about 400 m S of the northernmost house at Värsände, bottom of the ditch, *Kettnerites* (*K.*) *martinssonii, K.* (*A.*) *microdentatus*, annelid jaws; 82-44CB, loose slab, Mulde Beds, *Hindenites* sp., *Kettnerites* (*K.*) *martinssonii* var. mulde, *K.* (*A.*) *fjaelensis*, sp., annelid jaws.

VÄSTERBACKAR 1. Sundre Beds, middle upper part. *Age:* Latest Ludlow, Whitcliffian.
References: Laufeld 1974a, b*.
Samples/jawed annelids: 71-184LJ, top of the section, no annelid jaws; 75-2LJ, *Kettnerites polonensis*, annelid jaws.

VÄSTERBJÄRS 1. Klinteberg Beds. *Age:* Latest Wenlock – earliest Ludlow.
Reference: Frykman 1989*.
Samples/jawed annelids: 83-38LJ, from the road-side section north of the cross-roads SW of Västerbjärs. The sample was taken below the road sign 'Hörsne', from the lowest accessible bed in the bottom of the ditch, *Kettnerites* (*K.*) *bankvaetensis, K.* (*A.*) cf. *sisyphi*, annelid jaws.

VÄSTLAUS 1. Hemse Beds, Hemse Marl SE part. *Age:* Ludlow, probably Leintwardinian.
References: Laufeld 1974a, b*; Larsson 1979; Fredholm 1988.
Samples/jawed annelids: 82-31LJ, *Lanceolatites gracilis, Kettnerites* (*K.*) *martinssonii, K.* (*A.*) *sisyphi klasaardensis, K.* (*K.*) *huberti, K.* (*K.*) *burgensis*, annelid jaws.

VÄSTÖS 1. Högklint Beds unit b. *Age:* Early Wenlock.
References: Laufeld 1974a, b*, Larsson 1979.
Samples/jawed annelids: 77-10CB, 3 m below the top of the surface of the bridge, fragments.

VÄTE 1. 637264 165427 (CJ 4257 7107), ca 800 m NE of Väte church. Topographical map sheet 6 J Roma SV. Geological map sheet Aa 160 Klintehamn.

Ditch section immediately SE of the intersection road and the ditch of the ca 250 m SE of the cross-roads at Bäcks. Väte 1 comprises the distance 0–30 m SE of the intersection. *Reference point:* The small carbonate build-up, about 1 m south of the road and below the surface of the road.
Halla Beds, unite b(?). *Age:* Wenlock.
Samples/jawed annelids: 82-35CB, the bed which underlies the carbonate build-up, no annelid jaws; 82-36CB, about 0.50 m above and 4 m SE of the previous sample and 5 m SE of the road, fragments.

VÄTE 2. 637373 165540 (CJ 4375 7206), ca 2350 m NE of Väte church. Topographical map sheet 6 J Roma SV. Geological map sheet Aa 160 Klintehamn.

Surface exposure 2 m south of the path and 10–15 m west of a small wooden hut (not marked on the topographical map). About 900 m NE of the house, at the end of the field road, about 1.5 km NE of Väte church. *Klinteberg Beds*, lower part. *Age:* Late Wenlock.
Samples/jawed annelids: 85-40LJ, surface exposure, paulinitid fragments, annelid fragments.

VATTENFALLSPROFILEN 1. Lower Visby, Upper Visby, and Högklint Beds, units a–d. *Age:* Early Wenlock.
References: Hedström 1904, p. 93, line 11, to p. 96, line 17, 1923, p. 195, Fig. 2; Hede 1925, p. 15, line 18 from below, to page 16, line 3; Martinsson 1972, pp. 128–129; Laufeld 1974b*; Bassett & Cocks 1974, p. 5; Laufeld & Jeppsson 1976; Franzén 1977, pp. 223, 226; Larsson 1979; Jaanusson, Laufeld & Skoglund 1979*; Claesson 1979; Bengtson 1981; Jeppsson 1982, 1983; Brood 1982; Ramsköld 1983, 1984, 1985b.
Samples/jawed annelids: Lower Visby Beds unit e: 76-6LJ, 0.94–0.99 m a.s.l., *Kettnerites* (*K.*) *versabilis, Kettnerites* (*K.*) *abraham abraham, Lanceolatites gracilis*, annelid jaws; Upper Visby Beds: 70-14LJ, 0.99–1.06 m a.s.l., *Kettnerites* (*K.*) *martinssonii, K.* (*K.*) *abraham abraham*, annelid jaws; 76-8LJ, 2.58–2.62 m a.s.l., *Lanceolatites gracilis, Kettnerites* (*K.*) *abraham isaac*, annelid jaws; Högklint Beds unit a: 70-20LJ, 10.02–10.04 m a.s.l., *Lanceolatites gracilis, Kettnerites* (*K.*) *martinssonii, K.* (*A.*) *microdentatus, K.* (*K.*) *abraham isaac*, annelid jaws; Högklint Beds unit b: 70-8LJ, 13.33 m a.s.l., *Lanceolatites gracilis, Kettnerites* (*K.*) *martinssonii, K.* (*A.*) *microdentatus*, annelid jaws; 82-5CB, 13.77–13.80 m a.s.l., *Kettnerites* (*K.*) *martinssonii, K.* (*K.*) *abraham isaac, Lanceolatites gracilis*, annelid jaws; 82-6CB, 13.80–13.83 m a.s.l., *Kettnerites* (*K.*) *martinssonii, K.* (*A.*) *microdentatus, K.* (*K.*) *abraham isaac, Lanceolatites gracilis*, annelid jaws; 70-6LJ, 15.33–15.41 m a.s.l., *Lanceolatites gracilis, Kettnerites* (*K.*) *martinssonii, K.* (*A.*) *microdentatus, K.* (*K.*) *abraham abraham, K.* (*K.*) *abraham isaac, Hindenites gladiatus*, annelid jaws; 70-5LJ, 16.0–16.10 m a.s.l., *Lanceolatites gracilis, Kettnerites* (*K.*) *martinssonii, K.* (*A.*) *microdentatus, K.* (*K.*) *abraham isaac*, annelid jaws; 70-2LJ, 19.24–19.29 m a.s.l., *Lanceolatites gracilis, Kettnerites* (*K.*) *martinssonii, K.* (*A.*) *microdentatus, K.* (*K.*) *abraham isaac*, annelid jaws; Högklint Beds, unit d: Light shale RM 29.6–30.0 m a.s.l., *Kettnerites* (*K.*) *martinssonii, Hindenites angustus*, annelid jaws; *Valdaria testudo* RM 29.6–30.0 m a.s.l., *Kettnerites* (*K.*) *martinssonii, K.* (*A.*) *microdentatus, K.* (*K.*) *abraham isaac, Hindenites angustus, H. gladiatus*; annelid jaws; *Herrmannina* RM 29.6–30.0 m a.s.l., *Kettnerites* (*K.*) *martinssonii, K.* (*A.*) *microdentatus, K.* (*K.*) *abraham isaac, Hindenites angustus, H. gladiatus*, annelid jaws; Light limestone RM 29.6–30.0 m a.s.l., *Kettnerites* (*A.*) *sisyphi sisyphi, Hindenites angustus*, annelid jaws; *Pterygotus* Marl RM 29.6–30.0 m a.s.l., *Kettnerites* (*K.*) *martinssonii, K.* (*A.*) *microdentatus, K.* (*K.*) *abraham isaac, Hindenites angustus, H. gladiatus*, annelid jaws.

VATTENFALLSPROFILEN 2. 639260 164823 (CK 3808 9181), ca 1100 m SW of Visby cathedral. Topographical map sheet 6 I Visby NO. Geological map sheet Aa 183 Visby & Lummelunda.

Temporary excavation for house construction, about 50 m south of the uppermost part of the locality Vattenfallsprofilen 1. The locality is no longer accessible. The exposure comprised many varied lithologies: almost black pyrite-rich calcarenite, red calcarenite, light yellow calcarenite, and reefal limestone, all found within an excavation roughly 10 m wide and 2–3 m deep. The lithologies formed bodies rather than restricted beds.
Högklint Beds, unit d. *Age:* Early Wenlock.
Samples/jawed annelids: 77-35CB, 77-36CB, and 77-37CB, no annelid jaws; 77-38CB, annelid jaws.

VIALMS 2. 642568 169043 (CK 8245 2117), ca 6820 m N of Bunge church. Topographical map sheet 7 J Fårösund SO & NO. Geological map sheet Aa 171 Kappelshamn.

Inland cliff section W of Buckhällar and 100 m N of the triangulation point 10,2 at Vialms.
Reference level: The contact between Högklint Beds and Slite Beds (Hede 1933, p. 33, lines 37–40.
Högklint and *Slite Beds*. *Age:* Early Wenlock.
Reference: Hede 1933, p. 33, line 22, to page 34, line 2.
Samples/jawed annelids: 84-70LJ, 1.40–1.25 below the reference level, fragments.

VIDFÄLLE 1. Hemse Beds, unit b. *Age:* Ludlow.
References: Laufeld 1974a, b*; Larsson 1979.
Samples/jawed annelids: 75-41CB, 0.15 m above the bottom of the ditch, *Kettnerites* (*K.*) *martinssonii, K.* (*K.*) *huberti, K.* (*A.*) *sisyphi*, annelid jaws.

VIKE 1. Slite Beds, *Pentamerus gothlandicus* or slightly older. *Age:* Late(?) Wenlock.
References: Laufeld 1974a, b*; Larsson 1979.
Samples/jawed annelids: 75-30CB, 0.30 m below the top of the section on the SE side of the bridge, *Gotlandites slitensis*, annelid jaws.

VIKE 2. 639466 167680 (CJ 6673 9129), ca 4990 m S of Boge church. Topographical map sheet 6 J Roma NV & NO. Geological map sheet Aa 169 Slite.

Drainage ditch section on the southern side of the ditch, 0–25 m W of the southwestern corner of the new concrete bridge for road 146.
Slite Beds, *Pentamerus gothlandicus* Beds or, more probably, slightly older. *Age:* Late(?) Wenlock.
Reference point: The intersection of the west side of the bridge and the southern side of the ditch.
Reference level: The bridge abutment which also forms the bottom of the ditch.
Samples/annelid jaws: 83-4CB, 0.20–0.40 m above the reference level, *Gotlandites slitensis, Kettnerites* (*K.*) *bankvaetensis, K.* (*K.*) *martinssonii, K.* (*A.*) *microdentatus, Hindenites gladiatus*, annelid jaws; 83-5CB, 1.2–1.3 m above the reference level, *Gotlandites slitensis*, annelid jaws.

VIKE 3. 639476 167666 (CJ 6660 9141), ca 4860 m S of Boge church. Topographical map sheet 6 J Roma NV & NO. Geological map sheet Aa 169 Slite.

Drainage ditch running NW–SE, exposure on the southwestern side.
Slite Beds, *Pentamerus gothlandicus* Beds or, more probably, slightly older. *Age:* Late(?) Wenlock.
Reference point: On the southern side of the ditch about 200 m W of the new concrete bridge, a protruding section of about 0.5 cubic metres.
Samples/jawed annelids: 83-3CB, about 0.5 m above mean water level, *Kettnerites* (*K.*) cf. *bankvaetensis, K.* (*K.*) *martinssonii, Gotlandites slitensis, Hindenites* sp., annelid jaws.

VIVUNGS 1. Klinteberg Beds, middle–upper part. *Age:* Wenlock–Ludlow boundary.
References: Laufeld 1974a, b*; Larsson 1979; Frykman 1989.
Samples/jawed annelids: 78-11CB, 0.2 m from the top of the section, annelid jaws.

YGNE 2. Upper Visby Beds, uppermost part. *Age:* Early Wenlock.
Reference: Laufeld 1974a, b*; Larsson 1979.
Samples/jawed annelids: 76-1KL, *Kettnerites* (*K.*) *abraham* cf. *abraham*, annelid jaws.

YXNE 1. Slite Beds, unit d. *Age:* Wenlock.
References: Laufeld 1974b*; Larsson 1979.
Samples/jawed annelids: 76-29CB, excavated material, *Kettnerites* (*K.*) cf. *bankvaetensis*, annelid jaws.